A Manual of
Anatomy and Physiology

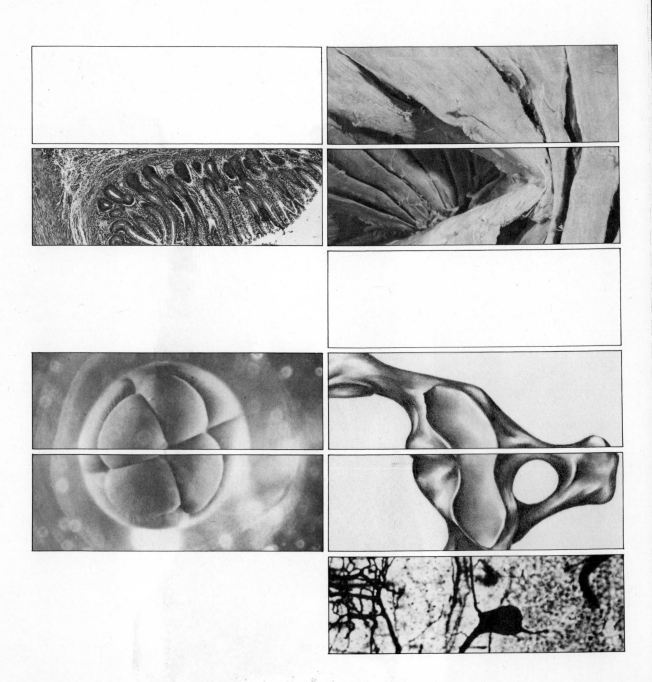

SECOND EDITION

A Manual of Anatomy and Physiology

Laboratory Animal: THE CAT

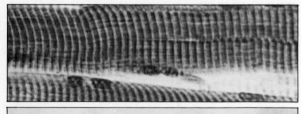

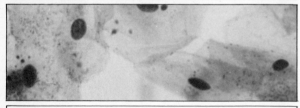
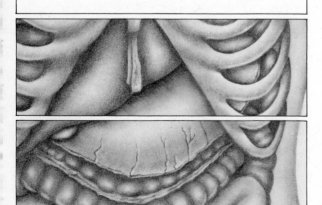

Anne B. Donnersberger
Moraine Valley Community College

Anne E. Lesak
Moraine Valley Community College

Michael J. Timmons
Moraine Valley Community College

D. C. HEATH AND COMPANY
Lexington, Massachusetts   Toronto

Direct gift 4-20-98 $25.00

A Note to the Student

A Manual of Anatomy and Physiology is designed for you, a beginning student enrolled in a health science or related program. It presents primary and practical information relevant to the following health science occupational areas: medical technology, nursing, respiratory therapy, medical record-keeping, operating room technology, dental hygiene, pharmacology, biomedical technology, physical and speech therapy, and special education. Studying anatomy and physiology will help you to see how systems serve the whole body and how the whole body serves its systems in an integrated manner. You will also better understand the relationship between anatomical structure and physiological functioning.

There are a number of special features in this manual that have been designed to aid your learning. Reinforcing the text are over 250 illustrations and photographs, 32 of which are in full color. The medical terminology exercises, the tables describing human bone markings with corresponding drawings, the tables and illustrations describing cat musculature and its correspondence to human musculature, and the photographs of the internal organs of the cat are all unique features. Throughout the manual, terms that appear in **boldface** type can be found in the Glossary at the end of the book.

The manual includes fifteen integrated units on medical terminology, the microscope, cells, tissues, the anatomy and physiology of the major body systems, and acid–base balance. Each unit consists of a purpose, objectives, materials, procedures, and a discussion (a self-test) related to the unit content. Within each procedure section, you will identify major organs, study the gross anatomy of selected organs, examine the histological features of selected tissue, and perform physiological exercises related to the organs or systems of study. As you examine the various drawings of the major body systems, you may find it useful to have a set of color pencils on hand and to actually color in the drawings. Doing so will help you see the dimensions of each of the systems more clearly. (This process is explained more fully in Exercise 4 of Unit V, "Identification of Bone Markings," and may be used other units as well.)

We have designed this manual to illustrate, through carefully chosen exercises and correlated photographs and illustrations, specific anatomical and physiological principles. Furthermore, as a self-testing device, we have partially labeled selected photographs, and you can complete the identification of unlabeled structures yourself. These unlabeled structures are identified in the Illustration Appendix.

This manual should provide you with a basic understanding of anatomy and physiology as well as provide the necessary background for further study and work in the health care delivery system. In view of the our desire to continue to improve the second edition of *A Manual of Anatomy and Physiology—Laboratory Animal: The Cat*, we invite constructive comments from those using it. A companion volume of *A Manual of Anatomy and Physiology* featuring the fetal pig as the anatomical subject is also available.

Anne B. Donnersberger
Anne E. Lesak
Michael J. Timmons
Palos Hills, Illinois

Acknowledgments

We wish to express our appreciation to the following individuals for their help and encouragement in the preparation of this laboratory manual: our students, who demonstrated the need for a manual of this design; Calvin C. Kuehner, Ph.D., for his contribution to the laboratory exercise on osmosis; Mary E. Bannon for her suggestions regarding laboratory design; Tom McCague for his helpful suggestions; Pat Oakes, who drew the finished three-dimensional illustrations; and Nancy Diepenbrock, for typing the manuscript of the second edition.

Photomicrographs were taken with a Carl Zeiss Research Microscope made available through the courtesy of James Sullivan at Eberhardt Instrument Company. Special appreciation is due Chuck Simak of Rayline, Inc., for technical photographic assistance.

We warmly acknowledge the editorial and production assistance of D. C. Heath and Company. Finally, a special note of thanks to our families, whose patience, encouragement, and understanding contributed significantly to the completion of this manual.

Contents

UNIT X **Nervous System**   263

UNIT XI **Special Senses**   295

UNIT XII **Urinary System**   313

Illustrations

*Drawing

†Micrograph or photograph

Color Plates

Medical Terminology

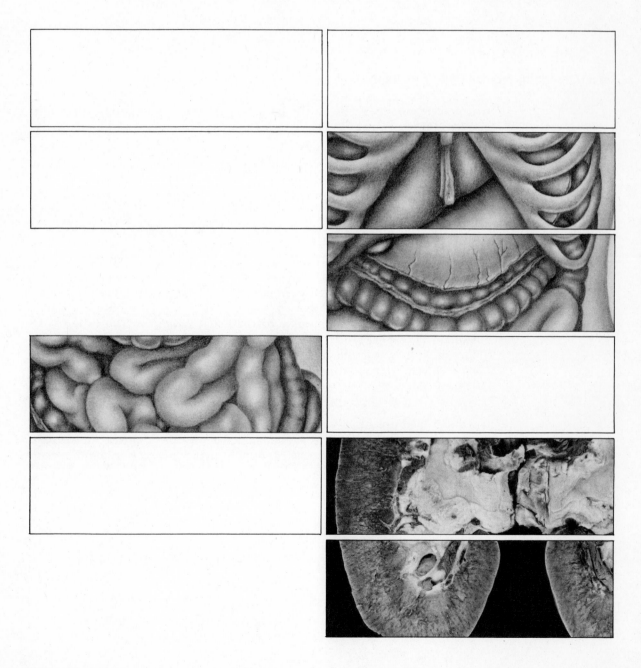

■ *Purpose*

The purpose of Unit I is to acquaint you with medical terminology with respect to body structure and medical procedures.

■ *Objectives*

In order to complete Unit I, you must be able to do the following:

1. Demonstrate proficiency in using terms describing body directions, planes, and abdominal regions.
2. Utilize prefixes indicating location, direction, tendency, and number.
3. Use prefixes that identify organs.
4. Use suffixes correctly that are commonly associated with various medical areas.
5. Construct medical terms from common descriptions.

■ *Procedure*

EXERCISE 1

Terms Describing Human Body Directions

Anatomical position refers to the human body in a standing position, with the palms facing forward. These terms are used in describing human body directions in anatomical position:

1. **Superior** (or **cranial**); Toward the head end of the body. *Example*—The shoulder is superior to the hip.
2. **Inferior** (or **caudal**): Away from the head. *Example*—The knee is inferior to the elbow.
3. **Anterior** (or **ventral**): Front. *Example*—The nose is anterior to the back of the head.
4. **Posterior** (or **dorsal**): Back. *Example*—The shoulder blades are on the posterior side of the body.
5. **Medial** (or **mesial**): Toward the midline of the body. *Example*—The great toe is medial to the little toe.
6. **Lateral:** Away from the midline of the body. *Example*—The little toe is lateral to the great toe.
7. **Proximal:** Toward or nearest the trunk or point of origin of a part. *Example*—The elbow is proximal to the wrist.
8. **Distal:** Away from or farthest from the trunk or point of origin of a part. *Example*—The foot is distal to the knee.

In a quadruped organism, such as the cat, the cranial portion of the body is referred to as *anterior* and the caudal portion *posterior*. Likewise, in a standing position, the quadruped dorsal, or upper, surface is *superior*, and the ventral, or belly, surface is *inferior*.

EXERCISE 2

Terms Designating Planes of the Body

These terms designate planes of the body (*see* Figures 1 and 2):

1. **Sagittal:** A lengthwise plane running from front to back that divides the body or any of its parts into right and left sides (not necessarily equal).

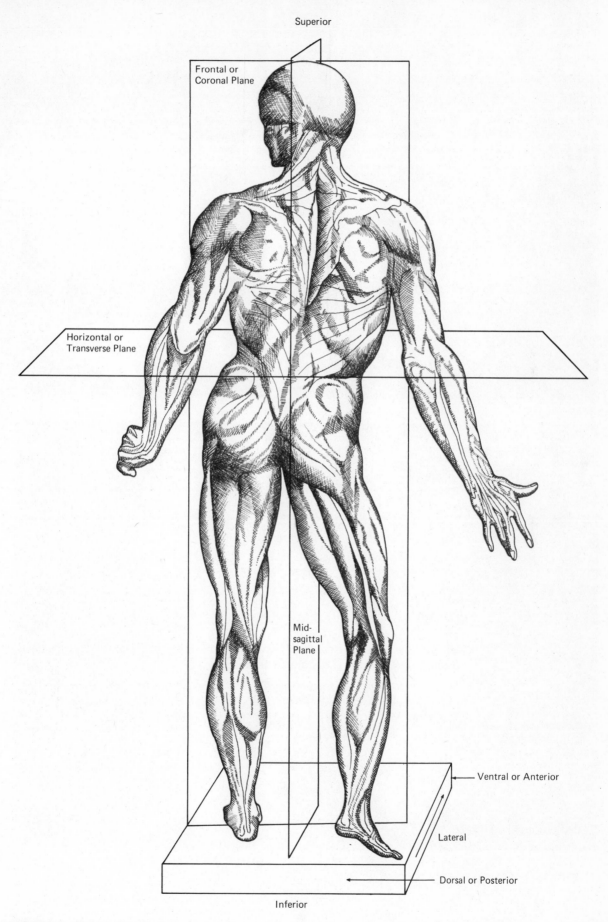

FIGURE 1 *Planes of the body*

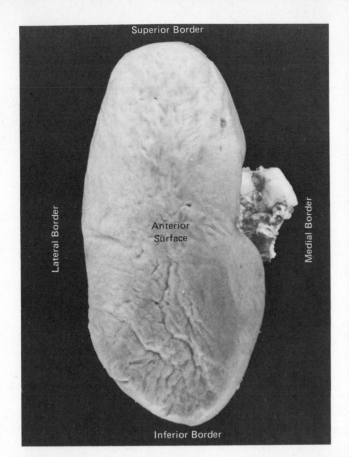

Superior Border

Lateral Border

Anterior
Surface

Medial Border

Inferior Border

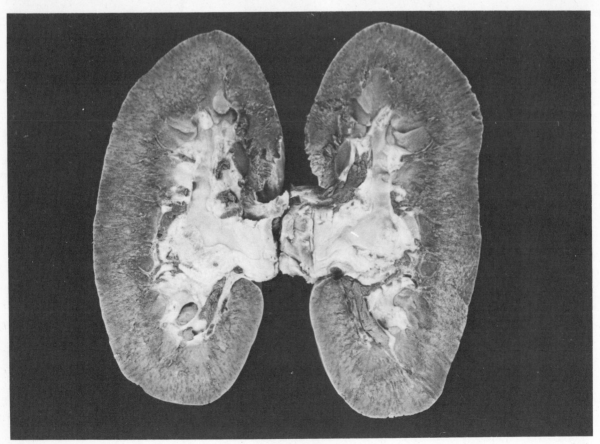

FIGURE 2 *Kidney as viewed in a coronal, or frontal, section*

4

Median Sagittal: A plane that divides the body or its parts into right and left halves.

2. **Coronal (or frontal):** A lengthwise plane running from side to side that divides the body or any of its parts into anterior and posterior portions, or, in the quadruped, into superior and inferior portions.

3. **Transverse (or horizontal):** A crosswise plane that divides the body or any of its parts into superior and inferior parts, or, in the quadruped, into anterior and posterior portions.

EXERCISE 3

Terms Designating Abdominal Regions

The following terms describe abdominal regions (*see* Figure 3):

1. Right **hypochondriac** region
2. **Epigastric** region
3. Left **hypochondriac** region
4. Right **lumbar** region
5. **Umbilical** region
6. Left **lumbar** region
7. Right **iliac** region
8. **Hypogastric** region
9. Left **iliac** region

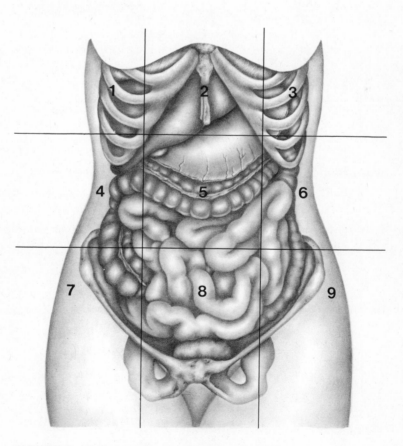

FIGURE 3 *Viscera in relation to abdominal regions*

EXERCISE 4

Prefixes Indicating Location, Direction, and Tendency

The prefixes given in the following table indicate location, direction, and tendency.

Prefix	Meaning	Example
AB-	From, away	Abnormal—away from normal
AD-	To, near, toward	Adrenal—adjoining the kidney
ANTE-	Before	Antepartum—before delivery
ANTI-	Against	Antiseptic—agent against infection
CIRCUM-	Around	Circumocular—around the eye
CO-	With, together	Coordination—to work together
CON-	With, together	Congenital—with birth
CONTRA-	Against	Contraindicated—not indicated
COUNTER-	Against	Counterirritant—nonirritating
DIS-	Apart from	Disarticulation—taking a joint apart
ECT-	Outside	Ectonuclear—outside the nucleus of a cell
END-	Within	Endocardium—membrane lining the inside of the heart
EPI-	Upon	Epidermis—upon the dermal layer of skin
EX-	Out from	Exhale—to breathe out
HYPO-	Under	Hypodermic—under the skin
IM-	Not	Immature—not mature
IN-	Not	Incurable—not curable
INFRA-	Under	Infrapatellar—under the kneecap
PERI-	Around	Pericardium—membrane around the heart
POST-	After	Postmortem—after death
PRE-	Before	Prenatal—before birth
PRO-	Before	Prognosis—a fore-knowing
SUPER-	Above	Superciliary—above the eyebrow
SUPRA-	Above	Suprapubic—above the pubic bone
SYM-	With, together	Symphysis—a growing together
SYN-	With, together	Synarthrosis—a bony union
TRANS-	Through	Transurethral—through the urethra

EXERCISE 5

Prefixes Denoting Number and Measurement

The prefixes in the following table denote number and measurement.

Prefix	Meaning	Example
UNI-	One	Unicellular—consisting of one cell
MON-	One	Mononuclear—one nucleus
BI-	Two	Bilateral—affecting two sides
BIN-	Two	Binocular—two-eyed
DI-	Two	Dicephalic—two heads
TER-	Three	Tertiary—the third stage
TRI-	Three	Trilobar—three lobes
QUADR-	Four	Quadriceps femoris—group of four muscles in the thigh
TETRA-	Four	Tetralogy of Fallot—heart anomaly having four features
POLY-	Many	Polydactyly—having (abnormally) many digits
MACR-	Large	Macrocephalic—having an exceptionally large head
MEGA-	Great	Megadontia—exceptionally large teeth
MICRO-	Small	Microscope—an instrument for viewing small objects
OLIGO-	Few	Oligophagous—eating only a few kinds of food

EXERCISE 6

Prefixes Denoting Organs and Structures

The prefixes in the following table refer to organs and structures of the body.

Prefix	Meaning	Example
ABDOMIN/O-	Abdomen	Abdominal—pertaining to the abdomen
ACR/O-	Extremity	Acromegaly—unusually large extremities
ADEN/O-	Gland	Adenitis—inflammation of a gland
ANGI/O-	Vessel	Angiogram—visualization of blood vessels
ARTHR/O-	Joint	Arthritis—inflammation of a joint
CARDI/O-	Heart	Cardiology—study of the heart
CHONDR/O-	Cartilage	Chondroma—a cartilaginous tumor
CYST/O-	Bladder	Cystoscopy—examination of the inside of the bladder

Prefix	Meaning	Example
CYT/O-	Cell	Cytokinesis—cytoplasmic division
DENT/O-	Tooth	Dental—referring to the teeth
DERMAT/O-	Skin	Dermatologist—physician specializing in skin diseases
DERM/O-	Skin	Dermatitis—inflammation of the skin
DUODEN/O-	Duodenum	Duodenal—having to do with the duodenum (first portion of the small intestine)
GASTR/O-	Stomach	Gastrointestinal—having to do with the stomach and intestines
HEPAT/O-	Liver	Hepatitis—inflammation of the liver
LARYN/GO-	Larynx	Laryngoscope—instrument for viewing inside the larynx
MY/O-	Muscle	Myocardium—heart muscle
NEPHR/O-	Kidney	Nephrology—study of the kidneys
NEUR/O-	Nerve	Neurologist—a physician specializing in diseases of the nervous system
OSTE/O-	Bone	Osteocyte—bone cell
OT/O-	Ear	Otology—study of the ear
PATH/O-	Disease	Pathological—relating to disease
PNEUMON/O-	Lung	Pneumonia—inflammation of the lung
RHIN/O-	Nose	Rhinitis—inflammation of the nasal passages
STOMAT/O-	Mouth	Stomatitis—inflammation of the mouth
THORAC/O-	Thorax or chest	Thoracentesis—puncture of the thorax for the removal of fluid

EXERCISE 7

Suffixes Denoting Relations, Conditions, and Agents

The following suffixes denote relations, conditions, and agents.

Suffix	Meaning	Example
-AC	Related to	Cardiac—relating to the heart
-IOUS	Related to	Contagious—communicable by touch
-IC	Related to	Pyloric—relating to the pyloric valve
-ISM	Condition	Mutism—inability to speak
-OSIS	Condition	Tuberculosis—infection by tuberculosis bacteria
-TION	Condition	Constipation—condition of passing infrequent dry, hard stools
-IST	Agent (one who practices)	Ophthalmologist—a medical specialist who treats eye disorders
-OR	Agent	Operator
-ER	Agent	Examiner
-ICIAN	Agent	Physician

EXERCISE 8

Suffixes Used in Operative Terminology

The following suffixes are used in operative terminology.

Suffix	Meaning	Example
-CENTESIS	To puncture	Amniocentesis—puncture of the amniotic sac
-ECTOMY	To cut out or excise	Appendectomy—excision of appendix
-OSTOMY	To cut into to form an opening	Colostomy—an opening cut in the large intestine, which drains to the outside
-OTOMY	To cut into	Tracheotomy—cut into the trachea
-PEXY	To fix or repair	Gastropexy—repair of the stomach
-PLASTY	To repair or reform	Rhinoplasty—repair of the nose
-(R) RHAPHY	To suture (a seam)	Arteriorrhaphy—suture of an artery
-SCOPY	To view	Otoscope—an instrument used to view in the ear canal

EXERCISE 9

Miscellaneous Suffixes

Suffix	Meaning	Example
-ALGIA	Pain	Neuralgia—nerve pain
-EMIA	Of the blood	Viralemia—viruses in the blood
-GRAM	Writing	Electrocardiogram—tracing of the electrical activity of the heart
-ITIS	Inflammation of	Appendicitis—inflammation of the appendix
-OLOGY	Study of	Ophthalmology—study of the eye
-ORRHEA	Flow	Amenorrhea—cessation of menstrual flow
-PHOBIA	Fear of	Claustrophobia—fear of confined spaces

UNIT I

Medical Terminology

DISCUSSION

Sentence Completion

Choose the correct response for Questions 1–4 from the answers listed below:

a. anterior	c. medial	e. distal	g. coronal
b. posterior	d. lateral	f. proximal	h. sagittal

_____ 1. The plane of the body dividing it into right and left sections is ....

_____ 2. The hip is .... to the ankle.

_____ 3. The navel is .... to the spinal cord.

_____ 4. The great toe is .... to the little toe.

Matching

Column A

_____ 5. Study of the heart

_____ 6. Inflammation of the skin

_____ 7. Removal of the gallbladder

_____ 8. Pain in a muscle

Column B

a. cardiology

b. cholecystectomy

c. dermatitis

d. myalgia

Multiple Choice

_____ 9. Which of the following would belong to the axial plane of the body?
   a. foot
   b. toe
   c. leg
   d. backbone

_____ 10. The kidneys are located primarily in the .... regions of the abdomen.
   a. hypochondriac
   b. hypogastric
   c. epigastric
   d. lumbar

_____ 11. When a surgeon amputates a leg, he or she makes a .... cut through the bone.
   a. sagittal
   b. coronal
   c. transverse
   d. longitudinal

_____ 12. A term describing a structure that surrounds an organ is:
   a. transurethral
   b. pericardium
   c. antepartum
   d. superciliary

13. In a quadruped organism, such as the cat, the cranial portion of the body is referred to as:
   a. anterior
   c. dorsal
   b. posterior
   d. superior

14. A cytological examination involves observing:
   a. the inside of the bladder
   c. tissues
   b. cells
   d. behavior

15. A person with pneumonitis would most likely have which of these symptoms?
   a. painful urination
   c. difficulty breathing
   b. inability to swallow
   d. nausea and vomiting

16. What is the medical term for inflammation of a kidney?
   a. neuritis
   c. cystorrhaphy
   b. nephrotomy
   d. nephritis

17. Viewing the inside of the duodenum is known as:
   a. duodenology
   c. duodenectomy
   b. duodenoscopy
   d. duodenitis

18. A person with otitis may have difficulty:
   a. hearing
   c. smelling
   b. seeing
   d. tasting

19. When puncturing the chest cavity with a needle, a surgeon performs a:
   a. pneumonectomy
   c. thoracentesis
   b. pneumonostomy
   d. thoracotomy

20. Which abdominal region is inferior to the umbilical region?
   a. left hypochondrium
   c. hypogastric
   b. epigastric
   d. right iliac

Word Relationships

Which is an incorrect word relationship in each of the following questions?

21. a. ante = before
   c. brady = slow
   b. bi = two
   d. cept = bag

22. a. cyto = cell
   c. epi = top
   b. entero = intestine
   d. hyper = below

23. a. algia = pain
   c. gram = tracing, mark
   b. ectomy = cut out
   d. itis = draining

24. a. centesis = puncture
   c. pexy = fixation
   b. ectasia = stretching
   d. gaster = kidney

25. a. derm = skin
   c. aden = gland
   b. arthr = joint
   d. opt = ear

DISCUSSION QUESTIONS (Optional)

1. List three advantages of using medical terminology.

    a.

    b.

    c.

2. List two disadvantages of using medical terminology.

    a.

    b.

3. From a recent newspaper or magazine list and define five medical or medically related terms. Attach newspaper clippings.

The Microscope

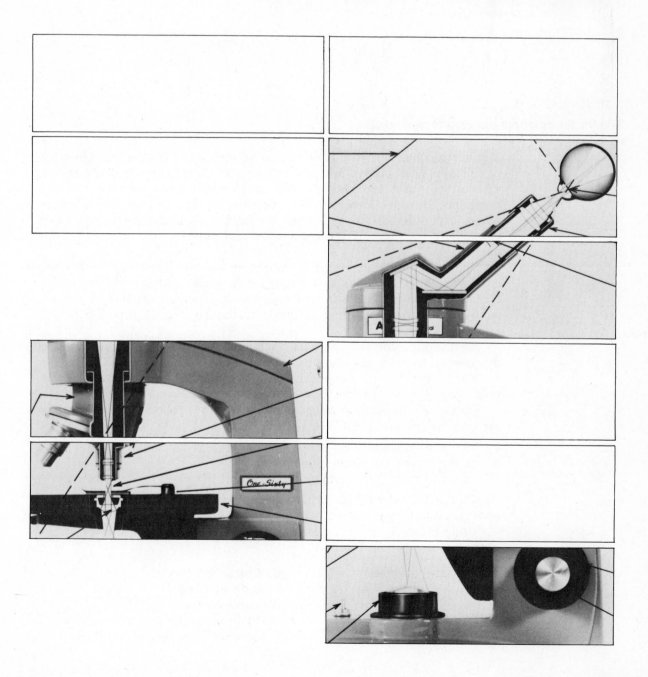

■ *Purpose*

The purpose of Unit II is to familiarize you with the structure and function of the microscope in order to better understand cytology and histology.

■ *Objectives*

In order to complete Unit II, you must be able to do the following:

1. Name and identify the major parts of the microscope.
2. State and demonstrate the functions of the microscope parts.
3. List and follow the directions for proper care of the microscope.
4. Demonstrate use of the microscope.

■ *Materials*

Compound microscope
Unprepared slides

Prepared slides of newsprint and other specimens

■ *Procedure*

EXERCISE 1

The Structure of the Compound Microscope

The microscope is an invaluable tool in your study of anatomy and physiology. Through its proper use you will be able to see very small structures, such as cells, that contribute to the total structure and function of living organisms. In order to maximize your benefits from the use of the microscope, it is essential that you know the basic facts about microscopy. You must know the parts of the microscope and how to use each of its components.

The **compound microscope** is commonly used in laboratory studies and consists of two lenses or lens systems—the ocular and the objective. The ocular, or eyepiece, magnifies an object that is again magnified by the objective. Total magnification is the product of ocular magnification × objective magnification.

Even though you may have used a microscope previously, work through this exercise to ensure its efficient use.

Your instructor will assign a microscope to you. You are to use this same microscope for your studies throughout this course.

Obtain the microscope assigned to you. Always follow these rules when carrying the microscope to your work area:

1. Put one hand beneath the scope for support.
2. Put the other hand around the curved arm.
3. Carry the microscope in an upright position.

Now place the microscope in front of you at your work area and plug the cord into the electrical outlet.

Using Figure 4 and the table on pages 18–19, identify the following parts of the microscope:

1. Ocular, or eyepiece
2. Revolving nosepiece
3. Objectives
4. Arm
5. Stage
6. Coarse adjustment
7. Fine adjustment
8. Condenser
9. Diaphragm
10. Base with illuminator

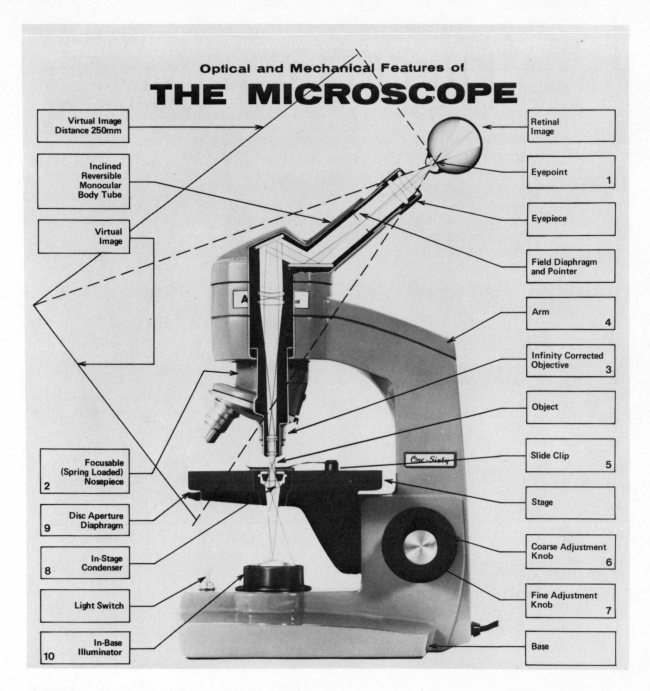

FIGURE 4 *Optical and mechanical features of the microscope (Courtesy American Optical Corporation)*

EXERCISE 2

Functions of the Microscope Parts

Reexamine the microscope parts that you identified in Exercise 1, noting the function of each part. Keep in mind that your primary purpose is to adjust the microscope to get the desired magnification and clarity (resolution) of the object viewed. Especially note the parts that allow you to achieve your primary purpose in focusing.

Microscope Parts and Their Functions (*see* **Figure 4**)

Parts	Function
1. **Ocular**	A lens of a given magnification, which is probably engraved on the rim, e.g., 10X. (You may also see a pointer embedded in the ocular. This is used to aid you in indicating specific locations within the microscopic field.)
2. **Revolving Nosepiece**	This plate, which is capable of rotation, allows you to utilize objectives of different magnifications.
3. **Objectives**	Lenses of varying magnifications. The values (e.g., 10X, 43X, and 100X) are usually engraved on the objectives.
a. Low power objective	Lens with the least magnification. (Often 10X.)
b. High power objective	Lens with greater magnification than low power objective. (Often 43X.)
c. Oil immersion objective	Lens with greatest magnification. (Often 100X. You will infrequently use this lens in this course. Proper technique must be demonstrated by your instructor prior to its use.)
4. **Arm**	Handle for holding and positioning microscope.
5. **Stage**	Platform upon which the slide is positioned for focusing. Note that your scope may have clips for anchoring the slide or it may have a mechanical slide holder. Obtain a clear slide and fit it into the holder. The slide can be moved forward, backward, and from side to side by rotating the small knob on the underside or on top of the stage.
6. **Coarse Adjustment**	Larger knobs on both sides of the base of the arm that allow for initial focusing of the object to be viewed.
7. **Fine Adjustment**	Smaller knobs on both sides of the base of the arm that allow for refinement of detail in focusing.
8. **Condenser**	A lens system that concentrates light from the illumination source so that a cone of light fills the aperture of the objective. After checking to see that the light bulb is on, move the condenser up and down and note the varying intensities of light visible in the ocular.

Parts	Function
9. Diaphragm	Plate with apertures (openings) of varying diameters from small to large. Open and close the diaphragm by adjusting it with its handle so that varying intensities of light are visible through the ocular.
10. Base with Illuminator	Platform upon which the microscope is structured, usually containing an electric light source.

EXERCISE 3

Proper Care of the Microscope

Care must be exercised when removing the microscope from its storage area and carrying it to your work area (*see* Exercise 1). Care must also be exercised when cleaning the lenses and returning the microscope to its storage area.

Do not remove the lenses to clean them. Gently wipe only the external surface of the lenses with lens paper. Do not use any other material for wiping because the lenses are damaged easily.

When returning the microscope to its storage area, make sure to:

1. Put the low power objective down toward the specimen stage.
2. Remove all slides from the stage.
3. Rewind the electrical cord.

EXERCISE 4

Use of the Compound Microscope

1. Using a prepared slide of newsprint, you will practice focusing. Obtain the slide from the slide tray and clean it if necessary. Then place the slide in the slide holder on the specimen stage. Make sure that the slide is properly anchored and movable in the slide holder.
2. Make sure that the light source is on, then adjust the diaphragm so that by looking into the ocular you see a bright field of light. You may also adjust the condenser to increase the intensity of light if necessary. Keep in mind that the intensity and brightness of light must be varied in relationship to the density of the specimen to be focused and the magnification used. Light must be transmitted through the specimen in order to be seen. Therefore, thinner specimens require less light than do thicker specimens.
3. Now make sure the *low* power objective is in place on the revolving nosepiece. It should be in a downward position toward the stage. By watching on the side of the stage and by rotating the coarse adjustment slowly away from you, you can bring the low power objective down to its limit.
4. With both eyes open, look into the ocular. At first you may be distracted with both eyes open, but eventually you will become unaware

of the surroundings. Only what is being focused will be seen. *Slowly* rotate the coarse adjustment toward you until you see a letter of newsprint. Adjust the slide position on the specimen stage if necessary.

5. Now focus with the fine adjustment knob. The edges of the letter of newsprint should be sharp and well defined. It may be necessary to readjust the light, allowing for more or less. Use the condenser and diaphragm for this adjustment.

6. In order to obtain greater magnification, it will now be necessary to rotate the nosepiece and place the high power objective in a downward position. Listen for the clicking sound to make sure it is in position. If your microscope is parfocal, the image under high power should be in focus. Many classroom microscopes, because of constant use, are not parfocal. In such cases, minor adjustments with both the coarse and fine adjustments must be made.

7. Under high power magnification, more light is often necessary. As the power of the objective increases, the diameter of the lower lens decreases, and less light enters the objective. Again readjust the condenser and diaphragm if necessary.

8. When using high power magnification, never rotate the coarse adjustment while looking into the ocular. If you must use the coarse adjustment, move your head so that you are looking at the side of the stage. Then slowly rotate the coarse adjustment until the objective is close to but *never* touching the slide. You may use the fine adjustment knob to focus while looking through the microscope under high power.

9. *Repeat* the preceding procedure using another demonstration slide from the microscope tray provided by your instructor.

10. Draw and label the images that you have viewed under both low and high power magnification in the spaces provided.

Newsprint Low power magnification	Newsprint High power magnification
Prepared slide Low power magnification	Prepared slide High power magnification

11. Using the guidelines set forth in Exercise 3, clean the microscope and replace it in the storage area. Replace the slides that you have used to their proper positions in the slide trays.

UNIT II

The Microscope

DISCUSSION

Matching

Column A

_____ 1. Condenser

_____ 2. Diaphragm

_____ 3. Fine adjustment

_____ 4. Ocular

_____ 5. Objective

Column B

a. a magnifying lens found on the revolving nosepiece

b. eyepiece

c. used for refinement of detail in focusing an image

d. a plate with apertures allowing for varying amounts of light

e. a lens system concentrating light from an illumination source

Multiple Choice

_____ 6. A scientific instrument that is used for visual study of tissues and cells is the:
   a. centrifuge
   b. pH meter
   c. microtome
   d. microscope

_____ 7. If the letter *P* is placed under the microscope in the normal reading position, which of the following orientations of the letter would the viewer see?
   a. b
   b. p
   c. q
   d. d

_____ 8. The part of the compound microscope that controls the amount of light penetrating the specimen is the:
   a. objective
   b. iris diaphragm
   c. eyepiece
   d. nosepiece

_____ 9. If a viewer wished to observe a particular region of a specimen in greater detail, which of the following objectives would provide the greatest magnification?
   a. 10X
   b. 20X
   c. 45X
   d. 100X

_____ 10. The objective on a compound microscope that needs the least amount of light is the:
   a. 100X
   b. 10X
   c. 20X
   d. 45X

_____ 11. Which of the following magnifications represents the "total magnification" that would be seen in the field of vision if a 5X eyepiece and a 20X objective were used?
   a. 100X
   b. 1000X
   c. 200X
   d. 20X

_____ 12. A . . . . microscope uses a lens system contained in each eye-piece and objective.
  a. simple
  b. complex
  c. compound
  d. magnifying

_____ 13. Which of the following operations should be performed first if the specimen image appears blurred to the viewer?
  a. turn the substage light on
  b. change to a higher power objective
  c. clean the eyepiece and objective lenses with lens paper
  d. adjust the stage

_____ 14. Which of the following procedures should be done just prior to viewing through the compound microscope?
  a. clean all lenses with lens paper
  b. turn the light on
  c. adjust the iris diaphragm
  d. all of the above

_____ 15. If the power of the ocular is 10X and total magnification is 1000, what is the magnification of the objective?
  a. 10X
  b. 100X
  c. 43X
  d. 1010X

_____ 16. The revolving nosepiece holds which microscope part?
  a. ocular
  b. condenser
  c. low power objective
  d. diaphragm

_____ 17. Which statement is *incorrect?*
  a. more light is necessary when using high power than low power
  b. when using the coarse adjustment the objective should touch the slide
  c. the compound microscope consists of two lens systems
  d. the oil immersion objective allows for greater magnification than low power

_____ 18. Two parts of the microscope that regulate the amount of light visible through the ocular are the:
  a. fine adjustment and coarse adjustment
  b. ocular lens and objective lens
  c. illuminator and revolving nosepiece
  d. condenser and diaphragm

_____ 19. Initial focusing of any slide is done under the:
  a. low power objective
  b. high power objective
  c. oil immersion objective
  d. objective demonstrating the greatest detail

_____ 20. Which of the following is the best method of regulating the amount of light that enters the specimen?
  a. condensing lens
  b. nosepiece
  c. fine adjustment knob
  d. iris diaphragm

_____ 21. Which of these represents poor technique when using a microscope?
  a. storing the microscope with the low power objective toward the specimen stage
  b. turning the coarse adjustment knob with the objective on high power while looking through the ocular
  c. looking through the ocular with both eyes open
  d. wiping the external surfaces of the ocular and objective lenses with lens paper prior to using the microscope

_____ 22. In which direction is the fine adjustment knob usually turned when it is used for focusing under high power?
a. clockwise   b. counterclockwise

_____ 23. Which of these objectives has the longest barrel length?
a. low power   c. oil immersion
b. high power

_____ 24. Under which magnification would there be the greatest field of vision?
a. 100X   c. 900X
b. 430X   d. 1000X

_____ 25. Under which magnification would the distance between the objective and the slide being observed be the least?
a. 100X   c. 900X
b. 430X   d. 1000X

DISCUSSION QUESTIONS (Optional)

1. If your field of magnification under low power appears too dark, what should you do to lighten the field?

2. What is meant by _parfocal?_

3. Does a diaphragm aperture of small diameter clarify or distort an image focused under high power? Explain.

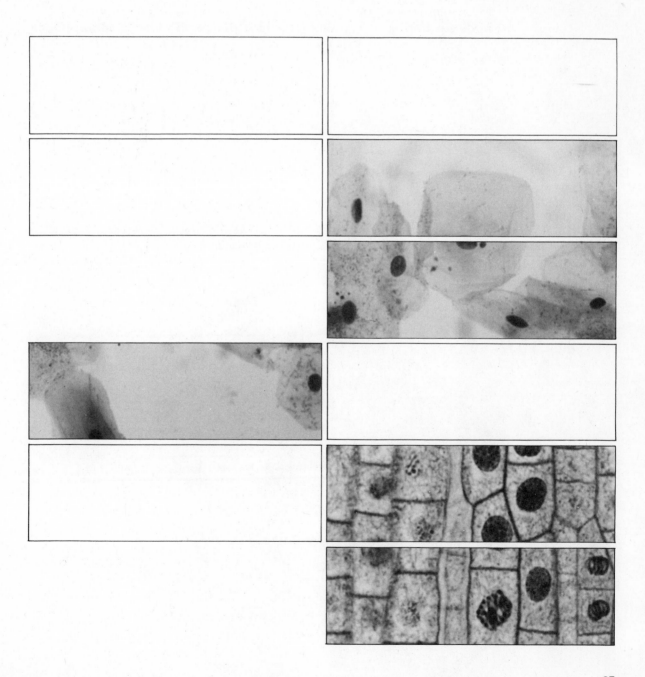

A. Cell Structure

- *Purpose*

  The purpose of Unit III-A is to acquaint you with cell structure and cell division.

- *Objectives*

  To complete Unit III-A, you must be able to do the following:

  1. Name and identify the organelles found in a **cell**.
  2. Prepare a smear of buccal mucosa and identify the **nucleus, cytoplasm,** and **plasma membrane** (cell membrane).
  3. Recognize the stages of **mitosis**.

- *Materials*

Clean microscope slides	Flat toothpicks	Prepared onion root tip slides
Medicine droppers	Methylene blue	Prepared whitefish mitosis slides

- *Procedure*

EXERCISE 1

Animal Cell Structure

Study the illustration of an animal cell as seen under an electron microscope (shown in Figure 5).

FIGURE 5 *Generalized animal cell drawn after electron micrograph (Adapted from* Biology Today, © *1972 by Ziff-Davis Publishing Company. Courtesy of CRM Books, a division of Random House, Inc.)*

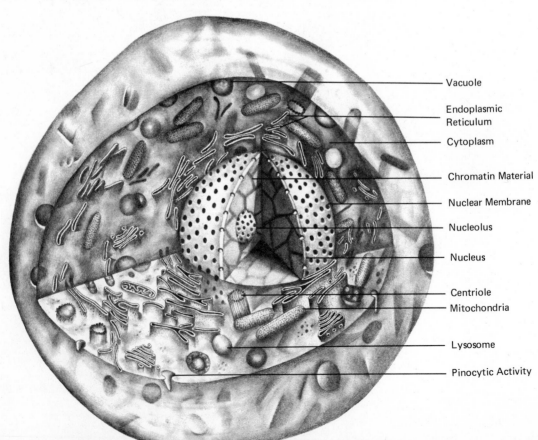

EXERCISE 2

Observation of Buccal Mucosa Cells

(*see* **Color Plate 2**)
Place a drop of water on a clean slide. Scrape the inside of your cheek (buccal mucosa) with a flat toothpick and mix with water on the slide. The thinner the *smear* the better. Allow the slide to air dry. Add one small drop of methylene blue to your smear and cover with a coverslip. Observe under a microscope and draw what you see. *LABEL:* **nucleus, cytoplasm, plasma membrane**, as shown in Figure 6.

FIGURE 6 *Human buccal mucosal cells*

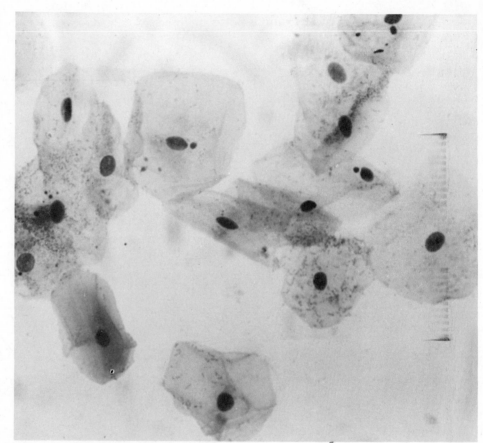

EXERCISE 3

Mitosis

(*see* **Color Plate 1**)

Obtain an onion root tip; locate the distal portion of the root tip under low power; then switch to high power. Look for the following stages of mitosis (see Figures 7a, 7b, and 7c):

1. **prophase**
2. **metaphase**
3. **anaphase**
4. **telophase**

When a cell is not actively dividing, it is in **interphase**. Notice that in interphase **chromatin** is not condensed into chromosomes.

FIGURE 7a *Mitosis in onion root cells*

1. _____

2. Interphase, Resting

3. Prophase

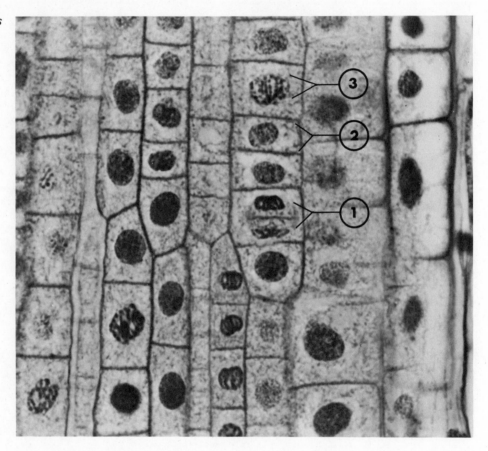

FIGURE 7b *Mitosis in onion root cells*

1. Daughter Cells

2. Anaphase

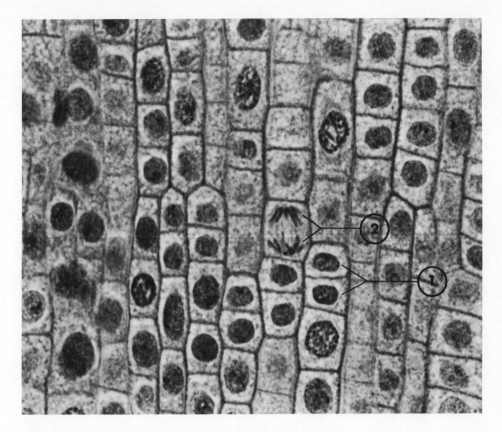

FIGURE 7c *Mitosis in onion root cells*

1. Metaphase

2. _____

3. Prophase, Late

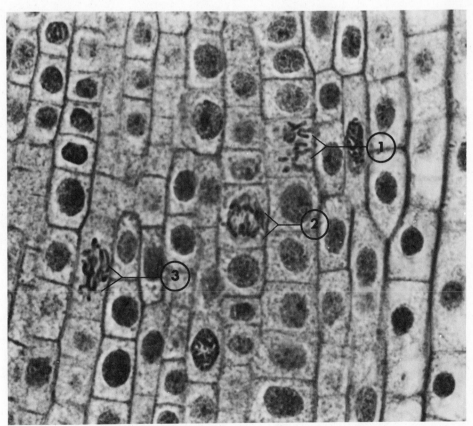

FIGURE 8 *Mitotic figures in whitefish eggs*

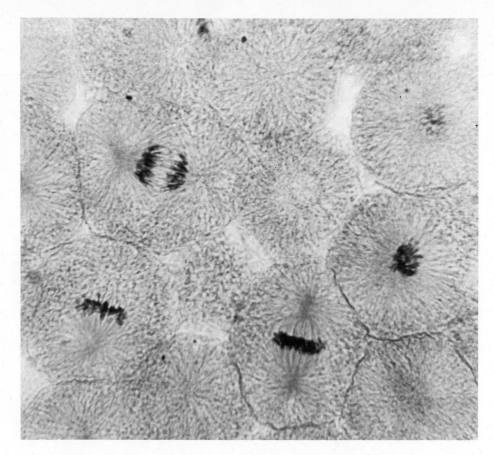

Observe a whitefish mitosis slide and look for similar mitotic figures (Figure 8). These cells are from the blastodisk portion of the whitefish embryo, which contains many rapidly dividing cells. Note that animal cells lack a cell wall.

Observe a prepared slide of human chromosomes as shown in Figure 9.

FIGURE 9 *Normal human somatic chromosomes (Courtesy Carolina Biological Supply Company)*

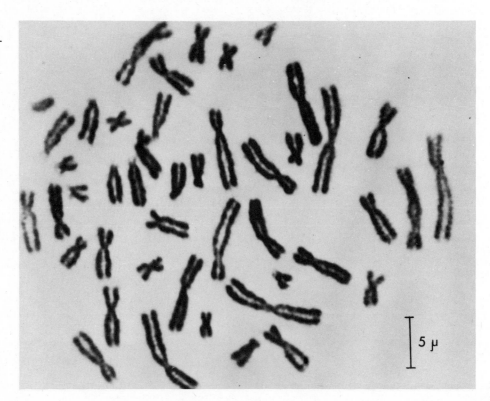

5 μ

B. Cell Physiology

- *Purpose*

  The purpose of Unit III-B is to give you an understanding of some basic concepts of cell physiology.

- *Objectives*

  To complete Unit III-B, you must be able to do the following:

  1. Observe various types of diffusion.
  2. Demonstrate the principles of osmosis and dialysis.
  3. Define a hypertonic, an isotonic, and a hypotonic solution.

- *Materials*

Ammonia or ether	1-ml and 5-ml pipettes
Distilled water	Dialysis membrane
Potassium permanganate crystals	Blood
Test tubes containing gelatin	Test tubes and racks
Probes	Dilute nitric acid
Sodium chloride (NaCl)	1% silver nitrate solution
Glucose	1.5% saline solution
Raw egg white or albumin	Benedict's solution
Molasses	Large beakers
Large white potato that has been soaked overnight	Capillary tubing, 60 cm long
½- to ¾-in. diameter drill bit	Rubber stopper
Methylene blue	Ring stands and test tube clamps
Bunsen burners or hot plate	Paper towels

- *Procedure*

EXERCISE 1

Diffusion

There are three major types of diffusion:

1. *Diffusion of gases.* Open a bottle of ammonia or ether at the front of the room. *Why may the odor soon be detected in all parts of the room?*

2. *Diffusion within a liquid.* Drop a crystal of potassium permanganate ($KMnO_4$) into a beaker of water. Observe during the laboratory period. *What happens?*

3. *Diffusion through a colloid.* Using a straight probe and medicine dropper, stab a tube of gelatin and add one to two drops of methylene blue. Observe during the laboratory period and at the beginning of the next period. *RESULT?*

EXERCISE 2
Osmosis

Bore a hole approximately two-thirds into the length of a large white potato using a drill bit. This exercise will give better results if the potato has been immersed in water overnight or for at least 8 hours. Fill the hole to within ¾ to 1 in. of the top with molasses. Insert a piece of glass tubing (capillary tubing is preferable) into a rubber stopper, and insert the stopper into the potato containing the molasses. The tubing should extend nearly to the bottom of the hole.

Place the prepared potato in a large beaker nearly full of water and place on a ring stand, anchoring the tubing with a test tube clamp. It is a good idea to wrap a paper towel around the tubing at the point where the clamp will be fastened to hold it more tightly.

Observe the apparatus during the laboratory period. The molasses should rise into the tubing as a result of water passing through the potato.*

EXERCISE 3
Dialysis

Place water, sodium chloride, glucose, and albumin or raw egg white from one egg into a bag of dialysis membrane or cellophane and suspend from a rod across a beaker of water. Allow the bag to be immersed in the water. Let stand for 1 to 2 hours; then run the following tests on the water in the beaker and the solution in the dialysis bag.

* The authors wish to thank Dr. Calvin Kuehner for contributing this exercise.

1. *Test for albumin.* Pour about 5 ml of fluid from the beaker into a test tube and add a few drops of nitric acid (HNO_3). If albumin is present, it will be coagulated by the HNO_3 and turn white. *CONCLUSION?*

2. *Test for NaCl.* Pour about 5 ml of fluid from the beaker into another test tube and add a drop of silver nitrate ($AgNO_3$). If the solution turns cloudy or white, silver chloride ($AgCl$) has been formed and the test is positive. *CONCLUSION?*

3. *Test for glucose.* Put 5 ml of Benedict's solution into a test tube and add four to five drops of the beaker water. Boil for 2 minutes over a Bunsen burner or in a water bath. Cool slowly. If a green, yellow, or red precipitate forms, the presence of glucose is indicated. *CONCLUSION?*

REPEAT the above tests using 5-ml aliquots of solution from within the dialysis bag.

EXERCISE 4

Hypertonic, Isotonic, and Hypotonic Solutions

If you are to understand this exercise, you should become familiar with the following concepts:

1. If cells are placed in a **hypertonic** solution, in which the solute concentration is greater outside the cell than inside, the cells will lose solvent (usually water) and shrivel. In erythrocytes (red blood cells), this process is known as *crenation.*

2. If cells are placed in an **isotonic** solution, in which the solute concentration outside is the same as inside, there will be no net change in the amount of water in the cell. Therefore, the cells will retain their original shape.

3. If cells are placed in a **hypotonic** solution, in which the solute concentration is greater inside the cell than outside, there will be a net gain of water in the cell, causing a rise in intracellular pressure, and the cells will burst. In erythrocytes, this is known as *hemolysis*.

Set six small test tubes into a rack and number left to right *1, 2, 3, 4, 5,* and *6*. First, pipette 1.5% saline into the tubes in the amounts shown in the following table. Then pipette the indicated amounts of water into each tube.

Test Tube No.	Add 1.5% Saline in Following Amounts (ml)	Add Distilled Water in Following Amounts (ml)	Final Concentration of Saline (%)
1	0	5.0	0
2	1.0	4.0	0.3
3	2.0	3.0	0.6
4	3.0	2.0	0.9
5	4.0	1.0	1.2
6	5.0	0	1.5

Carefully pipette 0.1 ml of blood into each tube and mix gently by tilting and rotating. Mix gently and thoroughly. In the space provided, describe the appearance of the tubes 5 minutes after the blood was added and mixed.

Tube 1:

Tube 2:

Tube 3:

Tube 4:

Tube 5:

Tube 6:

EXERCISE 5

Chemical Elements Found in the Human Body

Of the 105 known chemical *elements*, about 20 are needed by living organisms. Some of the more abundant elements found in the body are listed in Table 1, together with their representative symbols and ionic charges. In order to make the study of physiology easier and more meaningful, it is essential that you learn the names of these elements, their symbols, and ionic charges.

TABLE 1   Symbols and Ionic Charges for Common Chemical Elements

Element	Symbol	Ionic Charge
Hydrogen	H	usually +1
Carbon	C	varies, usually ± 4
Nitrogen	N	varies
Oxygen	O	usually –2
Sodium	Na	+1
Magnesium	Mg	+2
Phosphorus	P	varies
Sulfur	S	varies
Chlorine	Cl	–1 as chloride
Potassium	K	+1
Calcium	Ca	+2
Manganese	Mn	varies
Iron	Fe	+2 or +3
Cobalt	Co	+2
Copper	Cu	+1 or +2
Zinc	Zn	+2
Iodine	I	–1 as iodide

Chemicals found in cells and tissues (including blood) rarely occur as free elements. Rather, the invisible *atoms* comprising these elements are found as **ions** which are atoms that have gained or lost one or more *electrons* in their outer shells and thus carry either a positive or a negative electric charge. A positive ion is a **cation**; a negative ion is an **anion**. The major cation in extracellular fluid is sodium; in intracellular fluid it is potassium. The major anion in extracellular fluid is chloride; in intracellular fluid it is phosphate.

Ions are found in various body structures, such as nerve fibers, and circulate in body fluids, such as blood and interstitial fluid. Table 1 lists the ionic charges for the atoms of each element listed.

Ions of opposite charges (+ and −) have the tendency to combine with each other to form chemical *compounds*. Compounds such as NaCl (sodium chloride) and $NaHCO_3$ (sodium bicarbonate) are abundant in the body.

Certain ions have a variable charge. These ions tend to combine with ions of other atoms to form a charged particle known as a *complex ion*. Table 2 lists some of the common complex ions found in the body. Complex ions may also combine with other ions or complex ions to form chemical compounds.

TABLE 2   Symbols and Ionic Charges for Common Complex Ions

Complex Ion	*Symbol/Charge*
Hydroxide	OH^{-1}
Bicarbonate	HCO_3^{-1}
Carbonate	CO_3^{-2}
Phosphate	PO_4^{-3}
Sulfate	SO_4^{-2}
Ammonium	NH_4^{+1}

Chemical compounds are named by naming the positively charged ion first, then by naming the negatively charged ion or complex ion. Compounds comprised of a positive ion and a negative ion end in *-ide* (e.g., NaCl—sodium chloride). Compounds comprised of a positive ion and a negative complex ion usually end in *-ate* (e.g., $NaHCO_3$ —sodium bicarbonate). An exception to this rule is the hydroxide (OH^-) ion, which ends in *-ide*.

Procedure
You should practice naming compounds and writing symbols for compounds until it comes easily to you. Refer to Questions 1 and 2 below to test your understanding of this material. If you need further drill, consult an inorganic chemistry textbook.

1. Name the following compounds:

   a. KOH . . . .

   b. HCl . . . . (or hydrochloric acid)

   c. $H_2 SO_4$ . . . . (or sulfuric acid)

   d. HNO_3 . . . . (or nitric acid)

   e. $Na_2 CO_3$ . . . .

f. NaOH . . . .

g. H_2CO_3 . . . . (or carbonic acid)

h. H_3PO_4 . . . . (or phosphoric acid)

i. NH_4OH . . . .

2. Write the chemical formula for each of these compounds:

a. sodium chloride . . . .

b. sodium bicarbonate . . . .

c. magnesium sulfate . . . .

d. potassium chloride . . . .

e. calcium carbonate . . . .

f. potassium iodide . . . .

g. zinc oxide . . . .

h. ammonium chloride . . . .

i. sodium hydroxide . . . .

UNIT III
Cells

DISCUSSION

Multiple Choice

_____ 1. Diffusion rates of molecules may be controlled by which of the following?
   a. temperature
   b. solute/solvent ratio
   c. molecular state
   d. all of the above

_____ 2. A membrane is said to be .... if it will allow some but not all substances to diffuse through.
   a. impermeable
   b. permeable
   c. parapermeable
   d. differentially permeable

_____ 3. A solution of 0.9% sodium chloride, sometimes termed *physiological saline*, is said to be .... to the cells of humans and other mammals.
   a. hypertonic
   b. hypotonic
   c. isotonic
   d. osmotic

_____ 4. Protein synthesis in human cells occurs:
   a. on ribosomes
   b. in lysosomes
   c. in the nucleus
   d. in the Golgi apparatus

_____ 5. Of the following properties, which is *not* characteristic of water?
   a. universal solvent
   b. readily dissolves electrolytes and ionic substances
   c. has a high surface tension
   d. is lipophilic and fat soluble

_____ 6. The correct sequence of mitotic stages during cell division is:
   a. interphase, metaphase, anaphase, and telophase
   b. anaphase, metaphase, telophase, and prophase
   c. prophase, metaphase, anaphase, and telophase
   d. metaphase, anaphase, interphase, and telophase

_____ 7. A cell, during a normal cell cycle, spends most of its time in:
   a. prophase
   b. interphase
   c. anaphase
   d. telophase

_____ 8. During which stage of mitosis do the chromosomes line up across the equator of a cell?
   a. telophase
   b. prophase
   c. metaphase
   d. anaphase

9. If a dialysis bag containing 10% salt water were placed in a beaker of distilled water, what would be the net reaction?
   a. water would go in and salt would go out
   c. water would go out of the bag
   d. water would go into the bag
   b. nothing

10. The movement of water molecules (Question 9) will approach equilibrium when the concentration of water molecules on either side of the cellophane bag reaches . . . . concentration.
    a. 80%
    c. 90%
    b. 85%
    d. 95%

11. Enzymes are:
    a. lipids
    c. proteins
    b. carbohydrates
    d. nucleic acids

12. DNA replication takes place during:
    a. interphase
    c. metaphase
    b. prophase
    d. anaphase

13. –NH₂ and –COOH are characteristic of:
    a. organic acids
    c. carbohydrates
    b. amino acids
    d. nucleic acids

14. How many chromosomes are in a normal, mature human sperm?
    a. 23 pairs
    c. 23
    b. 46
    d. 46 pairs

15. Which of these is *not* a mechanism that enables substances to enter cells?
    a. active transport
    c. pinocytosis
    b. phagocytosis
    d. karyokinesis

16. Cytokinesis occurs during:
    a. interphase
    c. anaphase
    b. metaphase
    d. telophase

17. During which stage of mitosis are the chromosomes migrating toward opposite poles of the cell?
    a. prophase
    c. anaphase
    b. metaphase
    d. telophase

18. Which of these is not found in animal cells?
    a. cell membrane
    c. centriole
    b. cell wall
    d. vacuole

19. Which organelles function as the major sites of ATP production within cells?
    a. centrioles
    c. mitochondria
    b. lysosomes
    d. ribosomes

20. Which of these would not pass through a dialysis membrane because of large molecular size?
    a. salt
    c. protein
    b. simple sugars
    d. water

21. Normal saline is . . . . to body cells.
    a. hypertonic
    c. isotonic
    b. hypotonic

22. In the osmosis exercise of this unit, what would be the concentration of the molasses in the tubing with respect to the molasses first put into the potato?
   a. it would be more concentrated
   b. it would be more dilute
   c. it would be of the same concentration
   d. concentration differences would be impossible to determine

23. The chemical symbol for magnesium sulfate is:
   a. $Mg_2 SO_4$
   b. MnS
   c. MgS
   d. $MgSO_4$

24. A positive ion is an atom that has.....................one or more electrons from its outer shell.
   a. gained
   b. lost
   c. neither gained nor lost

25. If a person were to drink seawater, what would happen to his body's cells?
   a. they would multiply more rapidly
   b. they would burst
   c. they would lose water
   d. there would be no effect

DISCUSSION QUESTIONS (Optional)

1. Using your textbook or other references, state the function of each of these cellular organelles:

   a. chromatin . . . .

   b. nucleolus. . . . .

   c. centrioles. . . . .

   d. plasma membrane . . . .

   e. endoplasmic reticulum . . . .

   f. ribosomes . . . .

   g. Golgi apparatus . . . .

   h. lysosomes . . . .

   i. mitochondria. . . . .

   j. nuclear membrane. . . . .

2. Using your textbook or other references, name the stage(s) of mitosis during which each of the following occurs in the cell:

   a. cell is not actively dividing . . . .

   b. cytokinesis occurs. . . . .

   c. chromosomes are lined up at the equator of the cell . . . .
   . . . .

   d. chromatin condenses into chromosomes . . . .
   . . . .

e. nuclear membrane is not present..

f. DNA is replicating..

3. What physical principle determines whether various types of molecules will pass through dialysis membrane?

Tissues

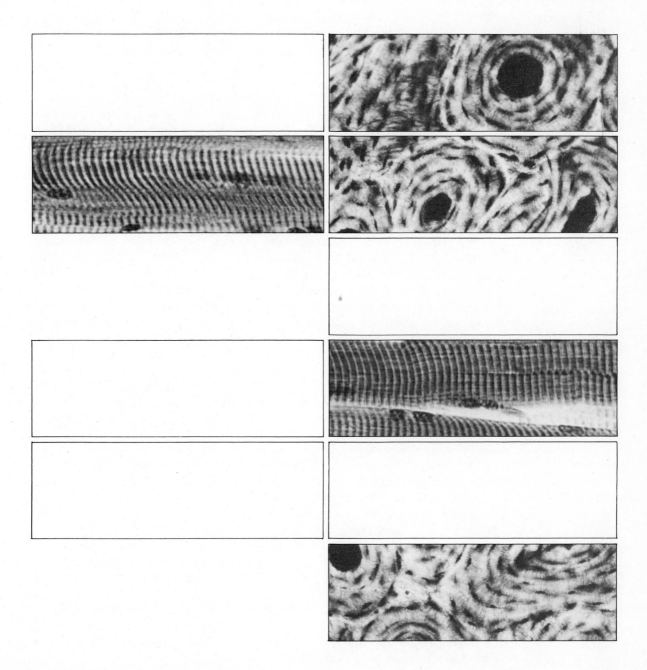

■ *Purpose*

The purpose of Unit IV is to enable you to recognize the various tissue types.

■ *Objectives*

In order to complete Unit IV, you must be able to do the following:

1. Name the general categories of tissues found in the body.
2. Identify general and specialized tissue types microscopically.
3. State the location of various tissue types in the body.
4. Identify the epidermal and dermal layers of skin.
5. Microscopically differentiate between Caucasian and Negroid skin.

■ *Materials*

Slides of the following tissue types:

Simple squamous epithelium	Dense fibrous connective tissue
Simple cuboidal epithelium	Areolar tissue
Simple columnar epithelium	Adipose tissue
Stratified squamous epithelium	Hyaline cartilage
Skin—Negroid and Caucasian	Osseous tissue
Pseudostratified ciliated columnar epithelium	Skeletal muscle
	Smooth muscle
Transitional epithelium	Cardiac muscle

Tissues are groups of cells that perform a common function. There are four primary tissue types—**epithelial** tissue that covers a surface and functions in protection, secretion, and absorption; **connective** tissue that binds and supports; **muscle** tissue that contracts; and **nervous** tissue that conducts impulses.

These four basic tissue types can be further subdivided as shown in the Classification of Tissues table. Epithelial tissue (**epithelium**) is comprised of many cells and very little intercellular material. Epithelium can be **simple**, meaning one cell layer thick, or **stratified**, which indicates several cell layers. If the cells comprising the epithelium are flat, they are known as **squamous**; if they are cube shaped, they are **cuboidal**; and if they are taller than they are wide, they are referred to as **columnar**. Ocasionally, the nuclei of a single cell layer are at different levels, giving the appearance of stratified epithelium. This type is referred to as **pseudostratified**. Another epithelial type consists of stratified balloon-shaped cells and possesses the property of stretching. This is known as **transitional** epithelium and is found in parts of the urinary tract.

Connective tissue, in contrast to epithelial tissue, contains relatively few cells and much intercellular material. It may be classified as **dense** and **fibrous**, as found in tendons and ligaments, or **loose, areolar**, which forms the "packing" in and around organs. There are several specialized types of connective tissue. **Adipose**, or fat, consists of cells in which most of each cell's volume is occupied by fat rather than cytoplasm. This feature imparts a characteristic "signet ring" appearance to adipose cells. Cartilage and bone are other specialized connective tissue types. Cartilage may be **hyaline**, **elastic**, or **fibrous**. Bone, or **osseous** tissue, may be *compact* or *spongy*. Most biologists classify *blood* as a specialized connective tissue, while others consider it a fifth basic tissue type. You will be studying blood in Unit VIII.

There are three types of muscle tissue—skeletal, **cardiac**, and **smooth**. The primary functional cells in nervous tissue are known as **neurons**. Muscle and nerve slides will be studied superficially in this unit and in greater depth in later exercises.

Using the labeled black/white photographs and color plates, examine the slides of tissue types your instructor has made available to you. By studying the slides and the summary table, you should be able to answer most of the questions in the Discussion.

CLASSIFICATION OF TISSUES

Primary Tissue	Types	Divisions		Example
Epithelium (*see* **Color Plates 2–8**)	A. Covering external body surface or lining internal surface	*Simple*	Squamous Cuboidal Columnar	Bowman's capsule (kidney) Collecting tubule (kidney) Gallbladder (nonciliated) Uterine tube (ciliated) Intestinal mucosa
		Pseudo-stratified	Columnar	Male urethra (nonciliated) Trachea (ciliated)
		Stratified	Squamous	Skin (keratinizing) Vagina (nonkeratinizing) Cornea
			Cuboidal Columnar Transitional	Sweat glands Male urethra Urinary bladder
	B. Multicellular glands	*Exocrine*	Simple Compound	Gastric, sweat Salivary
		Endocrine		Thyroid, adrenal
Muscle (*see* **Color Plates 17–20**)	A. Smooth (involuntary)			Intestinal tract, blood vessels
	B. Striated (voluntary)			Skeletal muscle
	C. Cardiac (involuntary)			Heart muscle
Connective tissue	A. General Loose (*see* **Color Plates 22–24**)	*Mesenchyme* *Mucoid* *Areolar*		Embryonic and fetal tissue Wharton's jelly (umbilical cord) Found in most organs and tissues
		Adipose *Reticular*		Subcutaneous tissue Bone marrow, lymph nodes
	Dense	*Irregular* *Regular*		Dermis, capsules of organs Tendon, cornea
	B. Special (*see* **Color Plates 9–14**)	*Cartilage*	Hyaline Fibrous Elastic	Costal cartilage, trachea Intervertebral disc External ear, epiglottis
		Bone	Cancellous Compact	Epiphyses of long bones Shaft of long bone
		Hemopoietic	Myeloid Lymphoid	Bone marrow Spleen, lymph node
		Blood *Lymph*		
Nervous tissue (*see* **Color Plates 25–29**)	A. Central nervous system	*Gray matter* *White matter*		Brain, spinal cord Brain, spinal cord
	B. Peripheral nervous system	*Nerves* *Ganglia* *Nerve endings*		
	C. Special receptors			Eye, ear, nose

COLOR PLATES

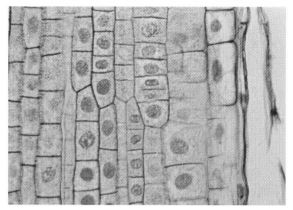

Plate 1 Mitosis in onion root

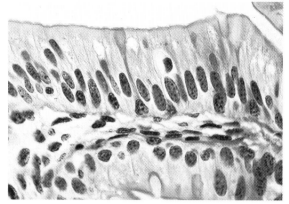

Plate 5 Simple columnar epithelium

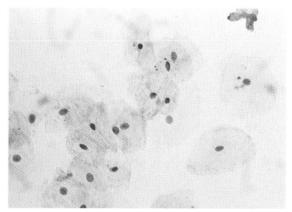

Plate 2 Squamous epithelial cells as seen in a smear from the oral mucosa

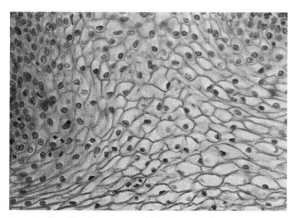

Plate 6 Simple columnar epithelium showing goblet cells and brush border

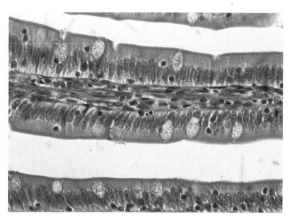

Plate 3 Stratified squamous epithelium, vagina

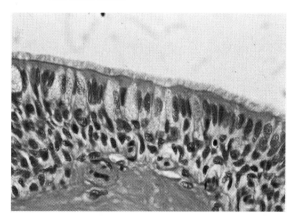

Plate 7 Pseudostratified columnar ciliated epithelium and goblet cells

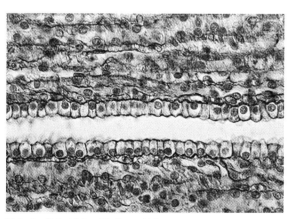

Plate 4 Cuboidal epithelium, kidney

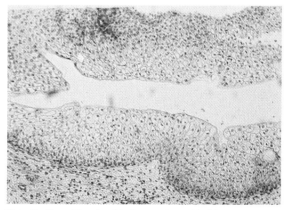

Plate 8 Transitional epithelium, urinary bladder

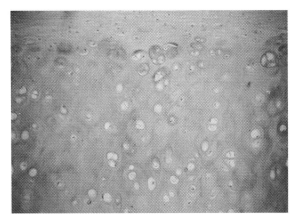

Plate 9 Hyaline cartilage as seen in the trachea

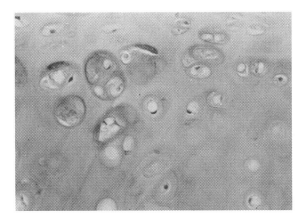

Plate 10 Hyaline cartilage showing chondrocyte and matrix detail

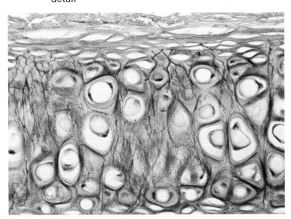

Plate 11 Elastic cartilage as seen in the epiglottis

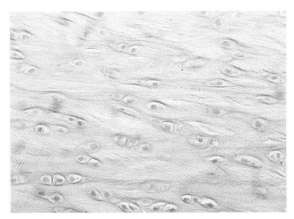

Plate 12 Fibrous cartilage as viewed in the intervertebral disc

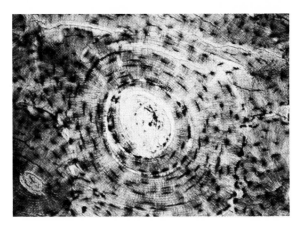

Plate 13 Human compact bone, unstained

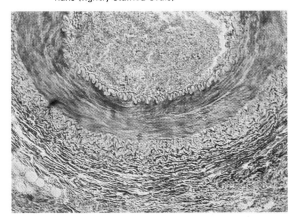

Plate 14 Human pancreas stained to show islets of Langerhans (lightly stained ovals)

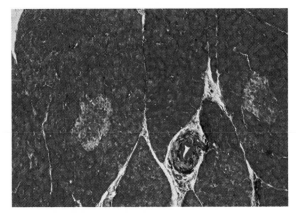

Plate 15 Large artery demonstrating tunica intima, media, and adventitia layers

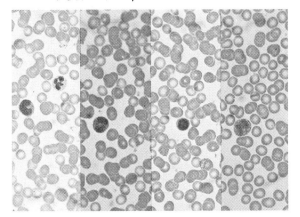

Plate 16 Composite picture of human blood, Wright's stain, showing from left to right, monocyte, neutrophil, eosinophil, lymphocyte, and basophil

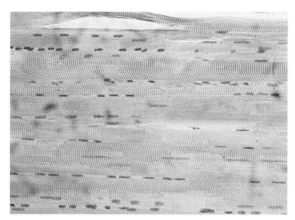

Plate 17   Skeletal muscle

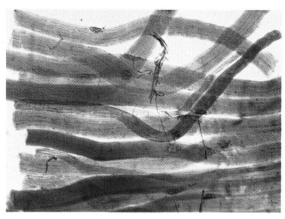

Plate 21   Myoneural junction showing motor end-plate detail

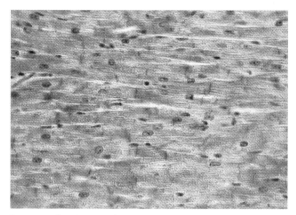

Plate 18   Cardiac muscle

Plate 22   Areolar connective tissue

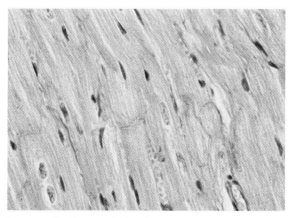

Plate 19   Cardiac muscle showing intercalated disc detail

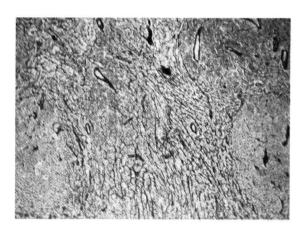

Plate 23   Reticular fibers as demonstrated in the spleen

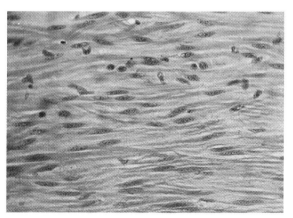

Plate 20   Smooth muscle

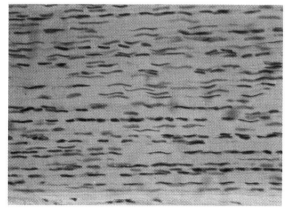

Plate 24   Collagenous fibers as seen in tendon of Achilles

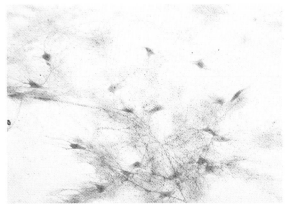

Plate 25 Motor neurons as seen in the spinal cord

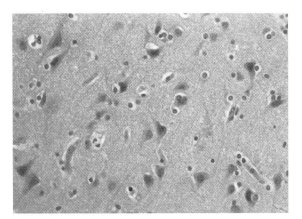

Plate 26 Human cerebral cortex stained to show pyramidal (larger) and glial (smaller) cells

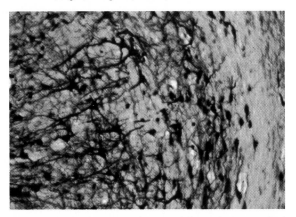

Plate 27 Cerebral cortex stained specially to show pyramidal cells

Plate 28 Fibrous astrocytes as seen in the brain

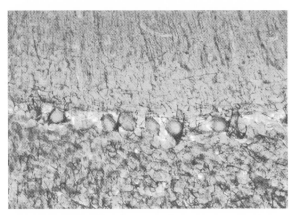

Plate 29 Purkinje cells (large ovals) as viewed in the human cerebellum

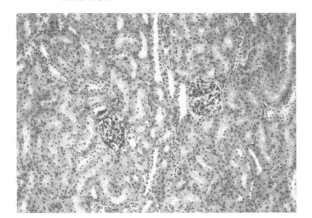

Plate 30 Glomeruli as seen in a human kidney

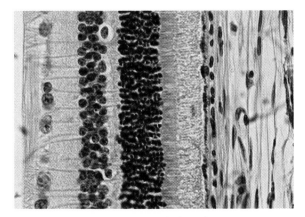

Plate 31 Human eye showing retinal details

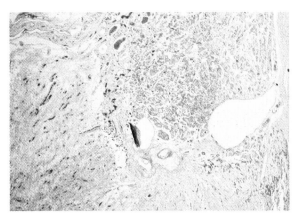

Plate 32 Panoramic view of human hypophysis showing the three lobes. Acidophilic (reddish), basophilic (bluish), and chromophobic (poorly staining) cells are also visible.

■ *Procedure*

EXERCISE 1

Microscopic Identification of Tissue Types

Examine each of the slides and identify each tissue type using the illustrations in this unit as a guide (see Figures 10–30 and **Color Plates 1–32**).

FIGURE 10 *Human squamous epithelial cells, smear from mouth*

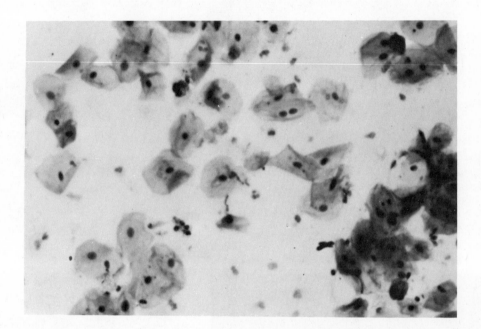

FIGURE 11 *Simple columnar epithelium, showing goblet cells and brush border*

1. Goblet Cells
2. Brush Border

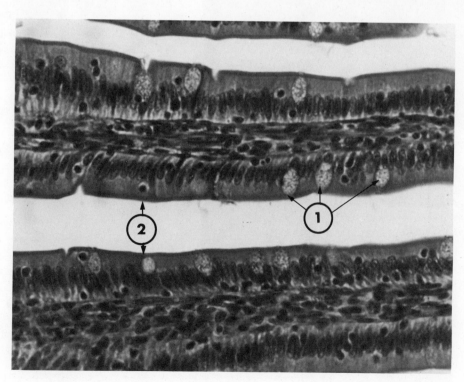

FIGURE 12 *Ciliated columnar epithelium*

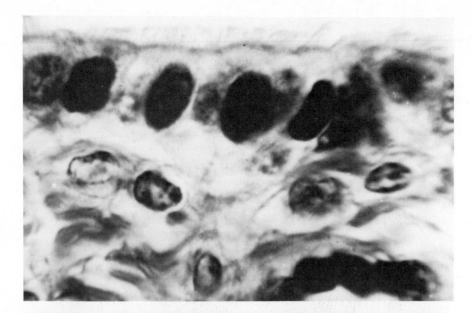

FIGURE 13 *Cuboidal epithelium as found in the collecting tubules of the kidney*

1. Cuboidal Cells

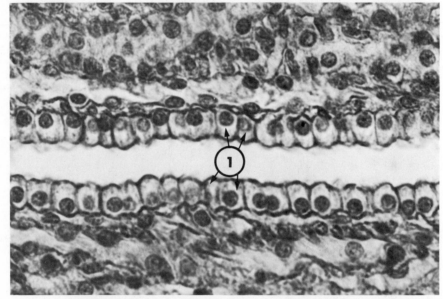

FIGURE 14 *Ciliated pseudostratified columnar epithelium*

1. Cilia
2. Pseudostratified Columnar Epithelium (Note appearance of nuclei.)

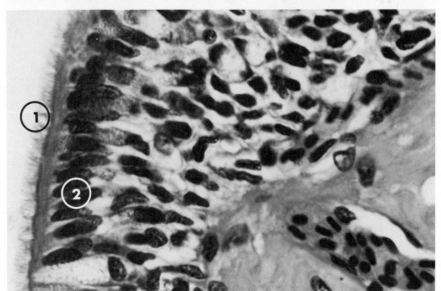

FIGURE 15 *Strat-ified squamous epithelium*

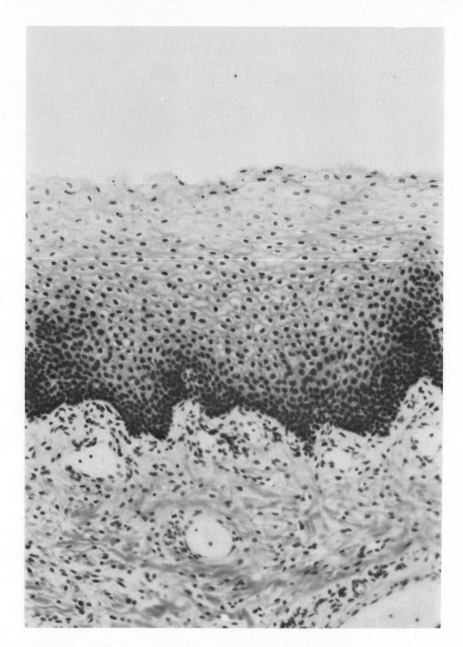

FIGURE 16 *Strat-ified columnar epithelium*

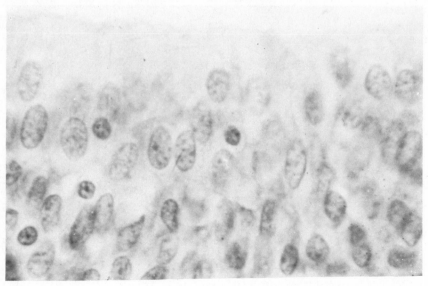

FIGURE 17 *Transitional epithelium*

1. Balloon-shaped Cells next to Lumen

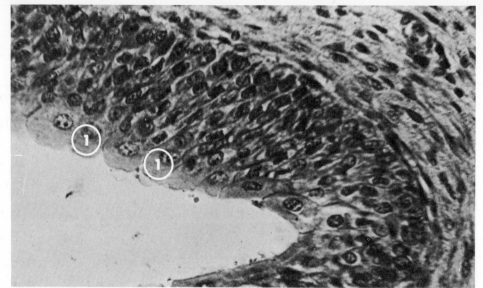

FIGURE 18
Smooth muscle fibers, longitudinal section

1. Fiber
2. Nucleus

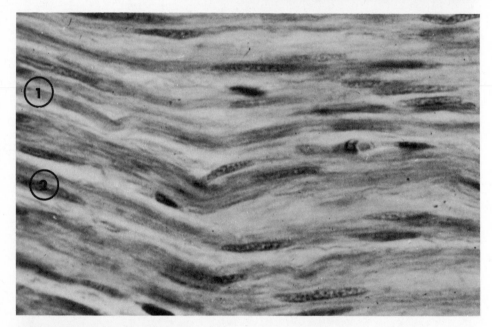

FIGURE 19
Smooth muscle

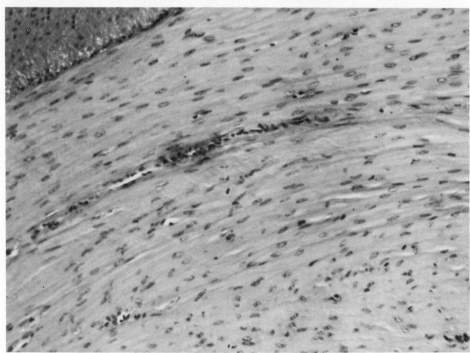

FIGURE 20 *Striated muscle fiber, longitudinal section*

1. Nuclei

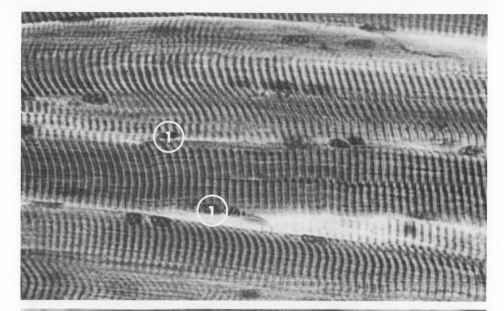

FIGURE 21 *Cardiac muscle, longitudinal section*

1. Nucleus
2. Intercalated Discs
3. Capillary with Blood Cells

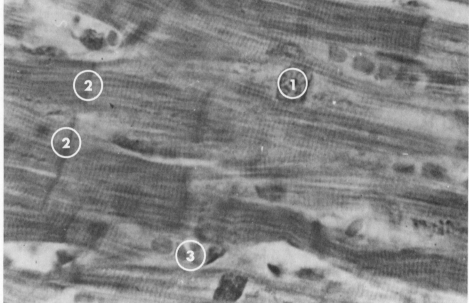

FIGURE 22 *Hyaline cartilage of the trachea*

1. Chondrocyte Cells
2. Intercellular Matrix
3. Perichondrium
4. Lacuna

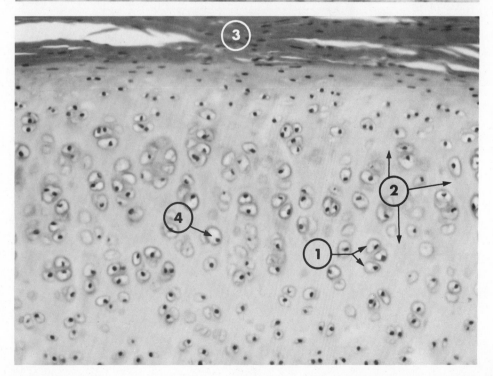

FIGURE 23 *Hyaline cartilage*

1. Intercellular Matrix
2. Nucleus of Chondrocyte Cell
3. Lacuna (Capsule), Containing Chondrocyte Cells

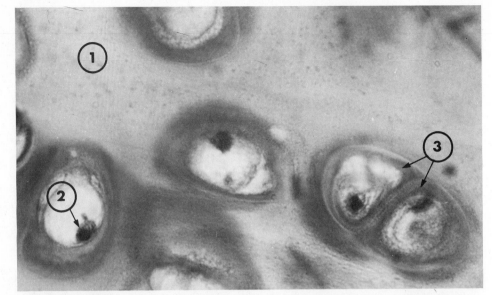

FIGURE 24 *Elastic cartilage and perichondrium*

1. Perichondrium
2. Intercellular Matrix
3. Elastic Fibers
4. Chondrocyte Cell

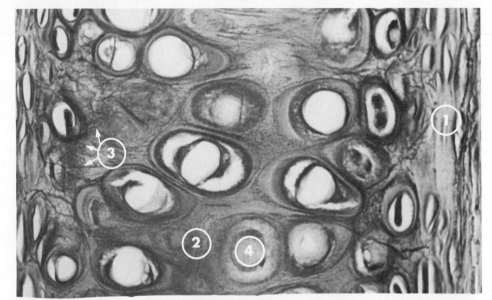

FIGURE 25 *Elastic cartilage showing elastic fiber and matrix detail*

1. Chondrocyte
2. Lacunae (Capsules)

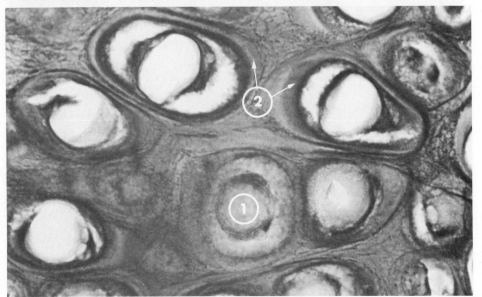

FIGURE 26
*Ground bone, pano-
ramic view, showing
Haversian systems*

1. Haversian Systems

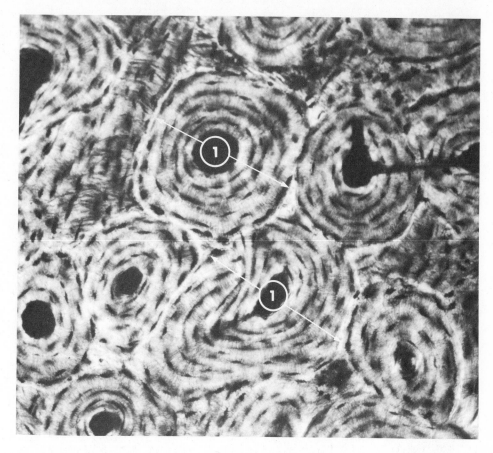

FIGURE 27
*Ground bone show-
ing detailed Haver-
sian system*

1. Bone Matrix
2. Haversian Canal
3. Osteocytes

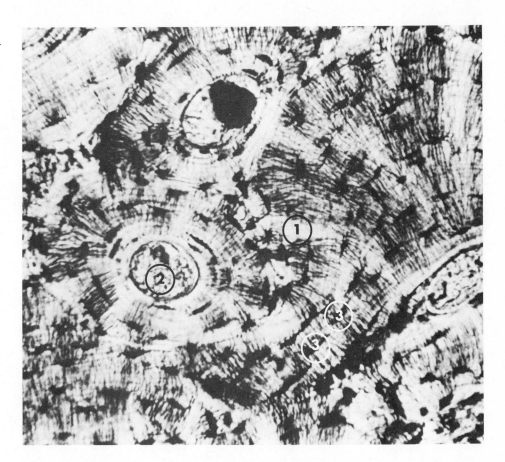

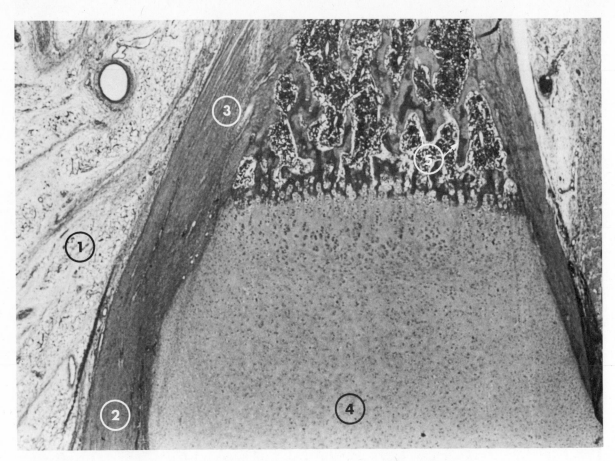

FIGURE 28 *Endochonral, or intracartilaginous, bone formation as seen in a metacarpal bone*

1. Subcutaneous Layer
2. Perichondrium
3. Periosteum
4. Proliferating Cartilage
5. Spicules of Bone

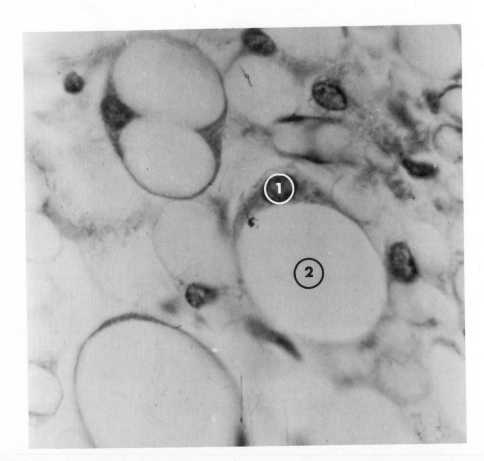

FIGURE 29 *Adipose tissue*

1. Signet-ring-shaped Cell
2. Fat

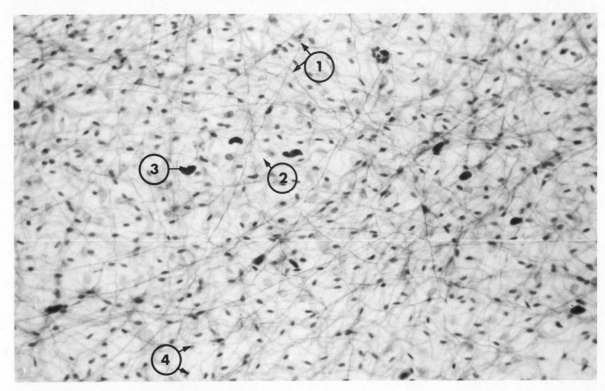

FIGURE 30 *Loose, irregular connective tissue*

1. Elastic Fibers
2. Collagenous Fiber
3. Nucleus of Macro-
phage Cell
4. Nuclei of Fibroblast
Cells

EXERCISE 2

Microscopic Identification of the Skin

The skin is comprised of an outer **epidermis** and a deeper layer, the **dermis**, or **corium**. The epidermis is composed of stratified squamous epithelium and consists of five layers:

1. **Stratum corneum:** This is the outermost layer of the epidermis, consisting of flattened, dead, keratinized (converted to protein) cells that are continuously shed. The stratum corneum helps serve as a barrier to light, heat, bacteria, and most chemicals.
2. **Stratum lucidum:** This layer is directly beneath the stratum corneum and is one or two cell layers thick. It may be difficult to see on your microscope slide.
3. **Stratum granulosum:** This is a thin layer lying beneath the stratum lucidum. It is thought to be the layer in which keratinization takes place. In this layer granules are numerous and tend to stain heavily.
4. **Stratum spinosum:** This layer may also be difficult to see. It is beneath the stratum granulosum and consists of "prickly" cells.
5. **Stratum germinativum:** This layer is the deepest layer of the epidermis. Cells in this layer actively undergo mitotic division and give rise to the four outer epidermal layers. **Melanin**, the principal pigment of the skin, is formed in this layer.

The dermis lies beneath the epidermis. It contains connective tissue fibers, blood vessels, nerves, sweat glands, sebaceous glands, and hair follicles.

Examine slides of Caucasian and Negroid skin, identifying the above structures (Figures 31–35).

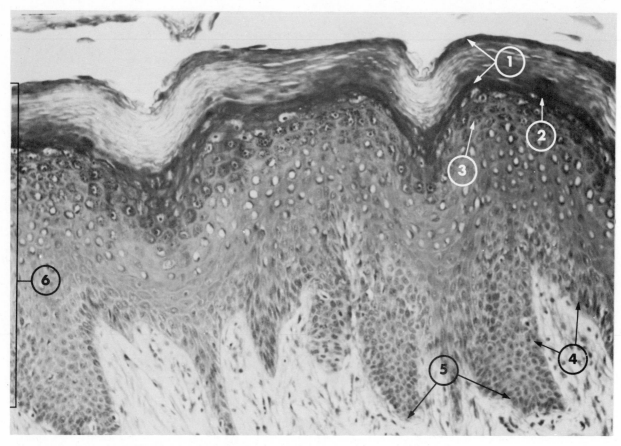

FIGURE 31 *Thick skin, palm, showing epidermal layers*

1. Stratum Corneum
2. Stratum Lucidum
3. _____
4. Stratum Germinativum
5. Dermal Papillae
6. Epidermis

FIGURE 32 *Skin, human, showing excretory duct of a sweat gland*

1. Stratum Corneum
2. Excretory Duct of Sweat Gland

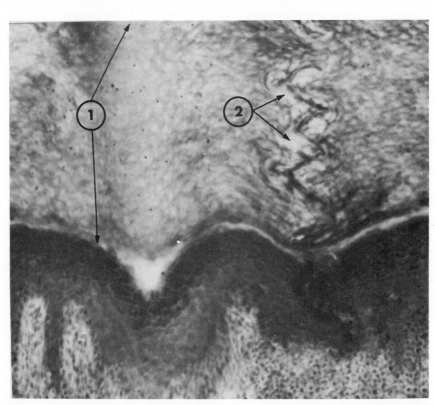

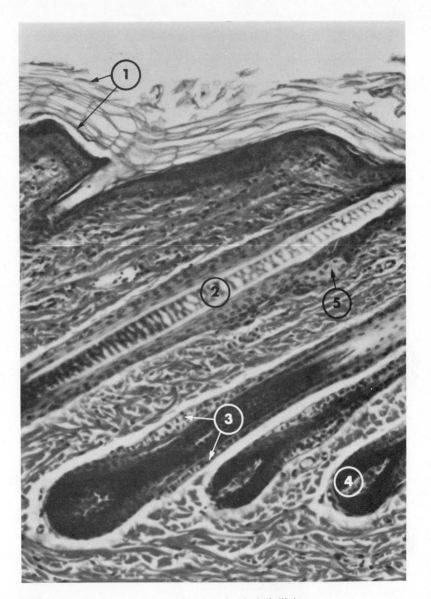

FIGURE 33 *Skin, human scalp, showing hair follicles*

1. _____

2. _____

3. Hair Follicle

4. _____

5. _____

FIGURE 34 *Skin showing sebaceous and sweat glands*

1. Sebaceous Glands
2. Ducts of Sweat Gland

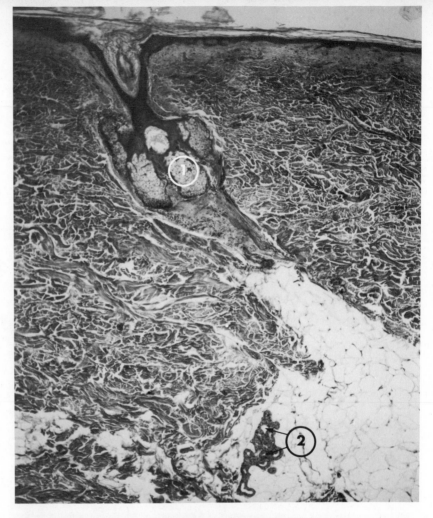

FIGURE 35 *Detail of sweat gland ducts*

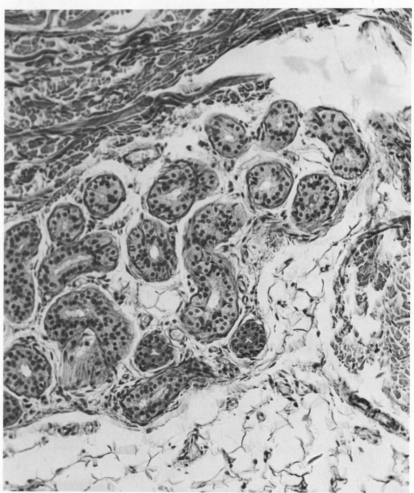

Tissues

DISCUSSION

Matching

Column A

_____ 1. This tissue type functions in impulse conduction.

_____ 2. This tissue type lines the stomach and small intestine.

_____ 3. This tissue type is found in the mouth.

_____ 4. This tissue type functions in contraction.

_____ 5. Blood is classified as this type of tissue.

_____ 6. This tissue functions in protection, absorption, and secretion.

_____ 7. Cartilage is of this tissue type.

_____ 8. This tissue includes striated and smooth types.

_____ 9. A glomerulus is comprised of this particular tissue type.

_____ 10. Adipose tissue is of this type.

Column B

a. epithelial

b. connective

c. muscular

d. nervous

Multiple Choice

_____ 11. Which of the following is *not* characteristic of epithelial tissue?
a. has no blood vessels
b. covers free surfaces of the body
c. typically involved in secretion, excretion, or absorption
d. contains an extensive nonliving matrix

_____ 12. Which is *not* a function of the skin?
a. protection against mechanical injury
b. protection against foreign invaders
c. regulation of body heat
d. all of the above are functions

_____ 13. The layer of skin that lacks blood vessels is:
a. the subcutaneous
b. the dermis
c. the integument
d. the epidermis

_____ 14. The integument of mammals and man does not contain . . . . glands.
   a. oil   c. mammary
   b. mucous   d. sweat

_____ 15. The color of human skin depends upon:
   a. the yellowish tinge of   c. the kind and amount of pigment
      epidermal cells
   b. the number of underlying   d. all of the above
      blood vessels

_____ 16. Through which epithelial type would diffusion most easily take place?
   a. simple squamous   c. cuboidal
   b. transitional   d. pseudostratified columnar

_____ 17. In which of these organs would striated involuntary muscle be found?
   a. small intestine   c. liver
   b. lung   d. heart

_____ 18. Which layer of epidermis would be gradually shed through bathing?
   a. stratum granulosum   c. stratum germinativum
   b. stratum corneum   d. stratum lucidum

_____ 19. In which tissue type are concentric rings of cells found?
   a. osseous   c. adipose
   b. hyaline cartilage   d. epithelial

_____ 20. The dermis is primarily comprised of which tissue type?
   a. nervous   c. connective
   b. muscle   d. epithelial

_____ 21. The amount of melanin produced in the skin is determined by:
   a. number of melanocytes   c. diet
   b. activity of melanocytes   d. proximity of blood vessels to the
                                          skin

_____ 22. When a person loses weight, adipose cells die.
   a. true   b. false

_____ 23. What tissue types are found in the intestinal tract?
   a. cuboidal epithelium, smooth   c. simple columnar epithelium, smooth
      muscle   muscle
   b. stratified squamous epithelium, d. pseudostratified columnar epithe-
      striated muscle   lium, reticular tissue

_____ 24. The criterion for categorizing epithelia as simple, stratified, or pseudostratified is:
   a. cell size   c. cell shape
   b. secretions of cells   d. arrangement of cells

_____ 25. Which epidermal layer has the greatest blood supply?
   a. stratum germinativum   c. stratum granulosum
   b. stratum spinosum   d. stratum corneum

DISCUSSION QUESTIONS (Optional)

1. For review, draw the following tissue types in the spaces provided:

Stratified squamous epithelium	Simple columnar epithelium	Adipose	Dense fibrous connective tissue
Hyaline cartilage	Osseous	Skeletal muscle	Smooth muscle

2. a. In which epidermal layer are melanocytes found?

b. Which layer(s) of epidermis is(are) comprised of dead cells?

c. Which epidermal layer(s) is(are) keratinized?

3. Which specialized tissue type(s) would you find in each of the following locations?

   a. collecting tubules of the kidney . . . .

   b. lining the urinary tract . . . .

   c. trachea . . . .

   d. muscle layer of the small and large intestines . . . .

   e. mesenteries . . . .

   f. biceps muscle . . . .

   g. tendon . . . .

   h. in the nasal septum . . . .

   i. lining the intestinal tract . . . .

   j. fat . . . .

Skeletal System

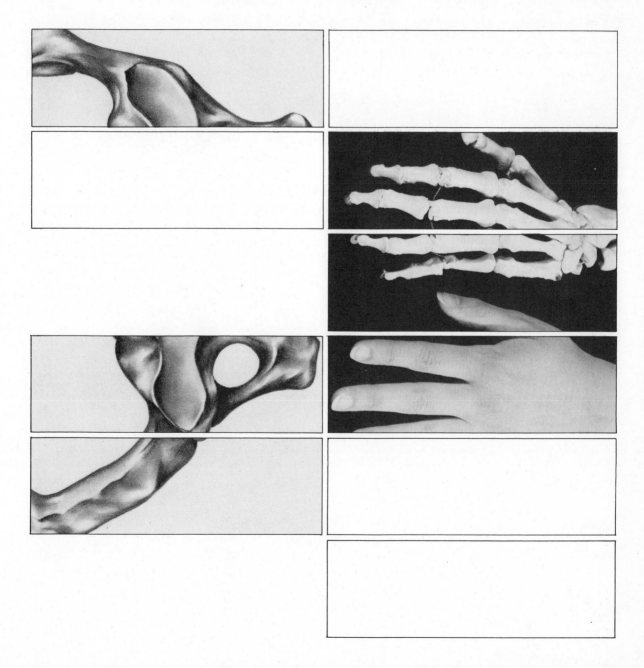

■ *Purpose*

The purpose of Unit V is to familiarize you with the various divisions of the human skeleton.

■ *Objectives*

To complete Unit V, you must be able to do the following:

1. Identify major components of a Haversian system microscopically.
2. Identify major anatomical structures of a long bone.
3. Identify bones of both axial and appendicular skeletons.
4. Identify major bone markings.
5. Identify locations of major sutures and fontanels.
6. Differentiate between diarthrodial and synarthrodial joint.

■ *Materials*

Articulated human skeleton
Slides of dry ground bone
   (osseous tissue)
Disarticulated human skeleton

Models: sagittal section of a long bone, adult and infant labeled skulls, Beauchene skull models (optional)

■ *Procedure*

EXERCISE 1

Microscopic Examination of Osseous Tissue

Since you studied osseous tissue previously (Unit IV), this should be a review. Examine a prepared slide of dry ground bone and identify the following: Haversian canal, lamella, lacuna, canaliculi, osteocytes, and interstitial material (*see* Figure 27 in Unit IV and **Color Plate 13**).

DRAW a Haversian system *and LABEL* the structures listed above to the right of your diagram:

EXERCISE 2

Examination of the Gross Anatomy of a Long Bone

You are to identify the gross anatomical features of a sagittal section of a long bone. Examine the available specimen and identify the following: articular cartilage(s), compact bone, cancellous bone, diaphysis, endosteum, epiphysis, medullary canal, and periosteum (*see* Figure 36).

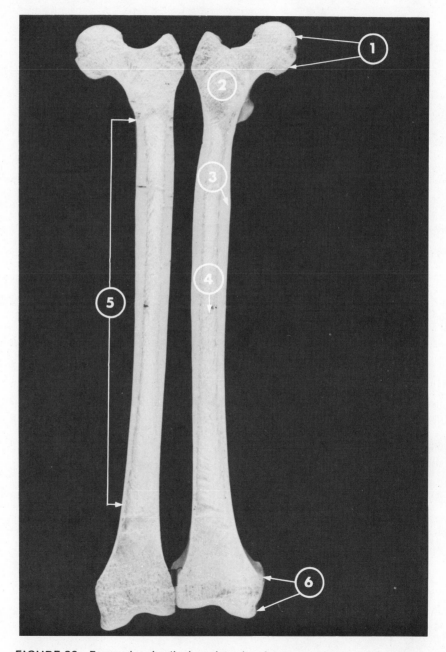

FIGURE 36   *Femur, longitudinal section, showing gross structure of the bone*

1. Articular Cartilage
2. Spongy (Cancellous) Bone
3. Compact Bone
4. Medullary Canal
5. Diaphysis
6. Epiphysis

FIGURE 37 *Anterior, posterior, and lateral views of the human skeleton showing the normal position of each bone*

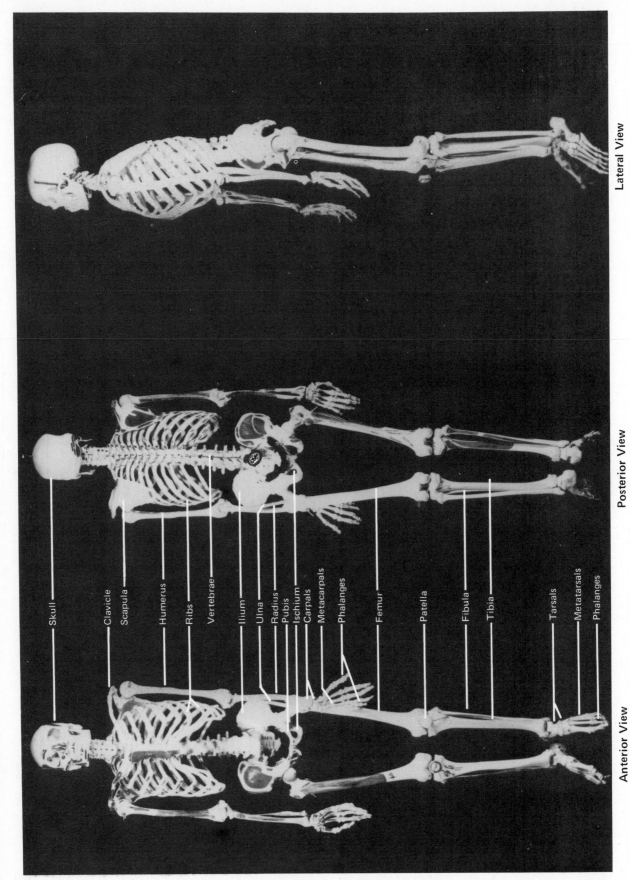

Skull
Clavicle
Scapula
Humerus
Ribs
Vertebrae
Ilium
Ulna
Radius
Pubis
Ischium
Carpals
Metacarpals
Phalanges
Femur
Patella
Fibula
Tibia
Tarsals
Metatarsals
Phalanges

Anterior View

Posterior View

Lateral View

EXERCISE 3

Identification of Bones of the Human Skeleton

You are to identify the bones of both the axial and appendicular skeleton. Identify the bones of the axial skeleton first. If you are working with a disarticulated (bones disjointed) skeleton, arrange the bones in the proper linear position on the laboratory table.

There are more than 200 bones in the human adult; some are paired, some are single. Other bones—such as the skull, sternum, and hip—are fusions of a number of bones. Be sure to identify the separate bones that comprise each fusion. (*See* Figure 37).

In infancy and early childhood, ossification has not been completed; therefore, bones are more cartilaginous. If you have a young skeleton, observe the delicate nature of the bones and obvious cartilage.

Identify all bones from the following table:

Skeletal Division	*Name of Bone(s)*	*Number of Bones*
A. AXIAL SKELETON		80 (total)
1. SKULL		28
a. **Cranium**		8
	Frontal	1
	Parietal	2
	Temporal	2
	Occipital	1
	Sphenoid	1
	Ethmoid	1
b. **Face**		14
	Mandible	1
	Nasal	2
	Lacrimal	2
	Vomer	1
	Inferior conchae (turbinates)	2
	Zygomatic (malar)	2
	Palatine	2
	Maxillary	2
c. **Ear Ossicles**		6
	Malleus	2
	Incus	2
	Stapes	2
2. HYOID		1
3. VERTEBRAE		26
	Cervical	7
	Thoracic	12
	Lumbar	5
	Sacrum	1 (fusion of 5)
	Coccyx	1 (fusion of 3-5

Skeletal Division	Name of Bone(s)	Number of Bones
4. STERNUM		1 (fusion of 3)
	Manubrium	1
	Gladiolus (body)	1
	Xiphoid process	1
5. RIBS		12 pairs (7 pairs true ribs, 5 pairs false ribs)

B. APPENDICULAR SKELETON		126 (total)
1. UPPER DIVISION		64 (total)
	Clavicle	2
	Scapula	2
	Humerus	2
	Ulna	2
	Radius	2
	Carpals (see Figure 59 in Exercise 4 of this unit)	16 (8 per hand)
	Scaphoid	1
	Lunate	1
	Triquetrum	1
	Pisiform	1
	Trapezium	1
	Trapezoid	1
	Capitate	1
	Hamate	1
	Metacarpals	10 (5 per hand)
	Phalanges	28 (14 per hand)
2. LOWER DIVISION		62 (total)
	Ossa Coxae	2 (each a fusion of 3)
	Ilium	1
	Ischium	1
	Pubis	1
	Femur	2
	Patella	2
	Tibia	2
	Fibula	2
	Tarsals (see Figure 66 in Exercise 4 of this unit)	14 (7 per foot)
	Talus	1
	Calcaneus	1
	Cuboid	1
	Navicular	1
	Cuneiforms	3
	Metatarsals	10 (5 per foot)
	Phalanges	28 (14 per foot)

EXERCISE 4

Identification of Bone Markings

Since the skeletal system serves as a basis for body topography, you should know more about the bones than simply their names and locations. A thorough knowledge of *bone markings*, specific identifiable projections or depressions on bones, is of great importance for it facilitates identification of other body structures, such as muscles, blood vessels, and nerves.

Listings of the major bone markings follow. Those classified as projections (also called **processes**) grow out from the bone, whereas those classified as depressions (also called **fossae**) are indentations in the bone. Locating specific bone markings will be easier if you first know the general names of the different types of markings. (*See* Figure 38.)

FIGURE 38 *Femur showing general types of bone markings*

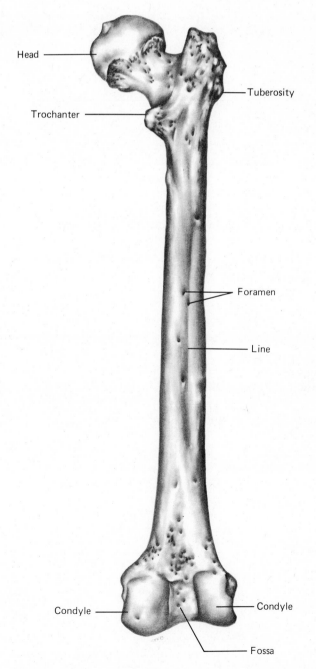

Head

Tuberosity

Trochanter

Foramen

Line

Condyle

Condyle

Fossa

BONE MARKINGS—GENERAL TYPES

Projections

1. **Trochanter:** a large irregularly shaped projection to which muscles attach.
2. **Tuberosity:** a large rounded projection to which muscles attach.
3. **Tubercle:** a small rounded projection to which muscles attach.
4. **Condyle:** a rounded convex projection.
5. **Head:** a rounded extension projecting from a tapering neck that forms a joint.
6. **Ramus:** an armlike branch extending from body of bone that forms a joint.
7. **Spine:** a sharp projection that serves as a point for muscle attachment.
8. **Crest:** an elevation or ridge serving as a point for muscle attachment.
9. **Line:** a lesser elevation or ridge that serves as a point of muscle attachment.
10. **Process:** term used to designate a general outgrowth of varying shape and size.

Depressions

1. **Fossa:** a hollow indentation, often rounded, often found in joint formations.
2. **Foramen:** a hollow opening that serves as a passageway for nerves and blood vessels.
3. **Sinus:** irregularly shaped space often filled with air and lined with mucosal tissue.
4. **Fissure:** a narrow slitlike depression often serving as a passageway for nerves.
5. **Meatus:** a canallike opening that sometimes serves as a passageway for nerves.

Now read through the table "Identification of Bone Markings" over the next several pages. Following the table sequentially, identify the bone markings in the accompanying figures. Use a set of color pencils (or felt-tip markers) and color each of the bone markings as you go.

The process of coloring will help you see the dimensions of the skeletal system more clearly. For example, locate the frontal bone of the skull and color it in. Selecting other pencils, locate and color in each bone marking on the frontal bone. Using a variety of colors, continue to identify and color in the remaining bones and bone markings. When you have finished your identification of the skull, review your work. Then go on to identify and color in the next segment of the skeleton, the vertebrae.

Identification of Bone Markings

A. AXIAL SKELETON

Bone	Bone Marking	Description
SKULL		
CRANIUM		
Frontal		A flat convex bone in the adult, formed by the union of two flat bones joined at frontal suture (suture partly visible in adult)
(Anterior view—Figure 39)	*Frontal eminences, or protuberances*	Protrusions above each optic orbit
	Supraorbital margins	Flat arched regions immediately inferior to eyebrows
	Supraorbital notches, or foramina	Irregularly shaped small openings in supraorbital margins toward nasal bones, serving as an opening for nerves and blood vessels

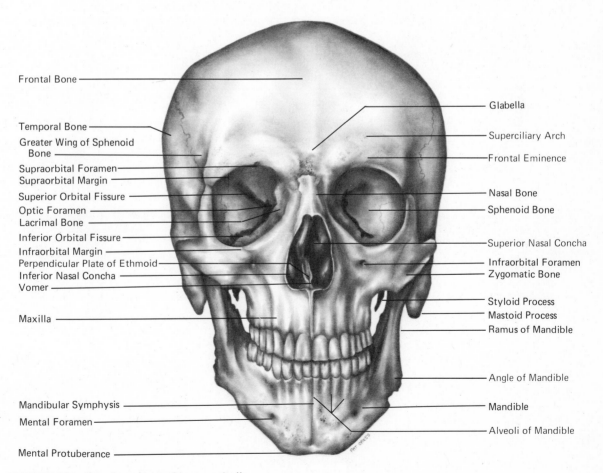

FIGURE 39 *Anterior aspect of human skull*

Bone	Bone Marking	Description
	Superciliary arches	Anterior ridges superior to optic orbits that form base for eyebrows; ridges formed by frontal sinuses that lie within them
	Glabella	Smooth area formed by the union of the superciliary arches above nasal bones
Parietal		Large irregularly shaped bones lateral and posterior to frontal bones; parietals merge at superior medial portion of the cranium to form sagittal suture
	Parietal foramen	Small rounded opening toward the posterior of the bone; immediately lateral to the sagittal suture; may be closed in some skeletons
	Parietal tuberosity	Slightly rounded protrusion on lateral surface of bone
Temporal		Bones form lateral; inferior sides of cranium; houses structures of middle and inner ear
(Lateral view—Figure 42)	*Mastoid process*	Rough projection immediately posterior to ear or external auditory meatus
	Squamous portion	Flat, thin superior portion of bone; transparent when held to light
	Zygomatic process (arch)	Thin, narrow bridgelike extension that articulates anteriorly with zygomatic bone
	External auditory meatus	Canal into ear running from exterior to interior of temporal bone
	Mandibular fossa	Rounded depression inferior to and partially formed by zygomatic process; immediately anterior to external auditory meatus; forms socket for mandibular condyle
(Inferior view—Figure 40) Skull viewed from below	*Styloid process*	Long needlelike projection anterior to mastoid process; serves as attachment for muscles and ligaments

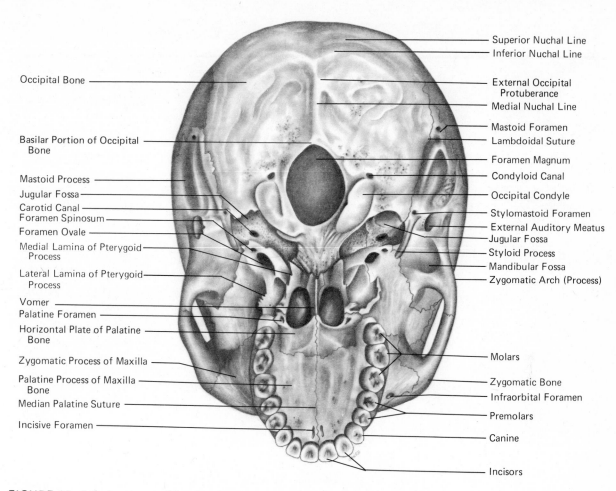

Occipital Bone

Basilar Portion of Occipital Bone

Mastoid Process
Jugular Fossa
Carotid Canal
Foramen Spinosum
Foramen Ovale
Medial Lamina of Pterygoid Process
Lateral Lamina of Pterygoid Process
Vomer
Palatine Foramen
Horizontal Plate of Palatine Bone
Zygomatic Process of Maxilla
Palatine Process of Maxilla Bone
Median Palatine Suture
Incisive Foramen

Superior Nuchal Line
Inferior Nuchal Line
External Occipital Protuberance
Medial Nuchal Line
Mastoid Foramen
Lambdoidal Suture
Foramen Magnum
Condyloid Canal
Occipital Condyle
Stylomastoid Foramen
External Auditory Meatus
Jugular Fossa
Styloid Process
Mandibular Fossa
Zygomatic Arch (Process)

Molars

Zygomatic Bone

Infraorbital Foramen

Premolars

Canine

Incisors

FIGURE 40 *Inferior aspect of human skull*

Bone	Bone Marking	Description
	Stylomastoid foramen	Small opening between styloid process and mastoid process; serves as a passageway for nerve
	Jugular fossa	Irregularly shaped depression lateral and anterior to occipital condyles; opening for internal jugular vein
	Jugular foramen	Fairly large oval-shaped opening at posterior end of jugular fossa; lateral to occipital condyle; serves as passageway for lateral sinus and IX, X, and XI cranial nerves
	Carotid foramen, or canal	Round openings medial and anterior to jugular foramen; serves as passageway for internal carotid artery

Bone	Bone Marking	Description
(Internal view—Figure 41) View of floor of cranial cavity; superior section of cranium removed	Petrous portion	Rocklike protrusion in center of cranial floor; flares from medial to lateral region; middle and inner ear structures are contained within
	Internal auditory meatus	Opening in medial side of petrous portion; it is the internal opening of canal running from external auditory meatus through which the acoustic nerve passes
Occipital		Large flaring bone forming inferior posterior section of cranium; articulates with parietal bones to form lambdoidal suture
(Inferior view—Figure 40)	External occipital protuberance	Protrusion, or eminence, that extends most posteriorly in medial portion; bumplike

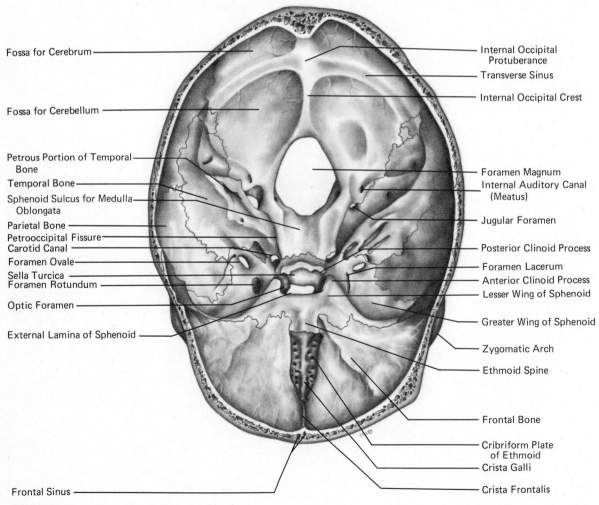

Fossa for Cerebrum

Fossa for Cerebellum

Petrous Portion of Temporal Bone
Temporal Bone
Sphenoid Sulcus for Medulla Oblongata
Parietal Bone
Petrooccipital Fissure
Carotid Canal
Foramen Ovale
Sella Turcica
Foramen Rotundum
Optic Foramen

External Lamina of Sphenoid

Frontal Sinus

Internal Occipital Protuberance
Transverse Sinus
Internal Occipital Crest

Foramen Magnum
Internal Auditory Canal (Meatus)
Jugular Foramen

Posterior Clinoid Process
Foramen Lacerum
Anterior Clinoid Process
Lesser Wing of Sphenoid

Greater Wing of Sphenoid

Zygomatic Arch

Ethmoid Spine

Frontal Bone

Cribriform Plate of Ethmoid
Crista Galli

Crista Frontalis

FIGURE 41 *Interior aspect of human skull*

Bone	Bone Marking	Description
	Superior nuchal line	Elevation projecting laterally from external occipital protuberance
	Inferior nuchal line	A lesser elevation inferior to superior nuchal line
	Condyles	Oval-shaped protrusions with flat surfaces lateral to foramen magnum; articulate with atlas
	Foramen magnum	Large round opening that serves as passageway for spinal cord in its connection with brain; medial to condyles
(Internal view—Figure 41) View of floor of cranial cavity	Internal occipital protuberance	Linear projection in internal medial region of bone
Sphenoid		Bat-shaped bone forming part of the floor of the cranial cavity; seen laterally anterior to the temporal bone
	Body	Medial cubelike region; hollow centrally
(Internal view—Figure 41) View of floor of cranial cavity	Greater wings	Laterally flaring portion forming floor of cranial cavity and outer wall of optic orbit
	Lesser wings	Thin flaring portions superior to greater wings and body; form superior posterior section of optic orbit
	Sella turcica	Saddle shaped indentation medial to wings; houses pituitary gland
	Optic foramen	Rounded opening inferior to lesser wings at medial point; can be seen anteriorly through optic orbit; transmits II cranial nerve
	Superior orbital fissure	Irregularly shaped furrow, immediately inferior to and covered by lesser wings; can be seen anteriorly through optic orbit; transmits III, IV, and V cranial nerves
	Foramen rotundum	Small rounded hole in greater wings; lateral to sella turcica; transmits maxillary division of V (trigeminal) cranial nerve

Bone	Bone Marking	Description
	Foramen ovale	Oval-shaped hole posterior to foramen rotundum; transmits mandibular division of trigeminal nerve
	Foramen lacerum	Oval-shaped opening in greater wing toward body; lateral to sella turcica at apex of petrous portion; transmits internal carotid arteries and branch of ascending pharyngeal artery
	Foramen spinosum	Small irregularly shaped opening posterior and lateral to foramen ovale; transmits mandibular nerves and meningeal arteries
	Pterygoid process	Wing-shaped downward projection posterior to maxilla and molar teeth; consists of medial and lateral lamina
Ethmoid		Irregularly shaped bone that forms anterior superior section of cranial floor; medial posterior walls of optic orbits; lateral walls, posterior septum, and roof of nasal cavity
	Crista galli	Flaglike structure pointing anteriorly; serves as a point of attachment for meninges
	Horizontal plate (Cribriform plate)	Porous flat section of bone lateral to crista galli; transmits olfactory nerves
	Perpendicular plate	Linear projection extending inferiorly from crista galli
	Lateral masses	Lateral portions of ethmoid; inner section forms walls of nasal cavities and conchae; contains many air cavities
	Ethmoid sinus	Air cavity within lateral masses
	Superior and middle conchae (Turbinates)	Extensions from lateral sides of nasal cavity pointing medially toward septum; contain spongy air-filled cavities

Bone	Bone Marking	Description
FACE BONES		
Mandible		Lower jaw; articulates with temporal bone; forms the only diarthrotic joint in cranium
(Lateral view—Figure 42)	*Body*	Rounded anterior section; convex posteriorly; forms chin
	Mental tubercles	Small rounded protrusions on inferior rim of body; on external surface
	Mental foramen	Small rounded opening on external surface below bicuspids; transmits nerves and blood vessels
	Angle	Junction of body and ramus on inferior side
	Ramus	Superior armlike projections on either side of body; posterior part of bone

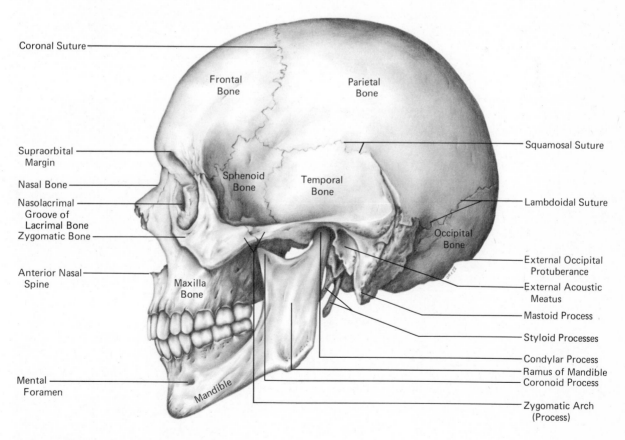

FIGURE 42 *Human skull, lateral view*

Bone	Bone Marking	Description
	Mandibular foramen	Opening on internal surface of ramus; passageway for nerves and blood vessels to teeth of mandible
	Mandibular condyle (Condyloid process)	Posterior rounded head that fits into mandibular fossa of temporal bone
	Mandibular notch	Depression between condyloid process and coronoid process
	Neck	Constricted projection at base of condyloid process
	Coronoid process	Thin pointed anterior projection that serves as point for muscle attachment
	Alveolar process	Sockets; process of bone into which teeth are anchored centrally
Nasal (Lateral view—Figure 42)		Small irregularly shaped bones inferior to glabella; form upper division of nasal bridge
Lacrimal (Lateral view—Figure 42)		Thin delicate bones lateral to nasal bones; form medial wall of optic orbit; form lateral wall of nasal cavity
	Groove	Slight indentation or furrow running in a superior to inferior direction; serves as a passageway for tears
Vomer		Single irregularly shaped bone found medially and posteriorly in the nasal cavity
Inferior conchae (turbinates)		Thin delicate bones flaring from lateral walls of nasal cavity; most posterior projections
Zygomatic (malar) (Figure 42)		Hard heavy bone; cheekbone
	Zygomaticofacial foramen	Small round opening anterior to zygomatic arch; inferior and slightly lateral to optic orbit
Palatine (Inferior view—Figure 40)		L-shaped bone that forms the posterior part of the hard palate; also forms lateral and inferior walls of posterior division of nasal cavity

Bone	Bone Marking	Description
	Horizontal plate	Horizontal junction with maxilla; forms most posterior section of hard palate
Maxilla		Upper jawbone, main bone of face; all other facial bones articulate with maxilla
	Antrum of Highmore (Maxillary sinus)	Largest of all sinuses within central cavity of maxilla; mucus lined and air filled
	Incisive foramen	Triangular opening between central incisors; located at midline of maxilla
	Palatine process	Forms most of hard palate; curves downward to form alveoli
	Alveolar process	Sockets containing teeth; archlike
	Anterior nasal spine	Projection inferior and anterior to nasal septum
	Infraorbital	
EAR OSSICLES		Small bones contained within the middle ear, which conduct vibrations from the eardrum to the inner ear
	Malleus	Largest of three ear ossicles, attached to eardrum, articulates with incus; also known as hammer
	Incus	The middle of three ossicles in middle ear; also known as anvil
	Stapes	Most medial of three ossicles, vibrates against oval window of inner ear; also known as stirrup
HYOID		U-shaped bone anteriorly located between larynx and mandible; serves as point of muscle attachment for some muscles of mouth and tongue
	Transverse body	Rounded anterior portion of bone
	Greater cornu	Posterior projections extending from transverse body
	Lesser cornu	Anterior projections on either side of the anterior portion of the transverse body

Bone	Bone Marking	Description
VERTEBRAE		Twenty-six blocklike bones comprising the vertebral column that protect the spinal cord and allow for anterior, posterior, and lateral movement
Typical Structures of All Vertebrae (See Figure 49)		
	Body	Rounded central bony portion facing anteriorly in column; points of attachment for intervertebral discs
	Vertebral arch	Junction of all posterior extensions from body
	Vertebral foramen	Central opening through which spinal cord passes
	Pedicle	Thick, bony lateral extension posterior to body; forms lateral portion of vertebral foramen
	Lamina	Flat, thinner bony extension; extends pedicle toward midline posteriorly; forms posterior portion of vertebral foramen
	Intervertebral foramen	Opening between vertebrae for emergence of spinal nerves
	Processes	Outgrowths of bone
	Transverse processes	Lateral extensions from junction of pedicle and lamina
	Superior articulating surfaces	Paired projections on superior surface lateral to vertebral foramen; slightly indented centrally; form articulation with immediately superior vertebra (or, in case of atlas, articulated with occipital condyles)
	Inferior articulating surfaces	Paired projections on inferior surface lateral to vertebral foramen; slightly indented centrally; form articulation with immediately inferior vertebra
	Spinous process	Single posterior and inferiorly extending process at midline of vertebra

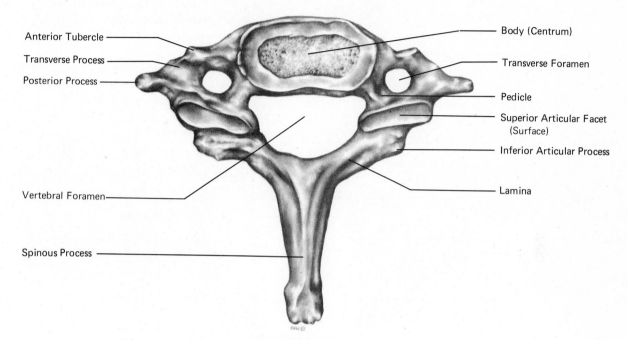

Anterior Tubercle

Transverse Process

Posterior Process

Vertebral Foramen

Spinous Process

Body (Centrum)

Transverse Foramen

Pedicle

Superior Articular Facet (Surface)

Inferior Articular Process

Lamina

FIGURE 43 *Superior aspect of typical cervical vertebra*

Bone	Bone Marking	Description
Specific Vertebral Structures		
Cervical *(Figure 43)*		All of the typical structures; smallest, lightest vertebrae; slightly larger, triangular vertebral foramen; spinous processes bifurcated and short; transverse processes contain foramen (transverse foramen)
Atlas (First cervical vertebra) *(Figures 44 and 45)*		No body; lateral processes contain large concave articulating surfaces

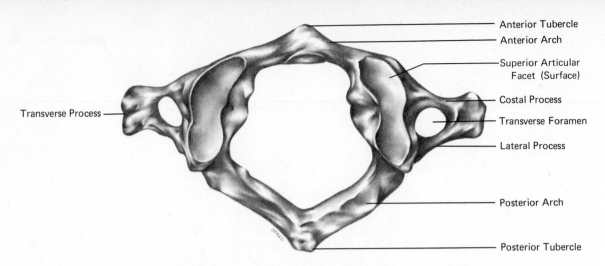

Anterior Tubercle

Anterior Arch

Superior Articular Facet (Surface)

Costal Process

Transverse Foramen

Lateral Process

Posterior Arch

Posterior Tubercle

Transverse Process

FIGURE 44 *Superior view of the atlas*

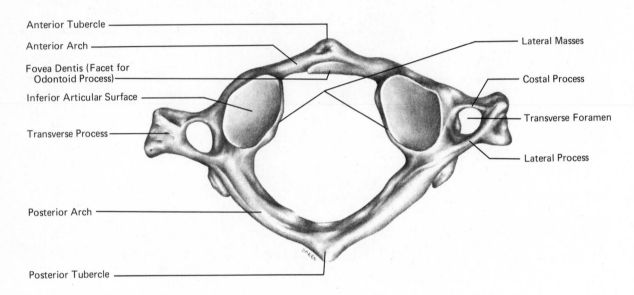

Anterior Tubercle

Anterior Arch

Fovea Dentis (Facet for Odontoid Process)

Inferior Articular Surface

Transverse Process

Posterior Arch

Posterior Tubercle

Lateral Masses

Costal Process

Transverse Foramen

Lateral Process

FIGURE 45 *Inferior aspect of the atlas*

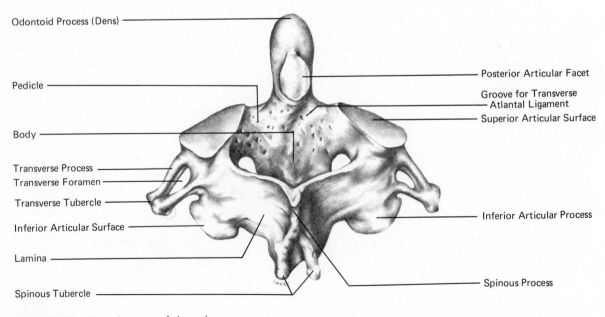

Odontoid Process (Dens)

Pedicle

Body

Transverse Process

Transverse Foramen

Transverse Tubercle

Inferior Articular Surface

Lamina

Spinous Tubercle

Posterior Articular Facet

Groove for Transverse Atlantal Ligament

Superior Articular Surface

Inferior Articular Process

Spinous Process

FIGURE 46 *Dorsal aspect of the axis*

Bone	Bone Marking	Description
Axis (Second cervical vertebra) (Figure 46)		Acts as a pivot for rotation of atlas and cranium; *odontoid process* is a superior vertical projection serving as pivot point and body
Thoracic (Figures 47 and 48)		All of the typical structures; larger body than cervical; oval vertebral foramen; sharp inferiorly projecting spinous process; contains lateral articulating surfaces, superior and inferior for rib articulation

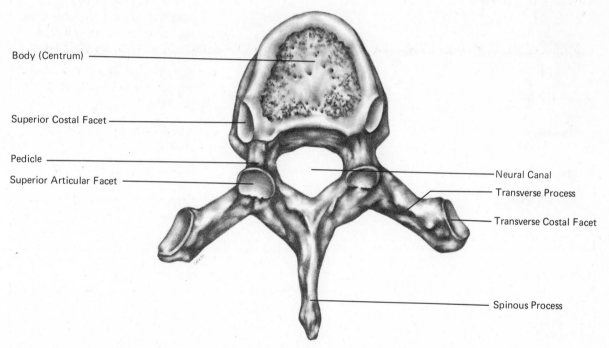

Body (Centrum)

Superior Costal Facet

Pedicle

Superior Articular Facet

Neural Canal

Transverse Process

Transverse Costal Facet

Spinous Process

FIGURE 47   *Superior aspect of typical thoracic vertebra*

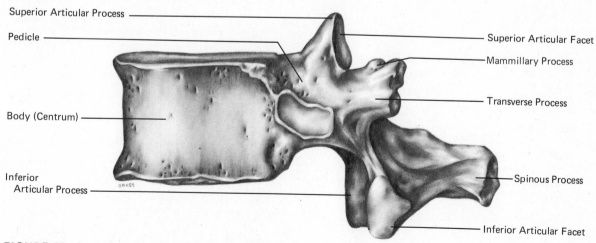

Superior Articular Process

Pedicle

Body (Centrum)

Inferior Articular Process

Superior Articular Facet

Mammillary Process

Transverse Process

Spinous Process

Inferior Articular Facet

FIGURE 48   *Lateral aspect of twelfth thoracic vertebra*

Bone	Bone Marking	Description
Lumbar *(Figure 49)*		All of the typical structures; large; thick blocklike body with superior articulating surfaces projecting inward and inferior articulating surfaces projecting outward; short, thick spinous process projecting posteriorly
Sacrum *(Figure 50)*		A fusion in the adult of the five sacral vertebrae that with the coccyx form the dorsal portion of the pelvic girdle
Coccyx *(Figure 50)*		Fusion of three to five small irregularly shaped vertebrae forming taillike projection inferiorly
STERNUM		Breastbone; fusion of three bones; T-shaped bone articulating with true ribs
Manubrium		Superior horizontal portion of bone
Gladiolus		Central vertical portion of bone
Xiphoid process		Cartilaginous inferior projection; pointed

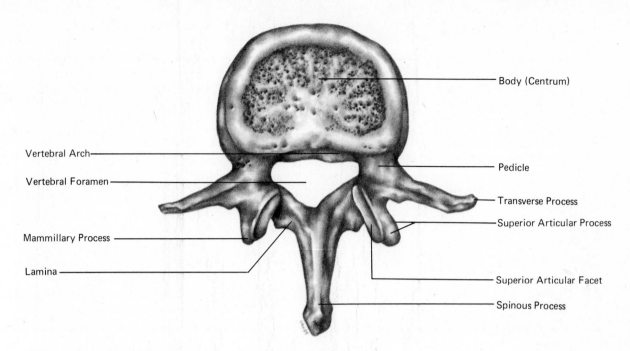

Body (Centrum)

Vertebral Arch

Vertebral Foramen

Pedicle

Transverse Process

Superior Articular Process

Mammillary Process

Lamina

Superior Articular Facet

Spinous Process

FIGURE 49 *Superior aspect of typical lumbar vertebra*

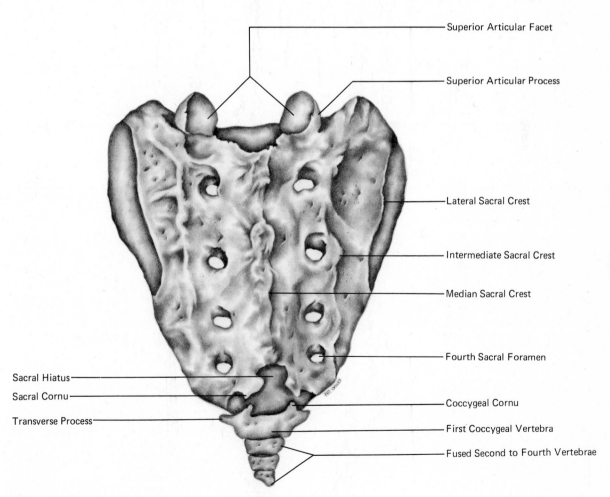

Superior Articular Facet

Superior Articular Process

Lateral Sacral Crest

Intermediate Sacral Crest

Median Sacral Crest

Fourth Sacral Foramen

Sacral Hiatus

Sacral Cornu

Transverse Process

Coccygeal Cornu

First Coccygeal Vertebra

Fused Second to Fourth Vertebrae

FIGURE 50 *Sacrum, posterior aspect*

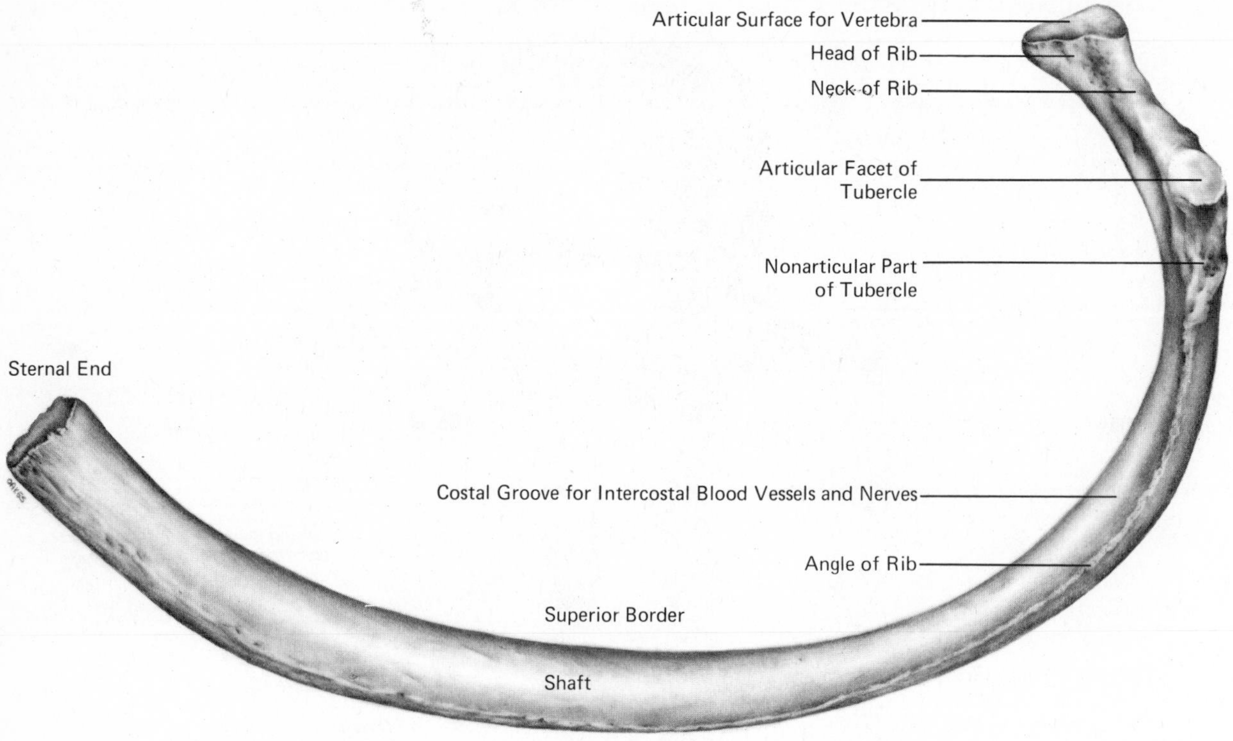

Articular Surface for Vertebra
Head of Rib
Neck of Rib

Articular Facet of
Tubercle

Nonarticular Part
of Tubercle

Sternal End

Costal Groove for Intercostal Blood Vessels and Nerves

Angle of Rib

Superior Border

Shaft

Inferior Border

FIGURE 51 *Sixth rib, inferior view*

Bone	Bone Marking	Description
RIBS (seven pairs true, five pairs false) *(Figure 51)*		Form walls of thoracic cavity; true ribs articulate with both vertebrae and sternum; false ribs articulate with vertebrae and have cartilaginous connections to sternum, except for last two pairs of "false" ribs—floating ribs that have no anterior connection to sternum
	Head	Posterior rounded projection, articulating with a thoracic vertebra
	Neck	Narrow portion below head and connecting head with shaft
	Shaft (body)	Long, thin curved portion
	Tubercle	Small rounded protrusion below neck on shaft
	Costal cartilage	Cartilaginous connection of ribs to sternum

B. APPENDICULAR SKELETON

Bone	Bone Marking	Description
UPPER DIVISION		
Clavicle *(Figure 52)*		Collarbone; S-shaped body articulating posteriorly with scapula and anteriorly with sternum
	Sternal end	Rounded end that articulates with sternum
	Acromial end	Flat, thinner end that articulates with acromion process of scapula

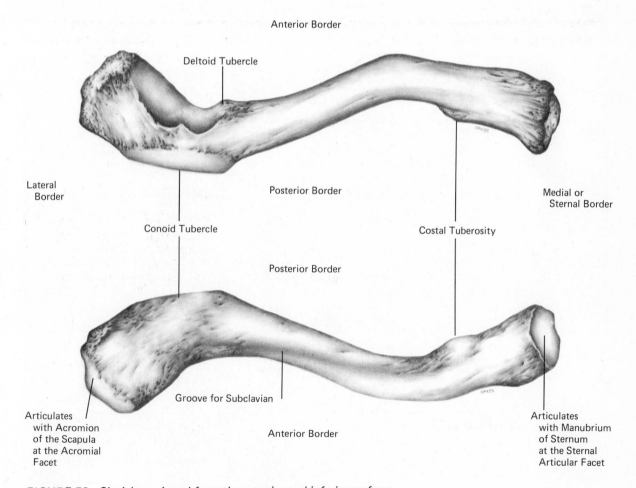

Anterior Border

Deltoid Tubercle

Lateral
Border

Posterior Border

Medial or
Sternal Border

Conoid Tubercle

Costal Tuberosity

Posterior Border

Groove for Subclavian

Articulates
with Acromion
of the Scapula
at the Acromial
Facet

Anterior Border

Articulates
with Manubrium
of Sternum
at the Sternal
Articular Facet

FIGURE 52 *Clavicle as viewed from the superior and inferior surfaces*

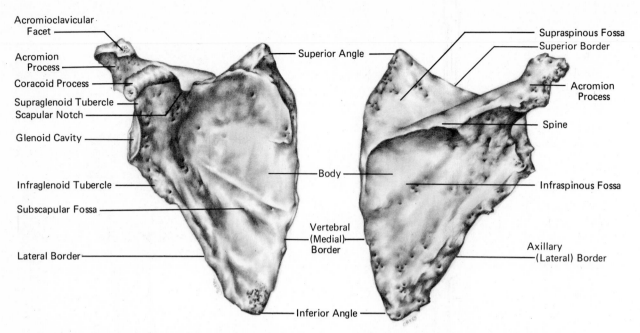

Acromioclavicular Facet

Acromion Process

Coracoid Process

Supraglenoid Tubercle

Scapular Notch

Glenoid Cavity

Infraglenoid Tubercle

Subscapular Fossa

Lateral Border

Superior Angle

Body

Vertebral (Medial) Border

Inferior Angle

Supraspinous Fossa

Superior Border

Acromion Process

Spine

Infraspinous Fossa

Axillary (Lateral) Border

FIGURE 53 *Anterior view of right scapula*

FIGURE 54 *Posterior view of right scapula*

Bone	Bone Marking	Description
Scapula *(Figures 53 and 54)*		Two wing-shaped bones on either side of body midline posteriorly; shoulder bones
	Body	Flat, thin central portion
	Superior border	Upper margin of body
	Vertebral border	Medial margin continuous with vertebral column
	Axillary border	Lateral margin
	Spine	Posterior ridgelike projection extending from vertebrae toward axillary border
	Acromial process, or acromion	Articulation with clavicle; rounded end of spine
	Coracoid process	Lateral rounded projection on superior border
	Glenoid cavity	Articulation point for humerus; rounded fossa on superior end of axillary border
Humerus *(Figures 55 and 56)*		Upper arm bone articulating with scapula proximally and with ulna and radius distally

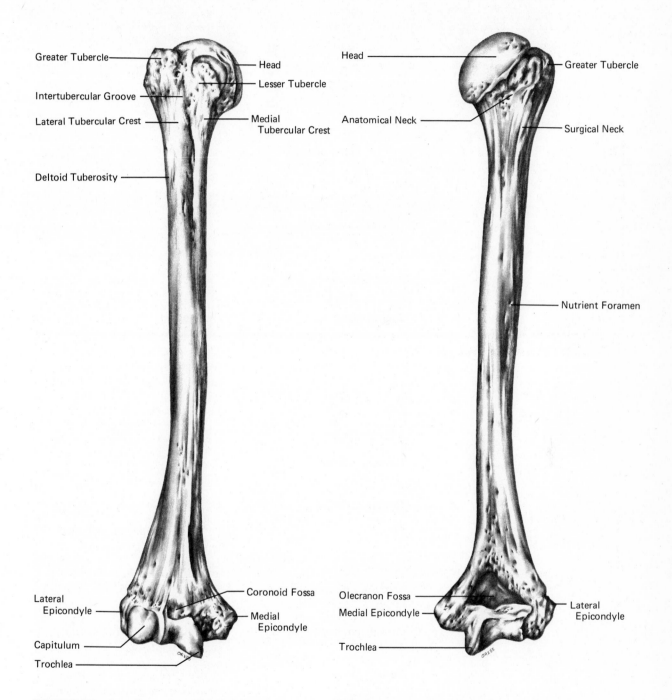

Greater Tubercle

Head

Intertubercular Groove

Lesser Tubercle

Lateral Tubercular Crest

Medial Tubercular Crest

Deltoid Tuberosity

Lateral Epicondyle

Coronoid Fossa

Capitulum

Medial Epicondyle

Trochlea

Head

Greater Tubercle

Anatomical Neck

Surgical Neck

Nutrient Foramen

Olecranon Fossa

Medial Epicondyle

Lateral Epicondyle

Trochlea

FIGURE 55 *Anterior aspect of right humerus*

FIGURE 56 *Posterior aspect of right humerus*

Bone	Bone Marking	Description
	Head	Rounded posteriomedial projection on proximal end
	Anatomical neck	Oblique constriction on anterior surface below head
	Greater tubercle	Larger anteriolateral projection from head
	Lesser tubercle	Smaller anterior projection below anatomical neck
	Intertubercular groove	Indentation between greater and lesser tubercles
	Surgical neck	Posterior constriction below tubercles; often fractured
	Deltoid tuberosity	Rough lateral protrusion on mid-diaphysis
	Medial epicondyle	Rough rounded projection on medial surface at distal end
	Lateral epicondyle	Rough smaller protrusion on lateral surface at distal end
	Capitulum (radial head)	Lateral rounded knob immediately anterior to lateral epicondyle that articulates with radius
	Trochlea	Pulley-shaped projection medial and inferior to medial epicondyle; articulates with ulna
	Olecranon fossa	Deep triangular depression on posterior surface between epicondyles; articulates with olecranon process during extension of forearm
	Coronoid fossa	Depression on anterior surface immediately superior to trochlea; articulates with coronoid process of ulna during flexion
Ulna (Figures 57 and 58)		Medial bone of forearm; longer than radius
	Olecranon process	Large round projection at proximal end; center concave (semilunar notch) resembles "open mouth"; forms elbow

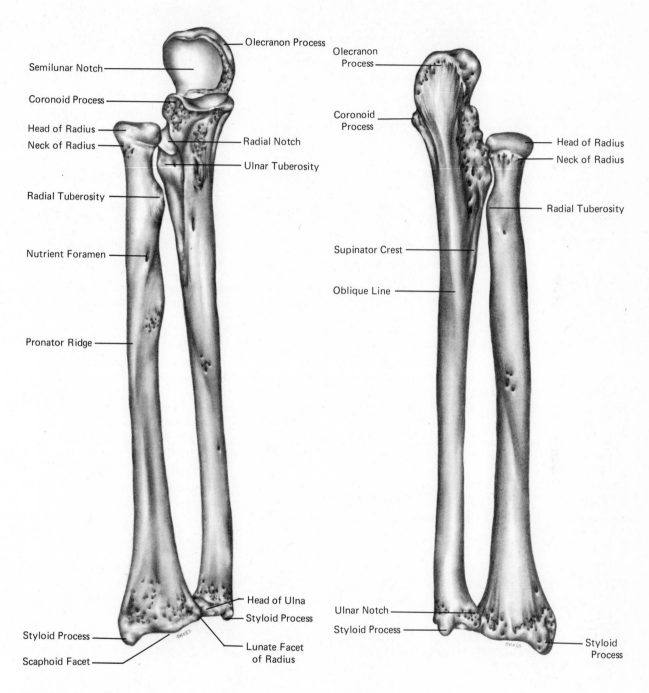

Semilunar Notch

Coronoid Process

Head of Radius

Neck of Radius

Radial Tuberosity

Nutrient Foramen

Pronator Ridge

Olecranon Process

Radial Notch

Ulnar Tuberosity

Styloid Process

Scaphoid Facet

Head of Ulna

Styloid Process

Lunate Facet
of Radius

Olecranon
Process

Coronoid
Process

Supinator Crest

Oblique Line

Ulnar Notch

Styloid Process

Head of Radius

Neck of Radius

Radial Tuberosity

Styloid
Process

FIGURE 57 *Anterior view of right radius and ulna*

FIGURE 58 *Posterior view of right radius and ulna*

Bone	Bone Marking	Description
	Coronoid process	Protrusion on anterior surface below olecranon process; lower half of "open mouth"; articulates with humerus
	Semilunar notch	Concave notch that articulates with trochlea of humerus
	Radial notch	Concave indentation that articulates with lateral aspect of radial head
	Styloid process	Small V-shaped projection at distal end of bone
Radius *(Figures 57 and 58)*		Shorter lateral bone of forearm
	Radial head	Rounded knoblike projection at proximal end of bone
	Radial tuberosity	A raised rough protrusion on the medial aspect near the proximal end
	Styloid process	A needlelike projection on the lateral aspect of the distal end
HAND *(Figures 59 and 60)*		
Carpals		Sixteen irregularly shaped bones; eight comprising each wrist
Metacarpals		Bones of the palm of the hand
Phalanges		Finger and thumb bones

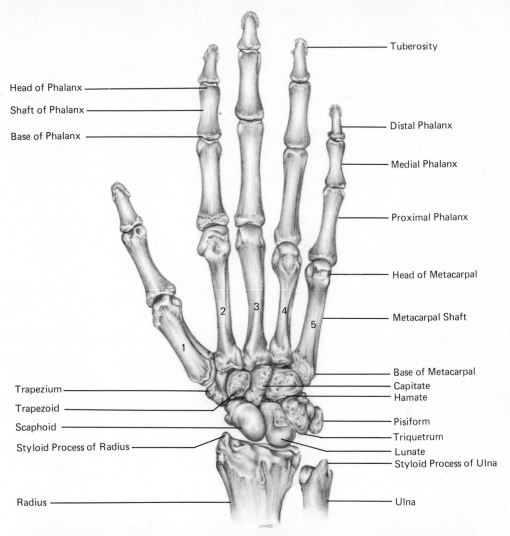

FIGURE 59 *Dorsal aspect of bones of right hand*

Tuberosity

Head of Phalanx

Shaft of Phalanx

Base of Phalanx

Distal Phalanx

Medial Phalanx

Proximal Phalanx

Head of Metacarpal

Metacarpal Shaft

1 2 3 4 5

Base of Metacarpal

Trapezium

Trapezoid

Scaphoid

Styloid Process of Radius

Capitate

Hamate

Pisiform

Triquetrum

Lunate

Styloid Process of Ulna

Radius

Ulna

FIGURE 60 *Human hand showing prone position and articulation*

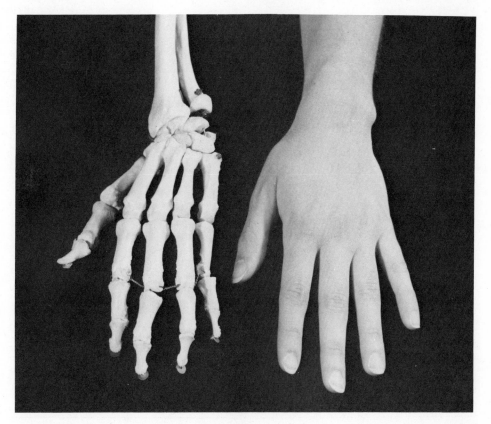

Bone	Bone Marking	Description
LOWER DIVISION		
Os coxa (Hip bone) *(Figures 61 and 62)*		Fusion of three smaller bones to form right and left pelvic bones
Ilium		Superior flaring portion of bone
	Iliac crests	Superior border of ilium
	Iliac spines	Projections on periphery of ilium (named by position)
	Anterior superior	Most anterior projection of hip
	Anterior inferior	Less prominent projection inferior to anterior superior spine
	Posterior superior	Most posterior projection on periphery of iliac crest
	Posterior inferior	Less prominent projection inferior to posterior superior spine
	Greater sciatic notch	Large irregularly shaped indentation inferior to posterior inferior spine
	Iliac fossa	Concave inner surface of ilium
Ischium		Inferior posterior bone that supplies support when sitting
	Ischial tuberosity	Irregularly shaped projection on inferior surface that is in contact with chair when sitting
	Ischial spine	Slightly pointed projection immediately superior to tuberosity
Pubis		Anterior irregularly shaped bones; two fuse to form anterior part of pelvic girdle
	Pubic symphysis	Anterior, medial cartilaginous articulation between pubic bones
	Obturator foramen	Very large rounded opening between pubis and ischium on anteriolateral surface

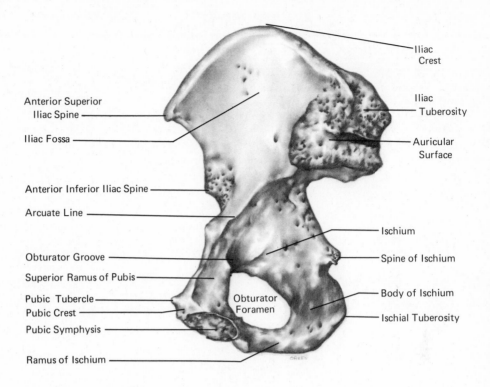

Iliac
Crest

Anterior Superior
Iliac Spine

Iliac Fossa

Iliac
Tuberosity

Auricular
Surface

Anterior Inferior Iliac Spine

Arcuate Line

Ischium

Obturator Groove

Superior Ramus of Pubis

Spine of Ischium

Pubic Tubercle

Pubic Crest

Obturator
Foramen

Body of Ischium

Ischial Tuberosity

Pubic Symphysis

Ramus of Ischium

FIGURE 61   *Medial aspect of right hipbone*

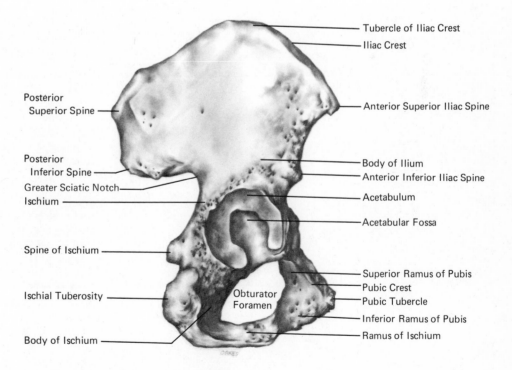

Tubercle of Iliac Crest

Iliac Crest

Posterior
Superior Spine

Anterior Superior Iliac Spine

Posterior
Inferior Spine

Greater Sciatic Notch

Ischium

Body of Ilium

Anterior Inferior Iliac Spine

Acetabulum

Acetabular Fossa

Spine of Ischium

Ischial Tuberosity

Superior Ramus of Pubis

Pubic Crest

Pubic Tubercle

Obturator
Foramen

Inferior Ramus of Pubis

Ramus of Ischium

Body of Ischium

FIGURE 62   *Lateral aspect of right hipbone*

Bone	Bone Marking	Description
	Pubic arch	Anterior arch formed inferior to pubic symphysis by inferior rami
	Pubic crest	Superior margin of superior rami above and lateral to pubic symphysis
	Pubic tubercle	Small rounded projections on superior rami; lateral and superior to pubic symphysis
	Pelvic girdle	Articulation of os coxa, sacrum, and coccyx; forms point of attachment for lower extremities
	Pelvic brim	Circumference of boundary formed by fusion of ilium, ischium, and pubis; oval shaped in male, rounded in female; of obstetrical importance in childbirth
	True pelvis	Space inferior to pelvic brim; contains pelvic organs
	False pelvis	Space superior to pelvic brim; medial to iliac bones; included in abdominal cavity
	Acetabulum	Lateral cavity formed by fusion of ilium, ischium, and pubis; hip socket articulates with femur head
Femur *(Figures 63 and 64)*		Large strong thigh bone
	Head	Rounded projection that articulates with acetabulum in ball-and-socket joint
	Greater trochanter	Large irregularly shaped process inferior and lateral to head
	Lesser trochanter	Small irregularly shaped process inferior and posterior to head
	Linea aspera	Ridge running approximately three-fourths the length of posterior bone surface
	Supracondylar ridges	Inferior division of linea aspera into two ridges—medial and lateral

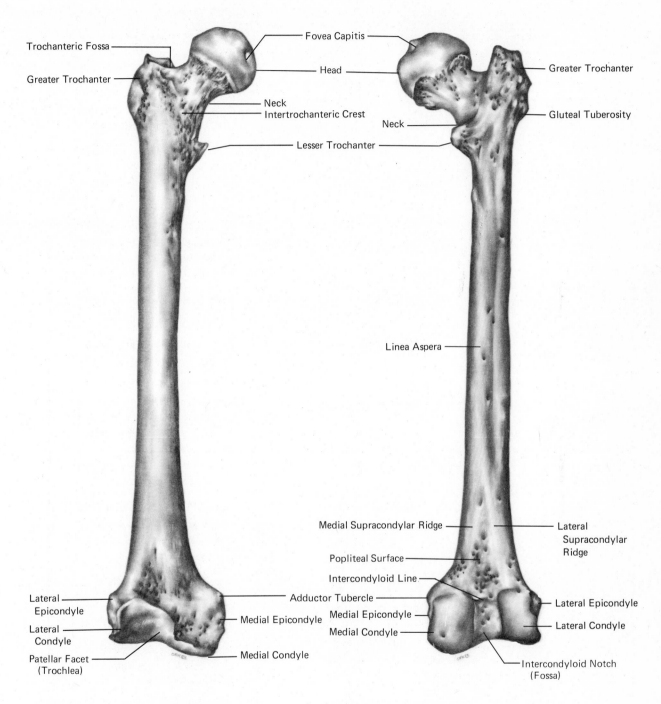

Trochanteric Fossa

Greater Trochanter

Fovea Capitis

Head

Neck

Intertrochanteric Crest

Lesser Trochanter

Greater Trochanter

Gluteal Tuberosity

Neck

Linea Aspera

Medial Supracondylar Ridge

Popliteal Surface

Intercondyloid Line

Adductor Tubercle

Medial Epicondyle

Lateral Supracondylar Ridge

Lateral Epicondyle

Lateral Condyle

Lateral Epicondyle

Lateral Condyle

Patellar Facet
(Trochlea)

Medial Condyle

Medial Epicondyle

Medial Condyle

Intercondyloid Notch
(Fossa)

FIGURE 63 *Anterior aspect of right femur*

FIGURE 64 *Posterior aspect of right femur*

Bone	Bone Marking	Description
	Gluteal tubercle	Small rounded projection inferior to greater trochanter
	Medial condyle	Large rounded protrusion on medial distal end
	Lateral condyle	Large rounded protrusion on lateral distal end
	Adductor tubercle	Very small rough projection above medial condyle marking termination of medial supracondylar ridge
	Trochlea	Depression on anterior surface between condyles
	Intercondyloid notch (fossa)	Deep depression between condyles on posterior surface
Patella		Irregularly shaped flat knee bone
Tibia (Figure 65)		Large strong bone of lower leg
	Lateral and medial condyles	Protuberance on lateral and medial surfaces at proximal end of bone
	Tibial tuberosity	Rough projection at midpoint on anterior surface at proximal end of bone
	Intercondylar eminence	Toothlike projection on superior surface between condyles
	Popliteal line	Slight ridge on posterior surface running from superior to inferior; extends approximately one-third the length of the bone
	Medial malleolus	Medial protrusion on distal end
Fibula (Figure 65)		Long thin bone of lower leg
	Lateral malleolus	Rounded projection on lateral surface of bone

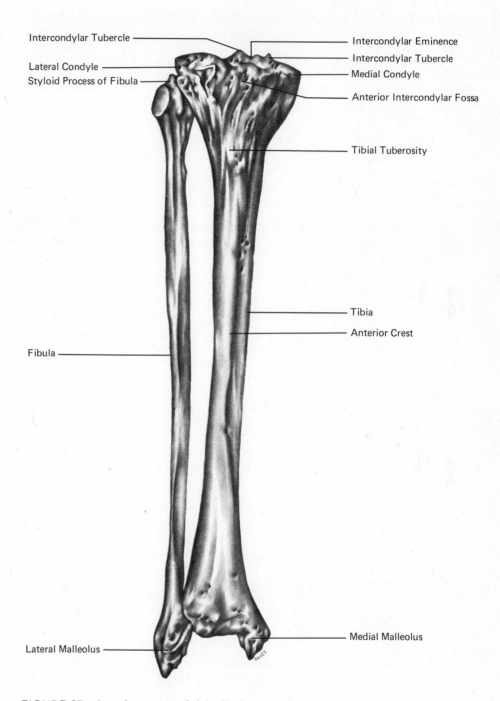

Intercondylar Tubercle

Lateral Condyle

Styloid Process of Fibula

Intercondylar Eminence

Intercondylar Tubercle

Medial Condyle

Anterior Intercondylar Fossa

Tibial Tuberosity

Tibia

Anterior Crest

Fibula

Medial Malleolus

Lateral Malleolus

FIGURE 65 *Anterior aspect of right fibula and tibia*

Bone	Bone Marking	Description
FOOT *(Figure 66)*		
Tarsals		Fourteen irregularly shaped bones of upper foot—seven per foot
Calcaneus		Large irregular bone that forms heel
Talus		Rounded bone with smooth superior surface that articulates with tibia and fibula
Metatarsals		Ten long bones articulating between the tarsals and proximal phalanges—five per foot
Phalanges		Toe bones

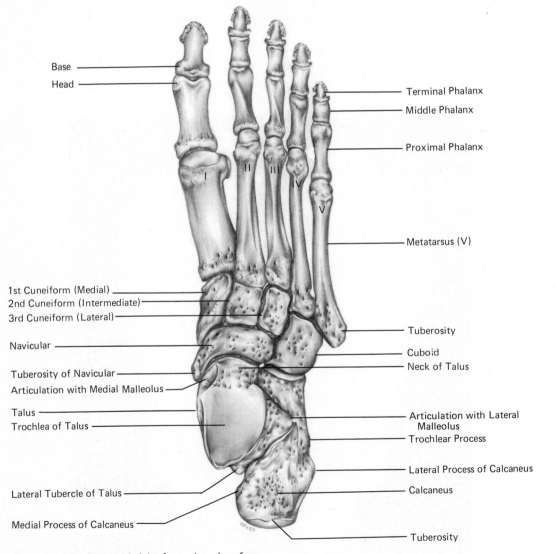

FIGURE 66 *Bones of right foot, dorsal surface*

EXERCISE 5
Identification of Joints

Most bones of the body join with one or more other bones. This union of bones is termed a **joint,** or **articulation.** According to the amount of movement allowed by the anatomical structures of these unions, joints are divided into two major groups—**synarthroses** and **diarthroses.**

a. Synarthroses

Synarthrodial joints are those lacking a joint cavity; the bones are tightly united by cartilage or fibrous tissue. Synarthroses can be further divided into two groups—those allowing *no movement* and those allowing *slight movement.*

1. **Immovable synarthrodial** joints are located in the skull between both cranial and facial bones except for the mandible, which is a movable joint. These joints exist between mature, ossified bones and are referred to as **sutures.** The major sutures observed in the articulated skull from a superior view are **coronal, sagittal,** and **lambdoidal.** The **coronal suture** joins the frontal bone with the parietals. The **sagittal suture** articulates the two parietal bones, and the **lambdoidal suture** unites the parietals with the occipital bone.

   The **squamous suture** formed by the union of the sphenoid, temporal, frontal and parietal bones can be observed from the lateral aspect as illustrated in Figure 42.

   Prior to maturation the four cranial bones are not completely formed and, therefore, fibrous areas, or **fontanels,** are observed in the fetal and newborn skull as shown in Figures 67 and 68.

FIGURE 67 *Fetal skull, full term, as viewed from the front and slightly above*

1. Half of Frontal Bone
2. Frontal Suture
3. Anterior Fontanel
4. Coronal Suture
5. Parietal Bone

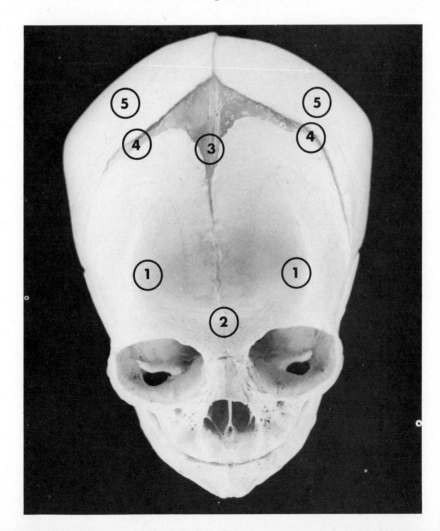

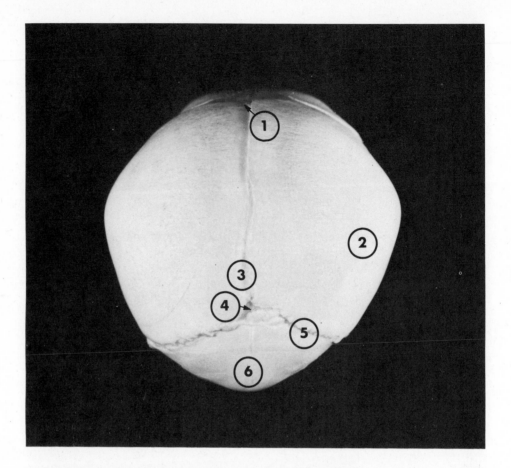

FIGURE 68 *Fetal skull, full term, as viewed from the posterior*

1. Anterior Fontanel
2. Parietal Bone
3. Sagittal Suture
4. Posterior Fontanel
5. Lambdoidal Suture
6. Occipital Bone

2. **Slightly movable synarthrotic,** or **amphiarthrodial,** joints are located in several places: between the ribs and sternum where these bones are united by cartilage, between the vertebrae, which also have cartilaginous connections, and between certain bones, such as the distal ends of the tibia and fibula, that are joined by ligaments.

b. Diarthroses

Diarthroses, the second major group of articulations, consist of those joints that have a **synovial cavity** and allow freedom of movement. (*see* Figures 69–71). The cavity, which is formed by a fibrous capsule, is lined with membranes that secrete a lubricating fluid called **synovial fluid.** Synovial fluid facilitates movement and reduces friction between articulating bones. Diarthrodial joints are further subdivided according to the shapes of the articulating bone surfaces or by the type of movement allowed by the joint. Table 1 lists the types of diarthrodial joints, movements allowed, and examples of joints in the human body.

TABLE 1 Diarthrodial Joints

Type of joint	Movement	Example
Hinge *(Figure 69)*	Flexion-extension	Elbow Knee
Gliding	Gliding	Carpals Tarsals
Pivotal joints	Rotation	Atlas-axis (atlantoaxial joint)
Spherodial, or ball-and-socket joints *(Figure 71)*	Wide range of movement Flexion-extension Abduction-adduction Rotation-circumduction	Shoulder Hip
Saddle joints	Flexion-extension Abduction-adduction Circumduction	Thumb between carpals and metacarpals
Ellipsoidal joints	Flexion-extension Abduction-adduction Circumduction	Wrist Ankle

FIGURE 69 *Left knee joint, anterior aspect*

1. Femur
2. Patellar Surface of Femur
3. Articular Surface of Patella
4. Medial Meniscus Ligament
5. Lateral Meniscus Ligament
6. Anterior Cruciate Ligament
7. Patellar Ligament
8. Tibia
9. Fibula
10. Interosseous Membrane
11. Fibular Collateral Ligament

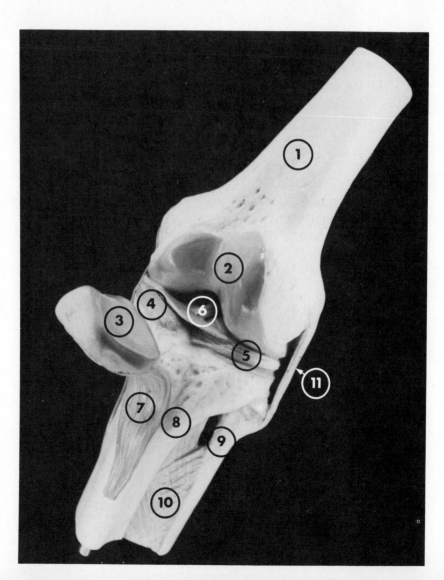

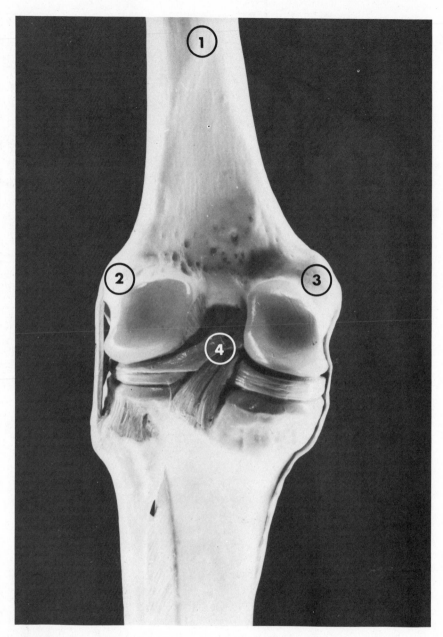

FIGURE 70 *Left knee joint, posterior aspect*

1. Femur, Posterior Surface
2. Lateral Condyle
3. Medial Condyle
4. Posterior Cruciate Ligament

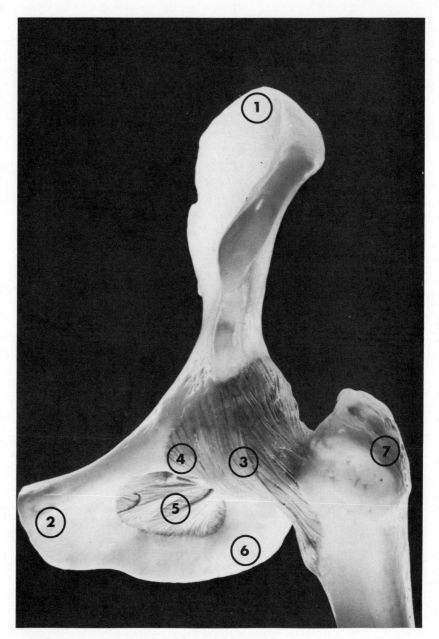

FIGURE 71 *Anterior view of left pelvis and femur*

1. Iliac Crest
2. Pubis
3. Iliofemoral Ligament
4. Ischiofemoral Ligament
5. Obturator Membrane
6. Ischium
7. Greater Trochanter

Procedure

1. On a fetal skull, observe and identify the anterior, posterior, anterolateral, and posterolateral fontanels.
2. On a mature articulated skull, identify the coronal, sagittal, and lambdoidal sutures.
3. On an articulated skeleton, identify a slightly movable synarthrosis between the sternum and the ribs.
4. With your partner as a guide, perform the listed movements and complete the following chart:

Movement	Name of Joint	Type of Movement
a. Turning the head to the right and then left		
b. Bending at the waist		
c. Speaking		
d. Shrugging the shoulders		
e. Kneeling		

5. Describe an action demonstrating each of the following types of movement:

   a. circumduction

   b. adduction

   c. extension

Skeletal System

DISCUSSION

Multiple Choice

_____ 1. Synovial joints are characterized by
    a. a joint cavity   c. lubrication by a thick fluid
    b. a synovial membrane   d. all of the above

_____ 2. Secondary centers of ossification in long bones are termed:
    a. diaphyses   c. compact bone
    b. epiphyses   d. cancellous bone

_____ 3. The articulation of the femur to the pelvic girdle is a good example of a . . . . joint.
    a. pivot   c. ball-and-socket
    b. hinge   d. pivot-condyloid

_____ 4. The human skeleton is divided into appendicular and axial components. Which of the following is *not* a part of the appendicular skeleton?
    a. skull   c. scapula
    b. pelvic girdle   d. pectoral girdle

_____ 5. The largest foramen in the skull is:
    a. foramen magnum   c. obturator foramen
    b. foramen lacerum   d. mental foramen

_____ 6. Intervertebral discs are composed of:
    a. hyaline cartilage   c. dense, white fibrous connective tissue
    b. bone   d. epithelial tissue

_____ 7. Which of the following is *not* an example of a diarthrotic joint?
    a. knee   c. atlas-axis
    b. hip joint   d. all of the above are diarthritic

_____ 8. Vertebrae are classified as . . . . bones.
    a. long   c. irregular
    b. sesamoid   d. short

_____ 9. A hollow opening in bone that serves as a passageway for nerves and blood vessels is:
    a. a sinus   c. a fissure
    b. a foramen   d. a tubercle

_____ 10. The odontoid process is found on which of the following bones?
    a. atlas   c. sternum
    b. axis   d. femur

_____ 11. Which of the following is a bone, *not* a bone marking?
     a. inferior turbinate   c. pedicle
     b. middle turbinate   d. acetabulum

_____ 12. An example of a synarthrosis is:
     a. elbow joint   c. hip joint
     b. coronal suture   d. wrist joint

_____ 13. An example of a pivotal joint is:
     a. wrist   c. femur head-acetabulum
     b. atlas-axis   d. knee

_____ 14. Another name for a ball-and-socket joint is:
     a. saddle   c. spheroidal
     b. ellipsoidal   d. hinge

_____ 15. An example of a joint type that does not allow for rotation is:
     a. condylar   c. spheroid
     b. pivot

_____ 16. Which of the following is a rounded extension that projects from a tapered neck and forms part of a joint?
     a. spine   c. tubercle
     b. head   d. trochanter

_____ 17. The shaft of a long bone is known as the:
     a. epiphysis   c. hypophysis
     b. diaphysis   d. metaphysis

_____ 18. Which of these is most centrally located in a Haversian system?
     a. Haversian canal   c. lamella
     b. osteocyte   d. osteocyte

_____ 19. Which of these bones is the smallest?
     a. hyoid   c. calcaneus
     b. incus   d. nasal

_____ 20. Which of these bone markings would be found at a hinge joint?
     a. spine   c. head
     b. trochanter   d. trochlea

_____ 21. Which of these bones would have the most compact bone in relation to spongy bone?
     a. lunate   c. thoracic vertebra
     b. scapula   d. tibia

_____ 22. Which of these bones is not the result of fusion of smaller bones?
     a. sternum   c. scapula
     b. os coxa   d. coccyx

_____ 23. The ilium, ischium, and pubis meet:
     a. at the pelvic brim   c. in the acetabulum
     b. at the pubic arch   d. inferior to the sciatic notch

_____ 24. If a patient is in a below-the-knee cast for a broken ankle, which of these movements could he *not* perform?
   a. inversion of the foot
   b. abduction of the leg
   c. hyperextension of the hip
   d. flexion of the knee

_____ 25. Which of these definitions is false?
   a. spine—a sharp slender projection
   b. trochanter—a large terminal enlargement
   c. fissure—a narrow slit
   d. tubercle—a small rounded process

DISCUSSION QUESTIONS (Optional)

1. Define the following bone markings and give an example of each:

   a. trochlea

   b. tubercle

   c. trochanter

   d. spine

   e. process

   f. fossa

   g. foramen

   h. condyle

   i. groove

2. Which bones are included in (a) the axial skeleton and (b) the appendicular skeleton?

3. Give the function of each of the following bone markings:

a. mandibular fossa . . . .

b. jugular foramen . . . .

c. carotid foramen . . . .

d. sella turcica . . . .

e. superior orbital fissure . . . .

f. foramen rotundum . . . .

g. crista galli . . . .

h. foramen magnum . . . .

i. superior articulating surface of a vertebra. . . . .

j. odontoid process . . . .

k. rib tubercle. . . . .

l. acromion of scapula . . . .

m. glenoid cavity. . . . .

n. coronoid fossa of humerus . . . .

o. semilunar notch of ulna . . . .

p. ischial tuberosity . . . .

q. acetabulum. . . . .

r. trochlea of femur . . . .

s. intercondylar eminence . . . .

t. trochlea of humerus . . . .

u. sternal end of clavicle. . . . .

v. costal cartilage of rib . . . .

Muscular System

A. Cat Musculature

- *Purpose*
  The purpose of Unit VI-A is to give you an understanding of the gross and microscopic anatomical relationships of cat muscles.

- *Objectives*
  In order to complete Unit VI-A, you must be able to do the following:
  1. Properly remove the skin and fascia from the cat.
  2. Identify selected muscles of the head, trunk, and limbs.
  3. Microscopically recognize skeletal, smooth, and cardiac muscle.
  4. Microscopically recognize a myoneural junction.
  5. Recognize similarities and differences in muscles found in the cat and human being.

- *Materials*
  Preserved cats
  Forceps
  Sharp scissors
  Dissecting pins
  Newspaper
  Slides of myoneural junction
  Sharp scalpel, preferably with removable blade
  Blunt probe
  Dissecting pan
  Slides of skeletal, smooth, and cardiac muscle

- *Procedure*

EXERCISE 1

Skinning the Cat

Your instructor will demonstrate the proper way to skin your specimen, to remove **fascia** (connective tissue), and to separate muscles. The usual procedure for skinning a cat is to make a midventral or middorsal incision with scissors and, taking care not to cut deeply into the underlying musculature, extend the incision from the neck to the external genitalia or base of the tail. Then cut around the neck, external genitalia, anus, and tail. Extend your initial incision down the lateral surface of each leg, then cut around the wrists and ankles. It is necessary to identify muscles on one side of your specimen only. The other side may be used for reviewing the musculature and locating blood vessels in a later unit.

EXERCISE 2

Identification of Muscles

You will be responsible for the separation and identification of muscles in the major body regions. Knowledge of the general description, origin, insertion, and action of each muscle will help in its identification. It is essential that fat and excess fascia be removed and each muscle separated so it can be readily identified. In order to aid in your identification of cat muscle origins and insertions, study the photograph of the cat skeleton in Figure 72.

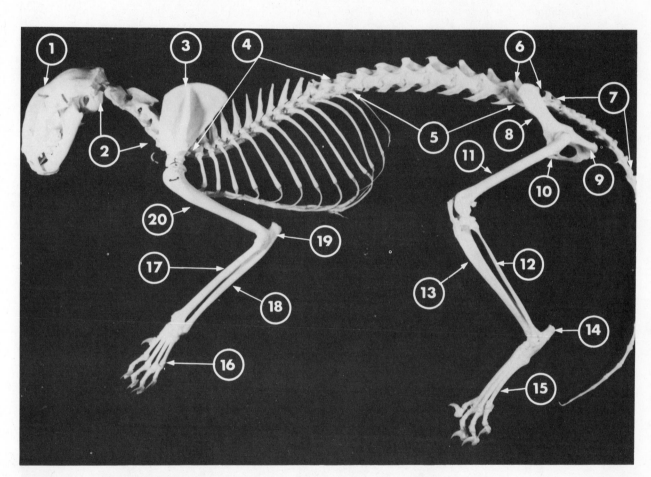

FIGURE 72 *Cat skeleton, lateral view*

1. Skull	6. Sacrum	11. Femur	16. Metacarpals
2. Cervical Vertebrae	7. Caudal Vertebrae	12. Fibula	17. Radius
3. Scapula	8. Ilium	13. Tibia	18. Ulna
4. Thoracic Vertebrae	9. Ischium	14. Calcaneus	19. Olecranon Process
5. Lumbar Vertebrae	10. Pubis	15. Metatarsals	20. Humerus

MUSCLES OF THE NECK AND THROAT (*see* Figures 73-76)

Name and Description	Origin	Insertion	Action
Sternomastoid: A superficial band of muscle extending cranially and laterally on the ventral surface of the neck	Manubrium of sternum and mid-ventral raphe at base of neck	Mastoid process and lambdoidal ridge of skull	Turns head when single muscle contracts; lowers head when both muscles contract
Sternohyoid: A thin, superficial longitudinal band of muscle covered posteriorly by the Sternomastoid; lies over the trachea	First costal cartilage and manubrium of sternum	Hyoid bone	Draws hyoid posteriorly

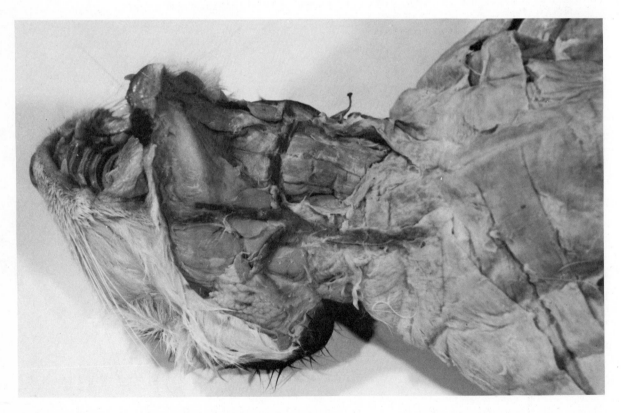

FIGURE 73   *Muscles, blood vessels, and glands of the neck*

Name and Description	Origin	Insertion	Action
Masseter: Cheek muscle	Zygomatic arch	Coronoid fossa of mandible	Elevates mandible
Digastric: A band of muscle under chin; medial to the Masseter	Mastoid process of skull	Mandible	Depressor of mandible
Mylohyoid: A thin, broad superficial sheet running transversely across the neck	Cricoid cartilage of larynx	Median raphe	Raises the floor of the mouth and draws hyoid bone forward

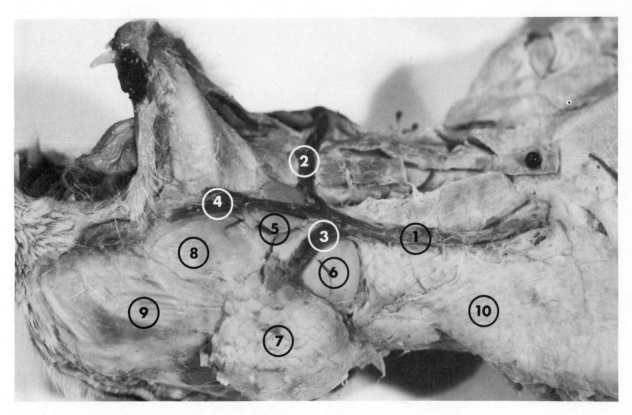

FIGURE 74 *Neck region, lateral view*

1. External Jugular Vein
2. Transverse Jugular Vein
3. Posterior Facial Vein
4. Anterior Facial Vein
5. Lymph Node
6. Submaxillary Gland
7. Parotid Gland
8. Parotid Duct
9. Masseter Muscle
10. Clavodeltoid

Name and Description	Origin	Insertion	Action
Stylohyoid: A thin, ribbonlike superficial muscle running transversely over the posterior portion of the Mylohyoid	Styloid process of hyoid bone	Hyoid bone	Elevates hyoid
Sternothyroid: A thin flat band of muscle lying under the Sternohyoid	Costal cartilage of sternum	Thyroid cartilage of larynx	Draws larynx posteriorly
Cleidomastoid: A flat thin muscle; fibers run beneath and lateral to the Sternomastoid	Clavicle	Mastoid process (reversible origin and insertions)	Together the muscles lower the head onto the neck; singly each turns head

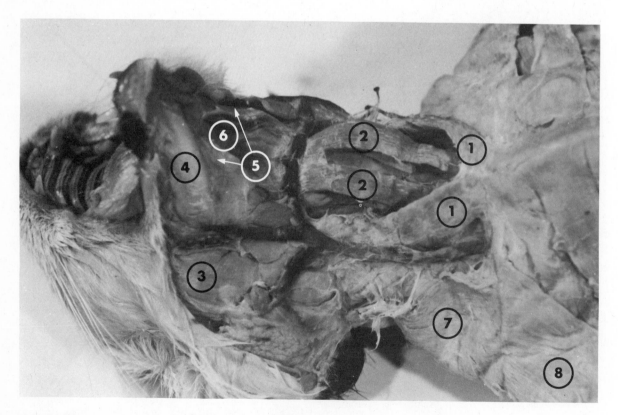

FIGURE 75 *Superficial muscles of the neck and throat, ventral aspect*

1. Sternomastoid
2. Sternohyoid
3. Masseter
4. Mandible
5. Digastric
6. Mylohyoid
7. Clavotrapezius
8. Clavodeltoid

Name and Description	Origin	Insertion	Action
Thyrohyoid: A narrow short muscle lying anterior to the Sternohyoid	Thyroid cartilage of larynx	Posterior of hyoid bone	Elevates the larynx
Cricothyroid: A thin narrow muscle that covers the ventral surface of the larynx	Cricoid cartilage of larynx	Thyroid cartilage of larynx	Causes vocal cords to become taut
Geniohyoid: A V-shaped muscle; fibers run parallel to and beneath the Mylohyoid muscles	Mandibular symphysis	Hyoid bone	Draws hyoid bone forward

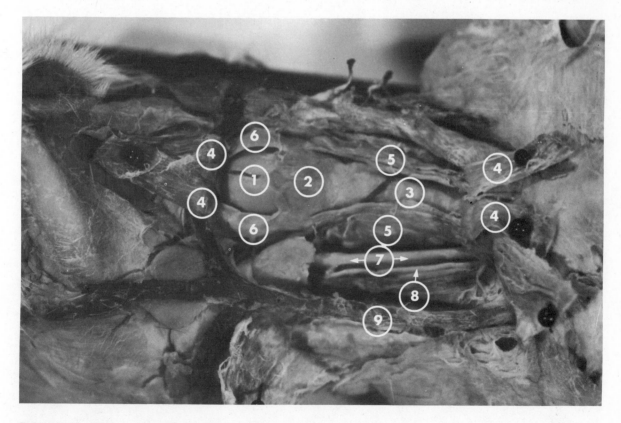

FIGURE 76 *Deep muscles of the neck, ventral view*

1. Thyroid Cartilage of Larynx
2. Cricoid Cartilage of Larynx
3. Trachea
4. Sternohyoid Muscles, Cut and Reflected
5. Sternothyroid
6. Thyrohyoid
7. Carotid Artery
8. Vagus Nerve
9. External Jugular Vein

MUSCLES OF THE FOREARM—LATERAL SURFACE (*see* Figures 77 and 78)

Name and Description	Origin	Insertion	Action
Extensor Carpi Ulnaris: A long muscle that forms the most inferior portion of the forearm when viewed laterally	Semilunar notch and lateral epicondyle of humerus	Proximal portion of fifth metacarpal	Extends the fifth digit and ulnar side of wrist
Extensor Digitorum Lateralis: A long wedge-shaped muscle that lies on top of the Extensor carpi ulnaris	Lateral surface of humerus	Attachment by means of tendons to third and fourth digits	Extends third and fourth digits

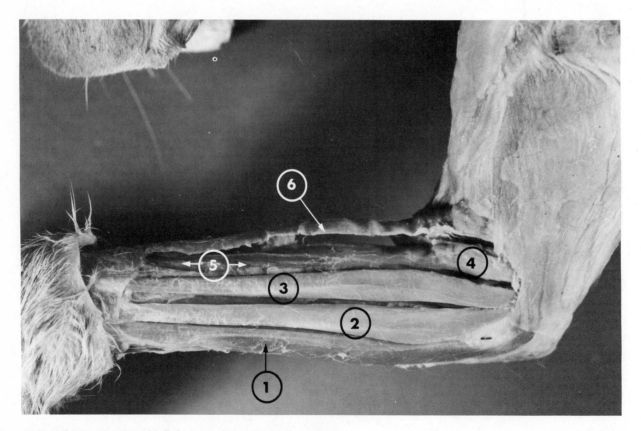

FIGURE 77 *Muscles of left forearm, lateral (dorsal) surface*

1. Extensor Carpi Ulnaris
2. Extensor Digitorum Lateralis
3. Extensor Digitorum Communis
4. Extensor Carpi Radialis, Long Head
5. Extensor Carpi Radialis, Short Head
6. Brachioradialis

Name and Description	Origin	Insertion	Action
Extensor Digitorum Communis: A thin muscle that lies on top of the Extensor digitorum lateralis	Lateral surface of humerus	Attachment by means of tendons to third and fourth digits	Extends third and fourth digits
Extensor Carpi Radialis: Narrow band of muscle that lies under Brachioradialis and on top of Extensor digitorum communis; has two heads—the brevis and longus	Middle of humerus	Attachment by means of tendons to second and third metacarpals	Extends the hand
Brachioradialis: The most superior ribbonlike muscle of the forearm group	Middle of humerus on the dorsal surface	Distal end of radius	Supinator of hand

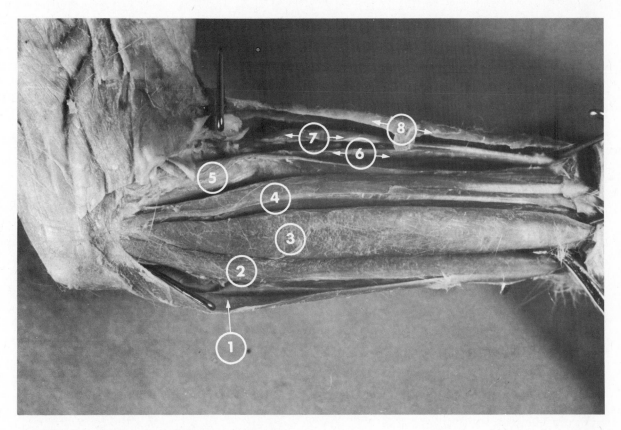

FIGURE 78 *Muscles of left forearm, medial surface*

1. Flexor Carpi Ulnaris, First Head
2. Flexor Carpi Ulnaris, Second Head
3. Palmaris Longus
4. Flexor Carpi Radialis
5. Pronator Teres
6. Extensor Carpi Radialis, Brevis
7. Extensor Carpi Radialis, Longus
8. Brachioradialis

MUSCLES OF THE FOREARM—MEDIAL SURFACE (*see* Figures 77 and 78)

Name and Description	Origin	Insertion	Action
Flexor Carpi Ulnaris: A large band of muscle that covers the medial surface of the humerus	Two heads—medial epicondyle of humerus and olecranon process	Pisiform bone of wrist	Flexor of digits
Palmaris Longus: Largest of the medial forearm group; lies above the Flexor carpi ulnaris	Medial epicondyle of humerus	Attachment by means of tendons to all digits	Flexor of digits
Flexor Carpi Radialis: A large narrow muscle that lies above the Palmaris longus	Medial epicondyle of humerus	Second and third metacarpals	Flexes second and third metacarpals
Pronator Teres: A short wedge-shaped muscle that lies above the Flexor carpi radialis	Medial epicondyle of humerus	Medial surface of radius	Pronator of the hand (rotates radius)
Flexor Digitorum Profundus: A long, thin ribbon-shaped muscle that lies between the Pronator teres and Extensor carpi radialis	Medial surfaces of ulna and humerus	Proximal end of phalanges	Flexor of digits
Extensor Carpi Radialis, Brevis and Extensor Carpi Radialis, Longus: Same muscle viewed from the medial aspect	*See* lateral surface description on page 121.		
Brachioradialis:	*See* lateral surface description on page 121.		

MUSCLES OF THE ABDOMINAL WALL (*see* Figure 79)

Name and Description	Origin	Insertion	Action
External Oblique: A superficial sheet of muscle forming the outermost muscle of the abdominal wall	Lumbodorsal fascia and posterior ribs	Linea alba by way of **aponeurosis**	Constricts abdomen

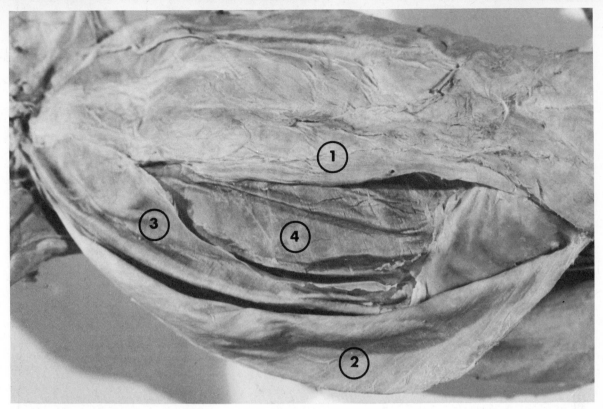

FIGURE 79 *Abdominal muscle detail*

1. Rectus Abdominus
2. External Oblique
3. Internal Oblique
4. Transverse Adbominis

Name and Description	Origin	Insertion	Action
Internal Oblique: A sheetlike muscle immediately beneath the External oblique; its muscle fibers run in a different direction from those of the External oblique	Lumbodorsal fascia and border of pelvic girdle	Aponeurosis to linea alba	Compressor of abdomen
Transversus Abdominis: The sheet of muscle beneath the fleshy portion of the Internal oblique; fibers run in a different direction than those of the Internal oblique	Similar to internal oblique	Similar to internal oblique	Compressor of abdomen
Rectus Abdominis: A superficial longitudinal band of muscle lying immediately lateral to the mid-ventral line of the abdomen	Anterior end of pubic symphysis	Sternum and costal cartilages	Retracts ribs and sternum and constricts abdomen

MUSCLES OF THE CHEST (PECTORALIS GROUP) (*see* Figures 80 and 81)

Name and Description	Origin	Insertion	Action
Pectoantibrachialis: A superficial band of muscle about 1 cm wide that extends laterally from the midline and caudally to the neck	Manubrium of sternum	Proximal end of forearm	Adducts arm
Pectoralis Major: A superficial triangular muscle caudal to the Pectoantibrachialis; a portion of the Pectoralis major lies under the Pectoantibrachialis	Sternum and median ventral raphe	Shaft of humerus	Adducts humerus

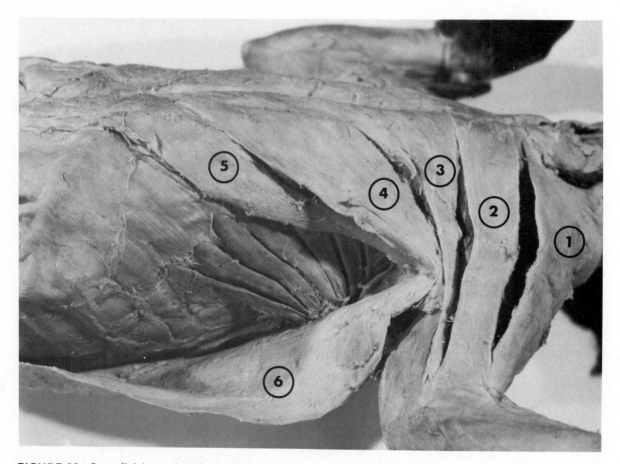

FIGURE 80   *Superficial muscles of the chest*

1. Clavodeltoid (Clavobrachialis)
2. Pectoantibrachialis
3. Pectoralis Major
4. Pectoralis Minor
5. Xiphihumeralis
6. Latissimus Dorsi

Name and Description	Origin	Insertion	Action
Pectoralis Minor: A superficial muscle caudal to the Pectoralis major; a portion lying deep to the Pectoralis major; Pectoralis minor is larger than the Pectoralis major in the cat	Sternum	Bicipital groove of humerus	Adducts humerus
Xiphihumeralis: A band of muscle caudal to the Pectoralis minor; fibers extend laterally and anteriorly and then pass deep to the Pectoralis minor	Xiphoid process of sternum	Bicipital groove of humerus	Adducts humerus

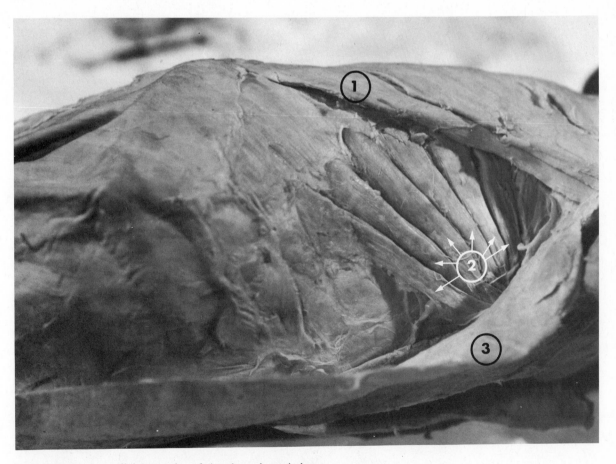

FIGURE 81 *Superficial muscles of the chest, lateral view*

1. Xiphihumeralis
2. Serratus Ventralis
3. Latissimus Dorsi

MUSCLES OF THE UPPER BACK, SHOULDER, AND BACK OF NECK (*see* Figures 82–86)

Name and Description	Origin	Insertion	Action
Latissimus Dorsi: A superficial, flat triangular muscle extending from the middorsal line to the medial surface of the humerus; a portion of this muscle is covered dorsally by a smaller triangular muscle, the Spinotrapezius	Neural spines of fourth thoracic to sixth lumbar vertebrae and lumbodorsal fascia	Medial surface of humerus	Retracts humerus dorsally and caudally
Spinotrapezius: A triangular superficial muscle overlying the Latissimus dorsi in the middorsal region	Spinous processes of thoracic vertebrae and spines of cervical vertebrae (1–4)	Fascia of scapula	Draws scapula medially and posteriorly toward the tail

FIGURE 82 *Detail of superficial muscles of left shoulder*

1. Acromiotrapezius
2. Levator Scapulae Ventralis
3. Spinodeltoid
4. Acrimodeltoid
5. Clavodeltoid
6. Clavotrapezius
7. Triceps, Long Head
8. Triceps, Lateral Head

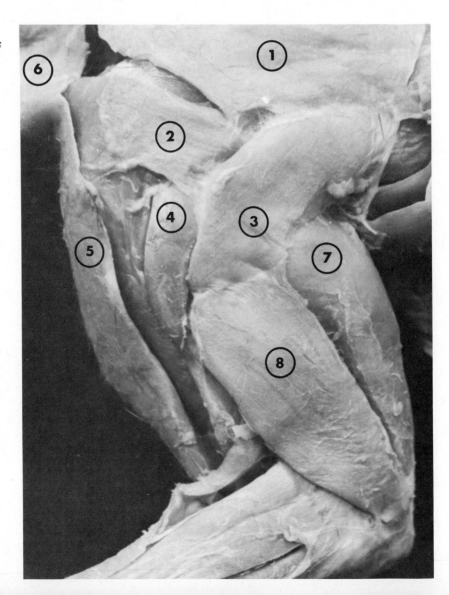

Name and Description	Origin	Insertion	Action
Acromiotrapezius: A flat superficial muscle covering the scapula	Spinous process of axis to third thoracic vertebra	Metacromion process and spine of scapula	Draws scapula medially
Clavotrapezius: A flat superficial muscle on the dorsal surface of the neck	Superior nuchal line and median dorsal line of neck	Clavicle	Draws clavicle dorsally and cranially
Levator Scapulae Ventralis: A longitudinal band of muscle on the side of the neck; between the Clavotrapezius and Acromiotrapezius	Transverse occipital bone and process of atlas	Metacromion process of scapular spine	Draws scapula cranially

FIGURE 83 *Muscles of scapular region after transection of the acromiotrapezius and spinotrapezius*

1. Acromiotrapezius, Cut and Reflected
2. Spinotrapezius
3. Supraspinatus
4. Infraspinatus
5. Rhomboideus
6. Clavotrapezius

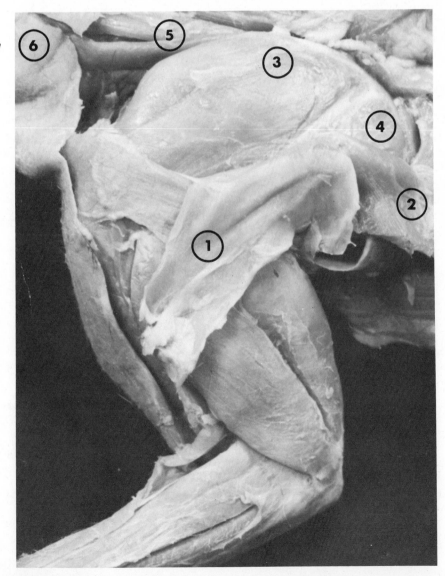

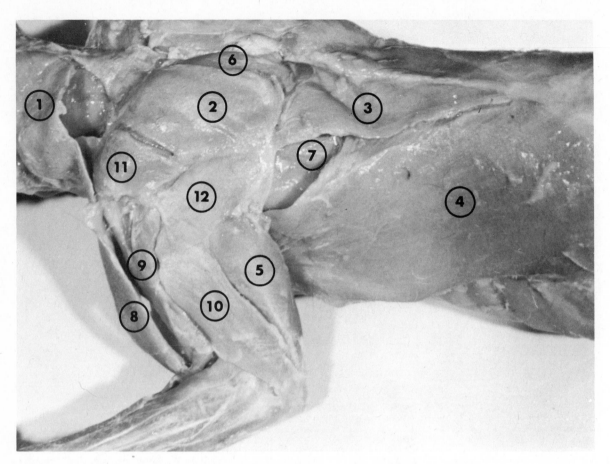

FIGURE 84 *Superficial muscles of the back, dorsal aspect*

1. Clavotrapezius	4. Latissimus Dorsi	7. Infraspinatus	10. Triceps Brachii Lateralis
2. Acromiotrapezius	5. Triceps Brachii Longus	8. Clavobrachialis (Clavodeltoid)	11. Levator Scapulae Ventralis
3. Spinotrapezius	6. Supraspinatus	9. Acromiodeltoid	12. Spinodeltoid

Name and Description	Origin	Insertion	Action
Supraspinatus: A thick muscle occupying the supraspinous fossa of the scapula; lies beneath the Acromiotrapezius	Supraspinous fossa of scapula	Greater tubercle of humerus	Draws humerus cranially
Infraspinatus: A muscle occupying the infraspinous fossa of the scapula	Infraspinous fossa of scapula	Greater tubercle of humerus	Rotates humerus laterally
Clavodeltoid (Clavobrachialis): A superficial muscle of the shoulder; continuous with the Clavotrapezius at its proximal end; extends into the arm	Clavotrapezius and clavicle	Proximal end of ulna	Flexes forearm

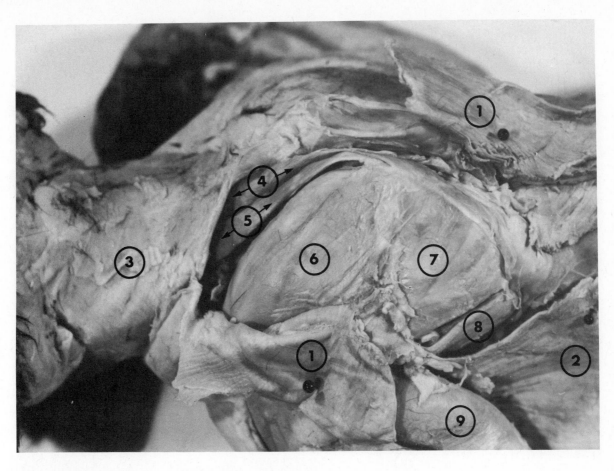

FIGURE 85 *Deep muscles of the shoulder, dorsal aspect*

1. Acromiotrapezius, Cut and Reflected
2. Spinotrapezius, Cut and Reflected
3. Clavotrapezius

4. Rhomboideus
5. Rhomboideus, Capitis
6. Supraspinatus

7. Infraspinatus
8. Teres Major
9. Triceps Brachii, Lateral Head

Name and Description	Origin	Insertion	Action
Acromiodeltoid: A muscle posterior to the Clavodeltoid; its distal portion lies under the Clavodeltoid	Acromion process of scapula	Deltoid ridge of humerus	Raises and rotates humerus with Spinodeltoid
Spinodeltoid: A band of muscle posterior to the Acromiodeltoid and Levator scapulae ventralis	Scapular spine	Deltoid ridge of humerus	Raises and rotates humerus
Serratus Ventralis: Fan-shaped slips of muscle deep to the Pectoralis group covering the ventro-lateral surface of the ribs	First nine or ten ribs and transverse processes of last five cervical vertebrae	Scapula near vertebral border	Draws scapula cranially, ventrally, and against thoracic wall

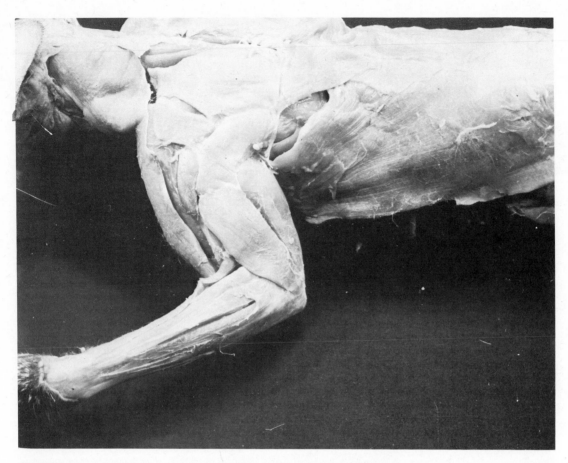

FIGURE 86 *Superficial muscles of the back, general view, dorsal aspect*

Name and Description	Origin	Insertion	Action
Serratus Dorsalis Cranialis: Fan-shaped slips of muscle in the thoracic and lumbar regions deep to the Latissimus dorsi; fibers run cranially	First nine ribs	Dorsal raphe between axis and tenth thoracic vertebra	Extends ribs forward and outward
Serratus Dorsalis Caudalis: Fan-shaped slips of muscle in the thoracic and lumbar regions deep to the Latissimus dorsi; fibers run caudally	Last five ribs	Spines of lumbar vertebrae	Draws ribs caudally

MUSCLES OF THE UPPER ARM (*see* Figures 85 and 87)

Name and Description	Origin	Insertion	Action
Epitrochlearis: A flat superficial muscle on the medial side of humerus	Lateral surface of Latissimus dorsi	Olecranon process of ulna	Extends forearm
Triceps Brachii Longus: A muscle covering the posterior surface of the humerus	Axillary border of scapula	Olecranon process of ulna	Extends forearm
Triceps Brachii Lateralis: A band of muscle covering the lateral surface of the humerus	Greater tuberosity of humerus	Olecranon process of ulna	Extends forearm
Triceps Brachii Medialis: A muscle lying deep to the Triceps brachii lateralis; transect the T. lateralis to see the T. medialis	Shaft of humerus	Olecranon process of ulna	Extends forearm

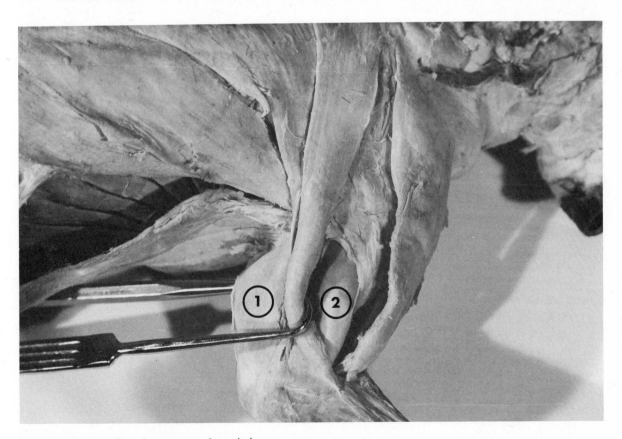

FIGURE 87 *Muscles of upper arm, lateral view*

1. Epitrochlearis   2. Biceps Brachii

Name and Description	Origin	Insertion	Action
Biceps Brachii: A muscle lying on the ventromedial surface of the humerus; it can be more easily seen by transecting and reflecting the Pectoralis muscles near the humerus	Glenoid fossa of scapula	Radius	Flexes forearm
Brachialis: A muscle lying along the outer lateral surface of the humerus; it can be exposed by transecting and reflecting the lateral head of the Triceps	Lateral surface of humerus	Ulna	Flexes forearm

MUSCLES OF THE THIGH—DORSAL ASPECT (*see* Figures 88-90)

Name and Description	Origin	Insertion	Action
Tensor Fasciae Latae: A triangular muscle on the lateral side of the hip medial to the tough white fascia (fascia lata) covering the anterior portion of the thigh	Ilium and fascia	Fascia lata	Tightens fascia lata, extends thigh
Biceps Femoris: A large thick muscle covering the lateral surface of the thigh, caudal to the T. fascia lata	Tuberosity of ischium	Patella to shaft of tibia	Abducts thigh and flexes shank
Caudofemoralis: A small muscle cranial to the dorsal portion of the Biceps femoris	Transverse processes of second and third caudal vertebrae	Patella	Abducts thigh, extends shank
Gluteus Maximus: A small muscle cranial to the Caudofemoralis	Fascia and transverse processes of last sacral and first caudal vertebrae	Fascia lata and slightly on greater trochanter	Abducts thigh
Gluteus Medius: A thick muscle dorsal to the Tensor fasciae latae	Crest and lateral surface of ilium	Greater trochanter	Abducts thigh

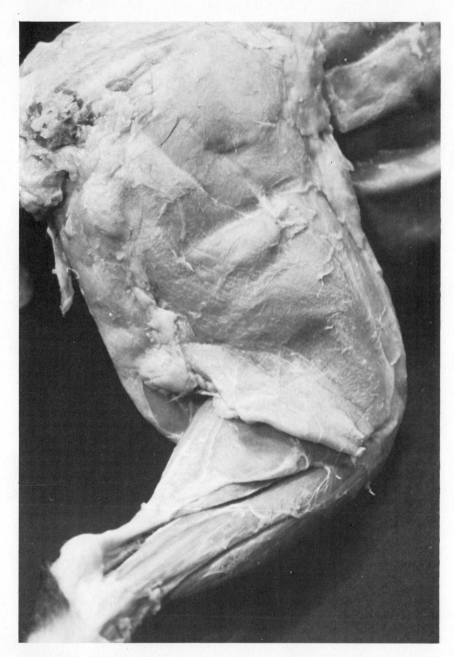

FIGURE 88   *Caudolateral aspect of superficial muscles of right thigh with fascia*

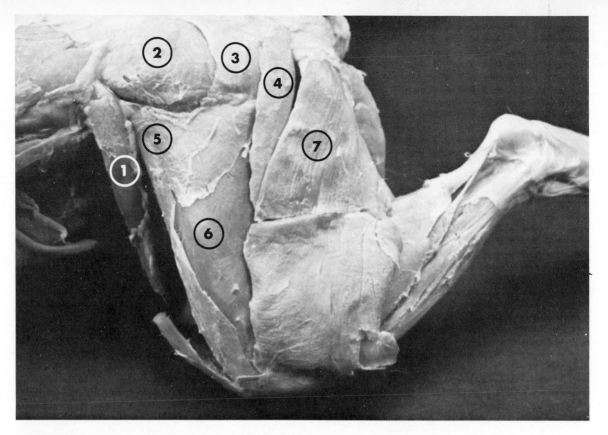

FIGURE 89 *Superficial thigh muscles, lateral aspect of left leg*

1. Sartorius, Cut
2. Gluteus Medius
3. Gluteus Maximus
4. Caudofemoralis
5. Tensor Fasciae Latae
6. Vastus Lateralis
7. Biceps Femoris, Cut

FIGURE 90 *Deep muscles of left thigh, lateral aspect*

1. Sartorius, Cut
2. Gluteus Medius
3. Gluteus Maximus
4. Caudofemoralis
5. Tensor Fasciae Latae
6. Vastus Lateralis, Exposed
7. Biceps Femoris, Cut
8. Semitendinosus
9. Tenuissimus
10. Sciatic Nerve
11. Gastrocnemius
12. Tendon of Achilles

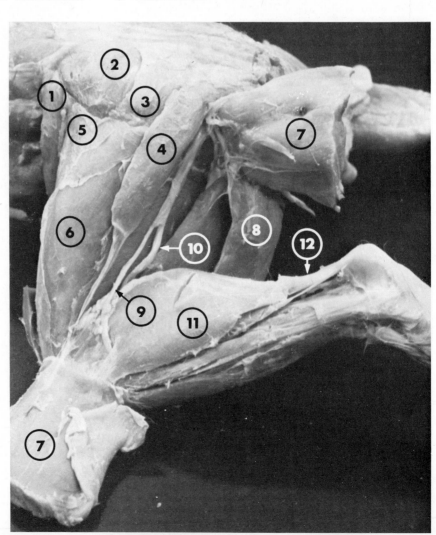

MUSCLES OF THE THIGH—VENTRAL ASPECT (*see* Figures 91-93)

Name and Description	Origin	Insertion	Action
Sartorius: A flat superficial band of muscle covering the lateral ventral portion of the thigh	Crest and ventral border of ilium	Proximal end of tibia	Adducts and rotates thigh, extends shank
Gracilis: A flat superficial band of muscle covering the medial ventral portion of the thigh	Pubic symphysis and ischium	Aponeurosis to tibia	Adducts and retracts leg
Vastus Lateralis (head of Quadriceps femoris): This large muscle can be seen better from the dorsal aspect; it is covered by the fascia lata of the Tensor fasciae latae muscle	Greater trochanter and shaft of femur	Patella and patellar ligament	Extends shank

FIGURE 91 *Superficial muscles of right thigh, ventral aspect*

1. Sartorius
2. Gracilis
3. Gastrocnemius
4. Tibialis Anterior
5. Flexor Digitorum Longus
6. Tibia

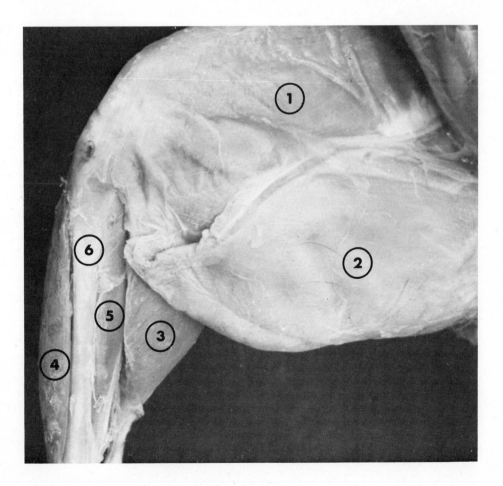

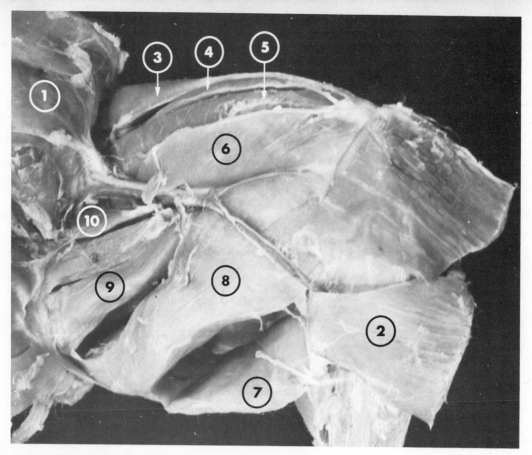

FIGURE 92 *Deep muscles of left thigh, ventral aspect*

1. Sartorius, Cut and Transected
2. Gracilis, Cut and Transected
3. Tensor Fasciae Latae
4. Vastus Lateralis
5. Rectus Femoris
6. Vastus Medialis
7. Semitendinosus
8. Semimembranosus
9. Adductor Femoris
10. Adductor Longus

Name and Description	Origin	Insertion	Action
Vastus Medialis (head of Quadriceps femoris): A band of muscle on the medial portion of the thigh, deep to the Sartorius	Femur	Patella	Extends shank
Rectus Femoris (head of Quadriceps femoris): An elongated muscle bordered laterally by the Vastus lateralis and medially by the Vastus medialis	Ilium anterior to acetabulum	Patella and patellar ligament	Extends shank
Vastus Intermedius (head of Quadriceps femoris): A flat muscle deep to the Rectus femoris	Shaft of femur	Patella and patellar ligament	Extends shank
Adductor Femoris: A muscle deep to the Gracilis, proximal to the trunk of the cat	Inferior rami of pubis and ischium	Shaft of femur	Adducts thigh

FIGURE 93 *Detail of deep muscles of left thigh, ventral aspect*

1. Semimembranosus
2. Adductor Femoris
3. Adductor Longus
4. Sartorius, Cut and Transected
5. Vastus Medialis
6. Rectus Femoris
7. Vastus Lateralis
8. Tensor Fasciae Latae

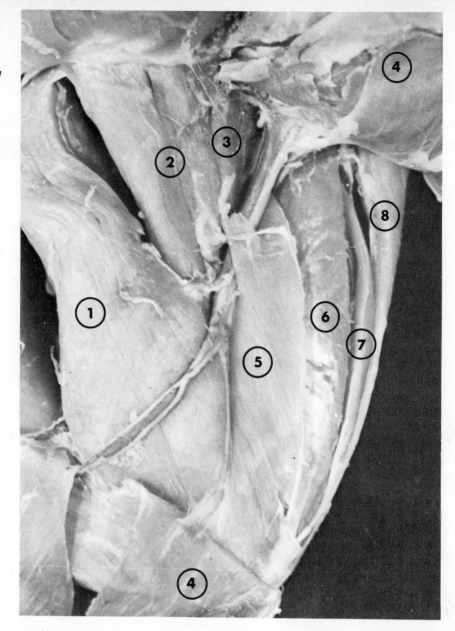

Name and Description	Origin	Insertion	Action
Adductor Longus: A narrow muscle band lying lateral to the Adductor femoris	Pubis	Femur	Adducts thigh
Semimembranosus: A large thick muscle deep to the Gracilis, caudal to the Adductor femoris	Ischium	Medial epicondyle of femur	Extends thigh
Semitendinosus: A stout band of muscle caudal to the Semi-membranosus; more easily seen from the ventral surface	Ischial tuberosity	Proximal end of tibia	Flexes shank

MUSCLES OF THE SHANK (see Figure 94)

Name and Description	Origin	Insertion	Action
Tibialis Anterior: A thin muscle lying along the lateral border of the tibia	Proximal portion of tibia	First metatarsal	Flexes foot
Extensor Digitorum Longus: A flat fusiform muscle lying ventromedial to the Tibialis anterior muscle	Tibia and fibula	All four digits	Extends phalanges

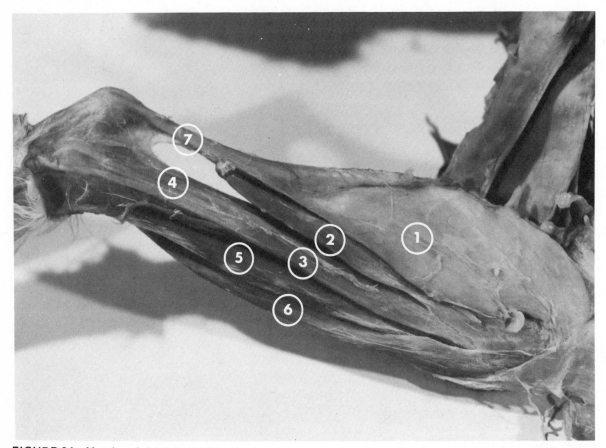

FIGURE 94 *Muscles of right shank, lateral aspect*

1. Gastrocnemius
2. Soleus
3. Peroneus Longus
4. Peroneus Brevis
5. Extensor Digitorum Longus
6. Tibialis Anterior
7. Tendon of Achilles

Name and Description	Origin	Insertion	Action
Peroneus Group (longus, brevis, tertius): A slender cylinder-shaped muscle located on the lateral surface of the leg	Lateral portion of fibula	Metatarsals and digits	Extends foot and flexes phalanges
Gastrocnemius: A large muscle mass on the dorsal surface of the foreleg	Lateral and medial condyles of femur	Tendon of Achilles on calcaneus bone	Extends foot
Soleus: A strip of muscle lying deep to the Gastrocnemius	Lateral portion of fibula	Tendon of Achilles on calcaneus bone	Extends foot

Now compare the muscles of the cat to those of the human, as can be seen in Figures 95 and 96.

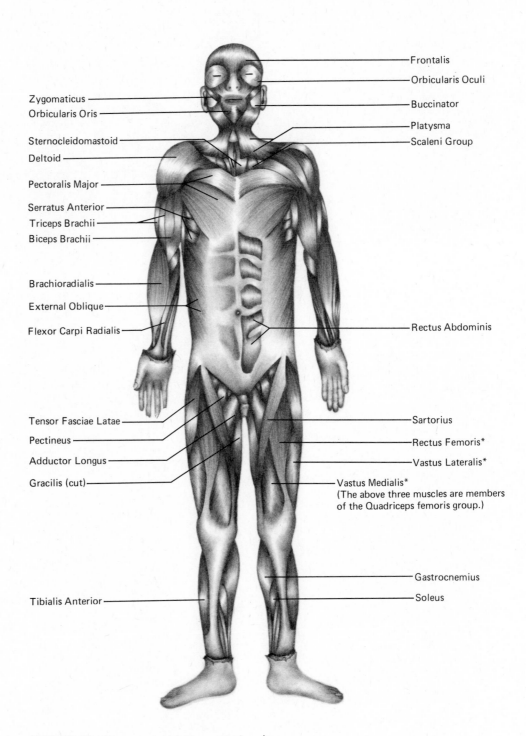

Frontalis

Orbicularis Oculi

Zygomaticus

Orbicularis Oris

Buccinator

Sternocleidomastoid

Platysma

Deltoid

Scaleni Group

Pectoralis Major

Serratus Anterior

Triceps Brachii

Biceps Brachii

Brachioradialis

External Oblique

Flexor Carpi Radialis

Rectus Abdominis

Tensor Fasciae Latae

Pectineus

Adductor Longus

Gracilis (cut)

Sartorius

Rectus Femoris*

Vastus Lateralis*

Vastus Medialis*
(The above three muscles are members
of the Quadriceps femoris group.)

Gastrocnemius

Soleus

Tibialis Anterior

FIGURE 95 *Human musculature, anterior view*

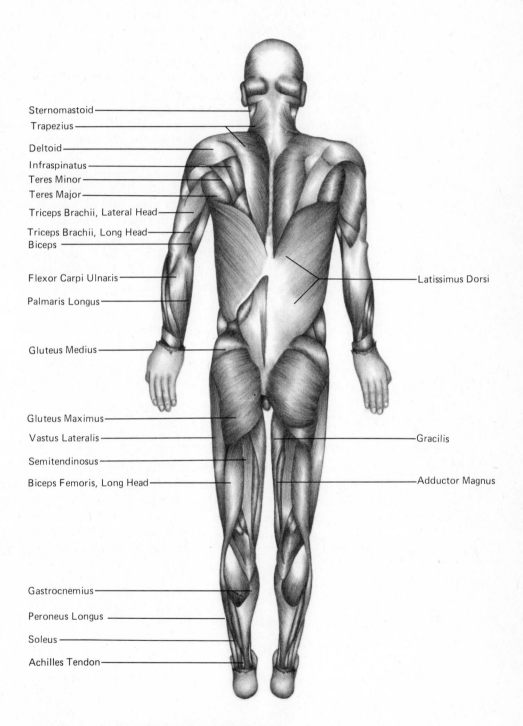

Sternomastoid

Trapezius

Deltoid

Infraspinatus

Teres Minor

Teres Major

Triceps Brachii, Lateral Head

Triceps Brachii, Long Head

Biceps

Flexor Carpi Ulnaris

Palmaris Longus

Gluteus Medius

Gluteus Maximus

Vastus Lateralis

Semitendinosus

Biceps Femoris, Long Head

Gastrocnemius

Peroneus Longus

Soleus

Achilles Tendon

Latissimus Dorsi

Gracilis

Adductor Magnus

FIGURE 96 *Human musculature, posterior view*

EXERCISE 3

Microscopic Identification of Muscle Types and Myoneural Junctions

a. *Skeletal Muscle:* (*see* Color Plate 17)

Skeletal muscle is voluntary striated muscle. Obtain a prepared slide of a longitudinal section of skeletal muscle and observe the **muscle fibers** (*see* Figure 20). Each fiber (muscle cell) is multinucleated. You should be able to distinguish these nuclei quite easily by their peripheral location. Each muscle fiber contains many **myofibrils,** that are made up of **myofilaments.** Careful observation under high power will reveal striations in the myofibrils. **A bands** (*anisotropic bands*) stain darkly and represent myosin and actin filaments. Between the A bands will be light areas known as **I bands** (*isotropic bands*) that represent actin filaments. Dividing the I bands are **Z bands,** or **Z lines.** The segment of a fiber between successive Z bands is known as a **sacromere,** the unit of structure and function of a muscle fiber.

Observe a cross section of skeletal muscle. An entire muscle is surrounded by connective tissue known as **epimysium.** The muscle is subdivided into **fasciculi** that are surrounded by **perimysium.** Look for muscle fibers that are contained in a fascicle in your slide. Under high power, you should be able to see the details of a muscle fiber, including nuclei and areas known as **Cohnheim's areas,** which contain myofibrils. In the space provided, sketch a longitudinal section and a cross section of skeletal muscle. *LABEL:* nucleus, A band, I band, and Cohnheim's area.

b. *Smooth Muscle:* (*see* Color Plate 20)

Smooth muscle appears nonstriated under the light microscope. It is usually considered involuntary and is present primarily in blood vessels and internal organs. Observe a slide of a longitudinal section of smooth muscle. Notice the "dovetailing" effect of the muscle fibers. In cross section, nuclei of smooth muscle fibers are located more centrally within the fiber rather than at the perimeter, as in skeletal muscle. *DRAW* several smooth muscle fibers. *LABEL:* nucleus and cytoplasm.

c. *Cardiac Muscle:* (*see* Color Plates 18 and 19)

Cardiac muscle is involuntary striated muscle. The muscle fibers bifurcate and connect with adjacent fibers to form a three-dimensional network (*see* Figure 21). Nuclei of cardiac muscle are situated deep within the fiber as in smooth muscle. Cardiac muscle can be recognized in longitudinal section on your prepared slide by the presence of **intercalated discs** that join the ends of cardiac muscle fibers together. The pattern of cross striations of myofibrils and the A and I bands and Z lines are the same as in skeletal muscle. *DRAW* a longitudinal section of cardiac muscle. *LABEL:* nucleus, A band, I band, Z line, and intercalated disks.

d.   *Myoneural Junction*

This slide represents the junction of a nerve ending and skeletal muscle fibers. Note the motor end-plate detail under the high power objective of your microscope. Figure 97 shows the motor end plate as seen under low power. *DRAW and LABEL* your high power observation.

FIGURE 97   *Motor end-plate detail as viewed in skeletal muscle*

B. The Physiology of Muscle Contraction

■ *Purpose*

The purpose of Unit VI-B is to enable you to understand the physiology of muscle contraction.

■ *Objectives*

In order to complete Unit VI-B, you must be able to do the following:

1. Name the ions that must be present for normal muscle contraction to occur.
2. Understand the importance of ATP in muscle contractions.
3. Calculate the degree of contraction in a Psoas muscle preparation.
4. Induce and interpret various responses to muscle stimulation by electrical and chemical methods.

■ *Materials*

Microscope slides	Living frogs
Sharp probes	Scissors
Psoas muscle preparation (kit*)	Forceps
ATP, KCl, MgCl₂ solutions (from kit)	Electronic stimulator and
Mechanical or electronic recording apparatus	electrodes
	Frog Ringer's solution

■ *Procedure*

EXERCISE 1

Chemistry of Muscle Contraction

In order for a muscle to contract, **ATP (adenosine triphosphate)** must be split and certain ions must be present.

1. Follow your instructor's directions for the preparation of needed solutions or use the solutions in your prepared kit.
2. Place several glycerinated Psoas muscle fibers in a small amount of glycerol on a microscope slide.
3. Measure the lengths of the muscle fibers in millimeters.
4. Flood the fibers with several drops of the magnesium and potassium solutions containing ATP, potassium chloride, and magnesium chloride.
5. After 30 seconds, measure the length of the fibers again. If there is a difference, calculate the percentage contraction.
6. *REPEAT* Steps 2-5, using (1) ATP alone and (2) potassium and magnesium salts alone.

---

*This kit may be obtained from Carolina Biological Supply Co., Burlington, N.C.

EXERCISE 2

Induction of Frog Skeletal Muscle Contraction

A single muscle contraction is known as a **muscle twitch**. A normal muscle twitch has three phases, as shown in the figure. Point *1* is the point of stimulation of the muscle by electrical, mechanical, or chemical means. Distance *1-2* is the **latent period**, during which electrical and chemical changes take place within the muscle prior to actual contraction. Distance *2-3* represents the **contraction period**, when the muscle contracts, and distance *3-4* is the **relaxation period**. Notice that the muscle twitch lasts approximately 0.1 second. The amount of stimulus necessary for a muscle to contract is referred to as the **threshold stimulus**, or **liminal stimulus**. A stimulus of less than threshold strength is known as **subthreshold**, or **subliminal**.

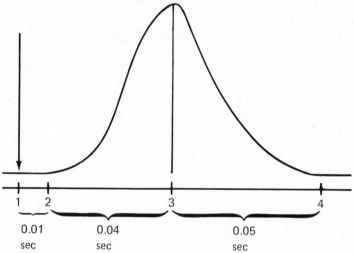

Your instructor will demonstrate the use of the mechanical or electronic physiological apparatus and the proper procedure for pithing a frog. First grasp the frog firmly with one hand, holding the head between your thumb and index finger. Incline the head forward about 90 degrees and feel the *occipital condyles* and *foramen magnum* between the condyles with your thumb. Insert a sharp probe into the foramen magnum and anteriorly into the brain. Move the probe from side to side in the brain, then remove the probe. The frog has now been **single pithed** and will feel no pain; the brain has been destroyed.

In order for the frog to lie limp for removal of the nerve-muscle preparation, the spinal cord must also be destroyed. This is done by inserting the tip of the probe into the depression between the occipital condyles and then pointing the probe inferiorly into the spinal cord. Work the probe up and down several times in this region. You should feel the frog relax in your hand. Remove the probe. The frog has now been **double pithed**; both the brain and spinal cord have been inactivated.

Expose the *Gastrocnemius muscle* and **sciatic nerve** in the leg of the frog by cutting the skin around the leg just inferior to the trunk with scissors. Do not cut too deeply. Peel the skin off the leg. From this point, it will be necessary to keep the nerve and muscles moist with frog Ringer's solution. Carefully insert a probe (perferably glass) under the Gastrocnemius and separate it from the other muscles. With scissors, cut the **Achilles tendon** at the heel. Insert the probe beneath the femur and separate the muscles from around it. Locate the sciatic nerve, which is silver white and stringlike and which should be dorsal to the femur. Sever the sciatic nerve at its proximal end (near the trunk). Remove the nerve-muscle preparation together with the femur. Your instructor will show you how to attach the nerve-muscle preparation to the recording apparatus and will demonstrate the operation of this equipment. *SAVE* your frog and maintain the intact leg with Ringer's solution for part "e. Muscle Fatigue."

a. *The Single Muscle Twitch*

In order to cause a muscle to twitch, it is necessary to achieve a threshold stimulus. Setting the voltage knob of the electronic stimulator at its lowest setting, stimulate the muscle itself with the electrodes with a single stimulus. If nothing happens, raise the voltage setting by 5 volts and again apply a single stimulus. Repeat the 5-volt increment until the muscle responds weakly and you see a response on the recording paper. The number of volts needed to produce this weak response is the threshold stimulus. What is the threshold stimulus in this exercise? . . . . volts. Remember to keep the nerve-muscle preparation moist with Ringer's solution. *SAVE and LABEL* the recording of the muscle twitch.

b. *Staircase Phenomenon (Multiple Motor Unit Summation, or Treppe)*

If single stimuli of constant intensity are applied to a muscle, each twitch will be slightly greater than the preceding one. Stimulate the muscle as rapidly as possible with single stimuli for 5 seconds with an intensity of 20 volts above threshold. *LABEL and SAVE* the tracing.

c. *Induction of Muscle Contraction by Chemicals and by Nervous Stimulation*

In this exercise you will be stimulating muscle contraction by using acetic acid and the sciatic nerve.

1. Using the same nerve-muscle preparation, touch the electrode to the sciatic nerve and apply electrical stimuli of increasing intensities, beginning with 0 volts. What is the threshold stimulus?

   . . . . volts

2. Apply several drops of 5% acetic acid to the sciatic nerve. Is there a twitch?

   . . . . *REPEAT* if necessary.

3. Apply several drops of 5% acetic acid to the muscle itself after rinsing the nerve with Ringer's solution.

   Result? . . . .

4. Rinse the acid off the muscle with Ringer's solution.

LABEL and SAVE the tracings.

d. *Tetanus*

Tetanus is a condition in which muscles are in a state of continual contraction Set the voltage at 20 volts above threshold and stimulate the muscle itself in the following manner:

1. A single stimulus at the rate of one per second for 10 seconds. Wait 1 minute.
2. A single stimulus at the rate of two per second for 10 seconds. Wait 1 minute.
3. A single stimulus as rapidly as possible for 10 seconds. This represents *incomplete tetanus.* Wait 1 minute.

4. Stimulate the muscle with multiple stimuli for 10 seconds. This represents *complete tetanus*.

LABEL and SAVE these recordings.

e. *Muscle Fatigue*

Muscle fatigue is caused by the accumulation of waste products, including CO_2, lactic acid, and acid phosphates.

1. Determine the threshold stimulus and record several normal muscle twitches.
2. Set the intensity of induction 20 volts above threshold stimulus.
3. With one person watching the time, apply multiple stimuli to the muscle itself and record the time that elapses between the initial application of stimuli and the beginning of complete tetanus.

    Time: . . . .
4. Determine the time between the beginning of complete tetanus and the beginning of fatigue (when the line begins sloping toward the baseline).

    Time: . . . .
5. Determine the time it takes for complete fatigue (when the line reaches the baseline).

    Time: . . . .
6. Allow the muscle to relax 1 minute, apply Ringer's solution, and then repeat multiple stimulation at 20 volts above threshold. Record times as in Steps 3, 4, and 5. *LABEL and SAVE* the recordings.

    Times: . . . ., ..., ...
7. Allow the muscle to relax 3 minutes, apply Ringer's solution, and then stimulate and record times as in Steps 3, 4, and 5.

    Times: . . . ., ..., ...

At this point, the condition of the muscle you are working with may necessitate dissection of the other Gastrocnemius from the frog to replace the first muscle on the recording apparatus. Please remember at all times to keep the muscle moist with Ringer's solution.

f. *Work Performed by Skeletal Muscle*

1. Keep the muscle moist with Ringer's solution. Attach a scale pan to the tendon and determine the threshold stimulus. Then add a 5-gram (5-g) weight to the pan and determine the threshold stimulus.

    . . . . volts
2. Set the intensity of induction 100 volts above the threshold stimulus determined above. Do a single muscle twitch. Add additional 5-g weights until the muscle can no longer contract. What weight was needed to achieve this point?

    . . . . g
3. As the weight of the load is increased, there is a proportionate increase in work accomplished to a certain point. After that, an additional load results in less work being performed. The amount of work done by the muscle can be calculated using this formula:

    Work = weight of load (in grams) $\times$ distance load is lifted (in millimeters)

    How much work was done by this Gastrocnemius muscle at its maximum point?

    . . . . g/mm

LABEL and SAVE this recording.

Label and save all recordings and attach to Question 3 of the Optional Discussion Questions at the end of this unit. If this series of exercises was done as a demonstration, then, in the space following Question 3, copy the tracings and hand them in with the other questions. A written interpretation of the various types of muscle contractions accompanying your tracings is required.

UNIT VI
Muscular System

DISCUSSION

Matching

Choose the correct response for Questions 1-3 from the following terms:
a. muscle twitch   c. ATP
b. tetanus   d. lactic acid

_____ 1. Prolonged contraction of a muscle

_____ 2. Single muscle contraction produced by a single stimulus

_____ 3. Muscle fatigue causes an accumulation of this substance

Choose the correct response for Questions 4-8 from the following muscles:
a. Sternocleidomastoid   c. Masseter
b. Gastrocnemius   d. Deltoid

_____ 4. Has clavicle and scapula as its origin and humerus as its insertion

_____ 5. Major muscle of chewing

_____ 6. Extends foot

_____ 7. Muscle used to turn the head

_____ 8. Calf muscle

Multiple Choice

_____ 9. Which of these muscles is present in the cat and absent in the human?
a. Pectoralis major   c. Epitrochlearis
b. Gluteus maximus   d. Sartorius

_____ 10. Which of these muscles is named according to its function?
a. Sternomastoid   c. Triceps brachii
b. Semimembranosus   d. Adductor femoris

_____ 11. The origin of a flexor muscle is located:
a. proximal to the insertion   c. lateral to the insertion
b. distal to the insertion   d. at its insertion

_____ 12. Which muscle type can be microscopically identified by the presence of prominent A bands and peripherally located nuclei?
a. cardiac   c. skeletal
b. smooth

_____ 13. Pointing the toes (plantar flexion) involves contraction of which of these muscles?
a. Tibialis anterior and Extensor   c. Peroneus longus and Semitendinosus
   digitorum longus   d. Calcaneus and Peroneus tertius
b. Gastrocnemius and Soleus

_____ 14. The longest phase of a skeletal muscle twitch is the:
   a. latent period
   b. contraction period
   c. relaxation period

_____ 15. Which of these represents largest to smallest in diameter?
   a. fasciculus, fiber, myofibril, myofilament
   b. fiber, fasciculus, fibril, actin
   c. actin, myofilament, fiber, fasciculus
   d. actin, myosin, myofilament, fibril

_____ 16. Which of these is not a muscle of the Quadriceps femoris group?
   a. Rectus femoris
   b. Vastus lateralis
   c. Vastus intermedius
   d. Rectus abdominis

_____ 17. In an adult, intramuscular injections are often given into which of these muscles?
   a. Gracilis
   b. Deltoid
   c. Latissimus dorsi
   d. Biceps femoris

_____ 18. The "all or none" principle provides for:
   a. graded contraction of single motor units
   b. a single contraction within single motor units
   c. summation
   d. maximum stimuli to contract motor units

_____ 19. Energy sources for skeletal muscular contraction include:
   a. phosphocreatine
   b. ATP
   c. glucose
   d. all of the above

_____ 20. The actual contractile elements of a skeletal muscle cell are the:
   a. tonofibrils
   b. sarcoplasm
   c. desmosomes
   d. myofibrils

_____ 21. Which of the following is associated only with cardiac muscle?
   a. intercalated discs
   b. the lack of distinct myofibrils
   c. reduced sacroplasmic reticulum
   d. sacromeres

_____ 22. The point of anatomical contact between a nerve and muscle is the:
   a. sarcoplasm
   b. motor end plate
   c. epineurium
   d. epimysium

_____ 23. Which of the following is *not* a phase of a simple muscle twitch?
   a. period of relaxation
   b. period of contraction
   c. latent period
   d. summation period

_____ 24. Tonus may be best described as:
   a. the response of a whole muscle to a single stimulus
   b. an increase in strength of response due to repeated strong stimuli
   c. a continuous tension maintained because the nervous system sends periodic nervous impulses to muscles even when they are at rest
   d. a single generalized muscle response due to a greater frequency of stimulation

_____ 25. During excessive muscular activity there is an accumulation and buildup of

   ...., which is a factor in muscle fatigue.
   a. oxygen
   b. glycogen
   c. motor end plate
   d. lactic acid

DISCUSSION QUESTIONS (Optional)

1. a. Why is ATP important for muscle contraction?

    b. Using your text or a reference book, write the equation for the breakdown of ATP:

2. Listed below are several cat muscles. Using reference books, compare them to human muscles with respect to presence or absence, location, number of heads, divisions, or size:

   a. Pectoralis major

   b. Pectoralis minor

   c. Pectoantibrachialis

   d. Xiphihumeralis

   e. Sternomastoid

   f. Clavotrapezius, Acromiotrapezius, Spinotrapezius

   g. Levator scapulae ventralis

   h. Clavodeltoid, Acromiodeltoid, Spinodeltoid

   i. Biceps brachii

   j. Triceps brachii

   k. Caudofemoralis

l. Gluteus maximus

m. Sartorius

n. Adductor femoris

o. Soleus

p. Epitrochlearis

3. If available, attach the following physiological tracings on this page. Use additional pages if needed. Below each recording, write your interpretation of the type of contraction.

   a. single twitch
   b. staircase phenomenon
   c. stimulation of the sciatic nerve by electricity and acetic acid
   d. tetanus
   e. fatigue
   f. work done by skeletal muscle

Digestive System

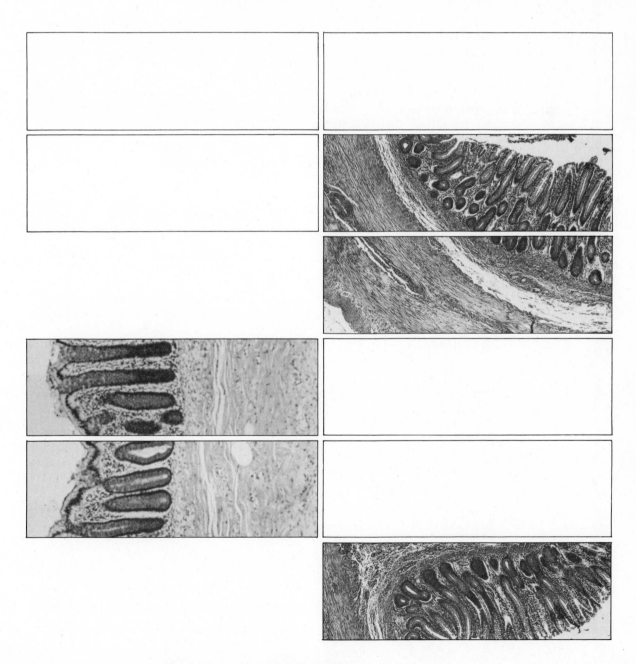

A. Digestive Anatomy

- *Purpose*

  The purpose of Unit VII-A is to enable you to locate and state the functions of the major digestive organs and accessory glands.

- *Objectives*

  To complete Unit VII-A, you must be able to do the following:

  1. Identify the major divisions and subdivisions of the digestive tract.
  2. State the functions of the structures of the alimentary tract.
  3. Briefly describe the histology of the digestive tract.

- *Materials*

  Dissecting instruments
  Preserved cats
  Display material
  Human skulls

  Prepared microscope slides of the following:

developing tooth	small intestine
dried tooth	large intestine
tongue	appendix
parotid gland	rectum
esophagus	anus
stomach	

- *Procedure*

EXERCISE 1

Digestive System

The digestive system is comprised of the following divisions and subdivisions as outlined in Table 1:

TABLE 1 Divisions of the Digestive Tract

MOUTH
Teeth
Incisors
Canines (cuspids)
Premolars (bicuspids)
Molars (tricuspids)
Tongue
PHARYNX
Nasopharynx
Oropharynx
Laryngopharynx

ESOPHAGUS	
STOMACH	
Cardiac sphincter	Fundus
Cardiac	Pylorus
Greater and lesser curvatures	Pyloric sphincter
SMALL INTESTINE	
Duodenum	
Jejunum	
Ileum	
LARGE INTESTINE (COLON)	
Illeocecal sphincter	Descending colon
Cecum	Sigmoid colon (not present in the cat)
Appendix (not present in the cat)	Rectum
Ascending colon	Internal anal sphincter
Hepatic flexure	Anus
Transverse colon	External and sphincter
Splenic flexure	
ACCESSORY ORGANS AND STRUCTURES OF THE DIGESTIVE TRACT	
Pancreas	Common bile duct
Pancreatic duct	Gastrohepatic ligament
Liver	Salivary glands
Gallbladder	Parotid
Cystic duct	Submaxillary
Hepatic duct	Sublingual

Identify on a human skull the four types of teeth—the **incisors**, the **canines** (cuspids), the **premolars** (bucuspids), and **molars** (tricuspids).

On your cat, identify the **hard** and **soft palates**, the **esophagus**, and **tongue.** The tongue is attached ventrally by the **frenulum linguae** and posteriorly by the hyoid bone. Note the papillae, or elevations, on the tongue.

The **pharynx** is the throat cavity and is divided into three regions. The area behind the nose is the **nasopharynx.** The region in which the larynx, or voice box, is situated is referred to as the **laryngopharynx.** The region behind the mouth, known as the **oropharynx,** can be identified by inserting your probe into the **isthmus of the fauces,** the aperture of the throat, and moving the probe anteriorly or posteriorly.

Observe the **parotid gland** beneath the ear in the neck. The **parotid duct,** or **Stenson's duct,** extends anteriorly and empties into the oral cavity near the maxillary fourth premolar. Beneath the parotid gland and posterior to the angle of the jaw is the **submaxillary gland.** Its duct, **Wharton's duct,** leads into the floor of the oral cavity. Last, identify the **sublingual glands** if possible. They are small and located at the base of the tongue and anterior to the submaxillary glands. The duct also empties into the floor of the oral cavity.

Place the cat on its dorsal side and make a midventral incision through skin and muscle with the blunt end of your scissors beginning at the pubic symphysis. Extend this incision cranially through the manubrium of the sternum. Avoid cutting too deeply or you will injure the **thoracic** and **visceral** organs. The smooth membrane lining the body cavity is *parietal peritoneum.* Extend your incision laterally at the level of the **diaphragm,** but do not cut into the diaphragm. Your instructor will demonstrate the proper procedure for exposing the organs in the thoracic region at a later time.

Using the photographs on the following pages, you will locate the **viscera** —that is, all the organs in the abdominal cavity listed in Table 1. Lying over most of the organs within the abdominal cavity is a connective tissue membrane containing a considerable amount of yellow adipose tissue—the **greater omentum.** The greater omentum is a large double fold of peritoneum. In order to see the organs beneath the greater omentum, carefully loosen it and move it to one side (*see* Figures 98 and 99).

FIGURE 98   *Abdominal organs with greater omentum*

Identify the esophagus as it enters the abdominal cavity through the diaphragmatic hiatus. It will be necessary to reflect the organs of the left side of the thoracic cavity gently to the right.

Observe the greater and lesser curvatures of the **stomach.** Cranially to caudally, the divisions of the stomach are **cardiac, fundus, body,** and **pylorus.** If your instructor so indicates, make an incision along the greater (convex) curvature of the stomach. Examine the inside surface for folded walls, **rugae,** that allow for distension and more surface area. The constriction between the stomach and small intestine is the **pyloric sphincter,** a circular smooth muscle that prevents the backflow of food from the small intestine to the stomach. The sphincter can be felt as a hard mass (*see* Figure 99).

The first portion of the small intestine is the **duodenum,** which begins at the pyloric sphincter and extends along the head of the **pancreas** for about 1 inch (12 inches in the human). Where the duodenum turns toward the left side of the body, it becomes the second portion of the small intestine—the

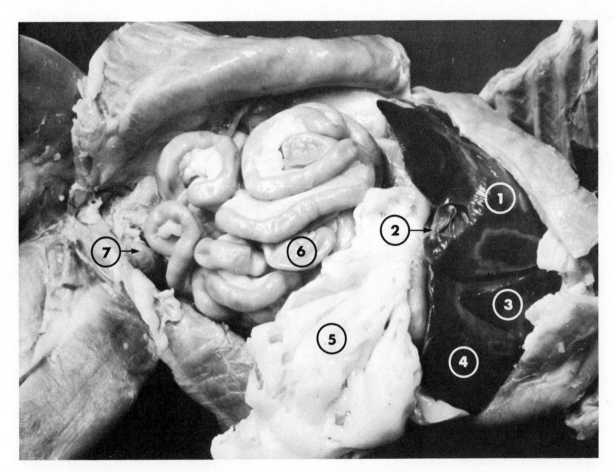

FIGURE 99 *Abdominal organs exposed after transection of greater omentum*

1. Liver, Right Medial Lobe
2. Gallbladder
3. Liver, Left Medial Lobe
4. Liver, Left Lateral Lobe
5. Greater Omentum, Transected
6. Small Intestine
7. Urinary Bladder

jejunum. The third portion of the small intestine is the **ileum** (*see* Figure 100). The jejunum and ileum are approximately equal in length in the cat. In the human being, the jejunum is approximately 8 feet and the ileum approximately 12 feet long. The ileum terminates on the right side of the body and joins the *large intestine*, or **colon.** The ileum and **cecum,** which is the first portion of the large intestine, are separated by the **ileocecal valve,** which you can feel in the cat. The large intestine is considerably larger in diameter than the small intestine and can also be recognized by the presence of sacculations, or **haustra.** The **cecum** is a blind pouch found posterior to the junction of the small and large intestines. In the human being, the appendix is inferior to the cecum. Anterior to the cecum in the cat are the **ascending colon, hepatic flexure, transverse colon, splenic flexure, descending colon,** and **rectum.** The rectum opens to the outside through the *anus.* In the human being, the anus is the terminal inch of the rectum.

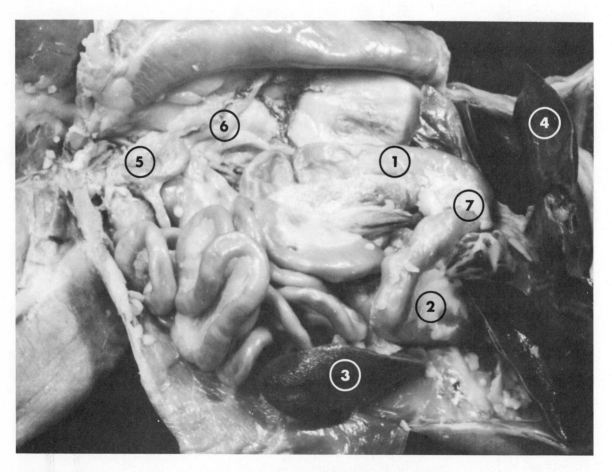

FIGURE 100 *Ventral aspect of deep abdominal organs after transection of small intestine*

1. Duodenum
2. Pyloric Portion of Stomach
3. Spleen
4. Liver, Right Medial Lobe
5. Uterus
6. Horn of Uterus
7. Pyloric Sphincter

The accessory organs to the digestive tract include the pancreas, liver, gallbladder, and salivary glands. The **pancreas** in the cat is composed of two parts: (1) the anterior head of the pancreas, which lies between the pylorus and duodenum; and (2) the body, which extends from the anterior head of the pancreas to the left side of the body dorsal to the stomach. It may be necessary to lift the stomach up toward the diaphragm to see this organ.

The **liver** is the most prominent organ in the abdominal cavity and overlies other organs. The cat liver has the following five lobes (*see* Figure 101):

Right lobe:

1. Right lateral lobe
2. Right medial lobe
3. Right caudate lobe

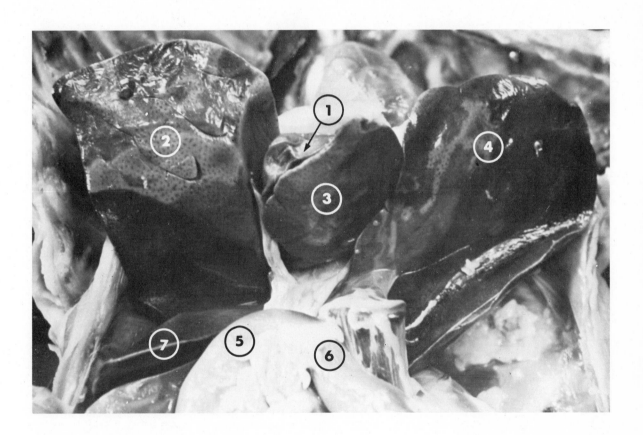

FIGURE 101 *Liver viewed from caudal aspect to show gallbladder*

1. Gallbladder
2. Liver, Right Medial Lobe
3. Liver, Right Medial Lobe
4. Liver, Left Medial Lobe
5. Duodenum
6. Region of Pyloric Valve
7. Caudate Lobe

Left lobe:

1. Left lateral lobe

2. Left medial lobe

The **falciform ligament** is part of the ventral mesentery between the liver and ventral body wall. It is also found between the left and right lobes of the liver. On the ventral surface of the right medial lobe is the **gallbladder,** which usually appears as a greenish sac tapering to form the **cystic duct.** This combines with the **hepatic ducts** to form the **common bile duct,** which penetrates the duodenum at the *ampulla of Vater.*

The **spleen** is an elongated brown organ located on the left side of the body. It is part of the lymphatic system. Supporting the visceral organs are **mesenteries,** which are comprised of simple squamous epithelium. The mesentery that suspends the small intestine from the dorsal body wall and stretches between the parts of the small intestine contains many blood vessels and a large lymph gland—the *pancreas of Aselli.*

EXERCISE 2

Microscopic Examination of Digestive Tissue

Observe the following slides and make a sketch of each:

BUCCAL CAVITY

a. *Developing Tooth*

In this slide, the developing deciduous tooth is embedded in the alveolar process of the jaw. Notice the bone of the **alveolus,** the central area of dental **pulp** that forms the core of the developing tooth, the broad layer of **dentin,** which may stain pink, surrounding the pulp, and the **enamel** immediately overlying the dentin. An area known as the *dental sac* surrounds the developing tooth. Also, notice the darkly staining *odontoblasts* around the outer margin of the dental pulp. *SKETCH and LABEL.*

b. *Dried Tooth*

The broad portion of this section represents the **crown** of the tooth, covered with enamel. The narrower portion represents the part of the tooth beneath the gum line and is referred to as the **root.** Notice that beneath the enamel and extending into the root is a thick layer containing wavy, closely spaced, parallel tubules. This layer is dentin. The junction of the root and crown of the tooth at the gum line is the **neck**. In the region of the root, the dentin is covered by a thin layer of **cementum** and a **periodontal membrane** that is adjacent to the alveolar bone. In the center of your section, you will see a clear area within the dentin. The broader end of this space represents the **pulp cavity**. The more constricted extension of the pulp cavity is the **root canal**. The pulp cavity and root canal contain fibroblasts, odontoblasts, blood vessels, and nerves embedded in connective tissue. *SKETCH and LABEL*.

c. *Tongue*

The mucosa of the tongue is stratified squamous epithelium. In some sections, you will be able to see various types of **papillae**, which contain taste buds (*see* Figures 102–105). Notice the longitudinal, transverse, and oblique planes of skeletal muscle fibers that occupy the interior of the tongue. Embedded in the muscle are lingual glands, blood vessels, and nerves. *SKETCH and LABEL*.

FIGURE 102 *Cleft of papillae showing numerous taste buds*

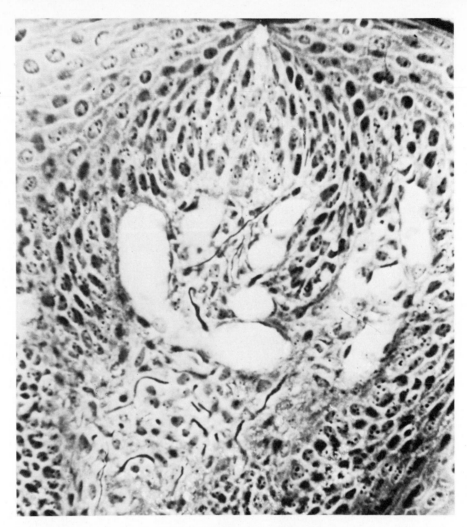

FIGURE 103
Tongue showing taste buds

1. Trench of Vallate Papilla

2. _____

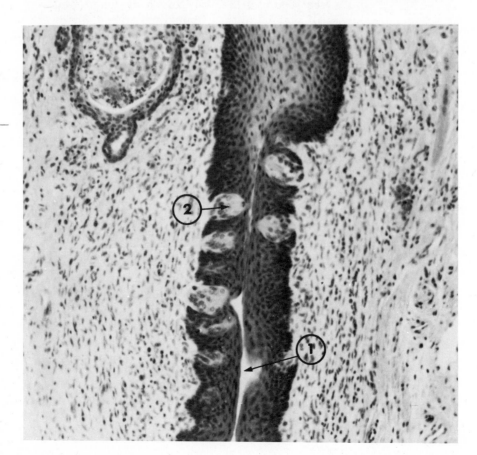

FIGURE 104
Tongue, filiform papilla

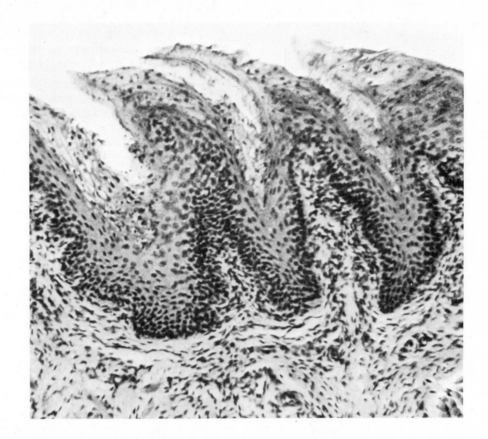

FIGURE 105
Tongue, fungiform papilla

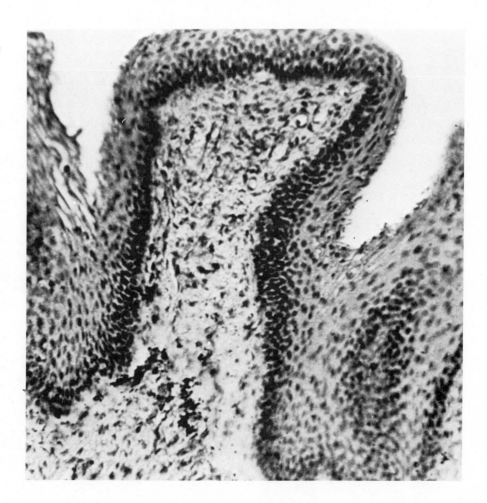

d. *Parotid Gland*

Parotid glands are one of three types of salivary glands (the others are the **sublingual** and the **submaxillary, or submandibular,** glands) located inferior and anterior to the ear. The predominant structures on this slide are *serous alveoli*, which secrete a watery saliva. These glands are comprised of pyramid-shaped cells arranged in a circular manner and contain darkly staining nuclei. In this section, there are also blood vessels, adipose tissue, connective tissue, and ducts that drain the glands. *SKETCH and LABEL.*

ALIMENTARY CANAL

The alimentary canal is essentially a long tube consisting of the esophagus, stomach, small intestine, and large intestine. The wall of the alimentary canal, or digestive tube, has four basic layers. The inner lining, **mucosa,** most often

FIGURE 106
Ileum showing plicae and general structural detail (Courtesy Carolina Biological Supply Company)

1. Serosa
2. Muscularis Externa
3. Submucosa
4. Mucosa

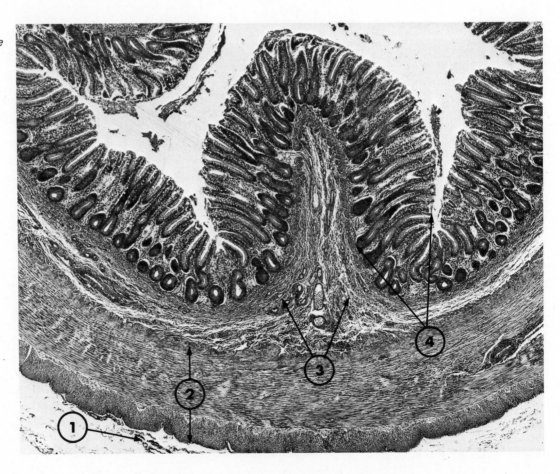

consists of a surface epithelium, an underlying connective tissue layer of **lamina propria**, and a thin layer of smooth muscle—**muscularis mucosae**. The second basic layer is a connective tissue layer beneath the mucosa known as **submucosa**. The third layer is the **muscularis externa** that is usually comprised of two layers of smooth muscle—an inner circular layer and an outer longitudinal layer. The outer layer, or coat, is **serosa**, which is a reflection of the peritoneum lining the walls of the body cavity. Some parts of the digestive tract have an **adventitia** rather than a serosa. Adventitia is a fibrous outer coat. Modifications of the four basic layers (mucosa, submucosa, muscularis externa, and serosa or adventitia) occur in different regions of the digestive tract (*see* Figures 106 and 107).

FIGURE 107
Stomach showing detail of mucosa, submucosa, and muscularis layers

1. _____

2. Muscularis Mucosae

3. _____

4. Mucosa Layer

5. _____

6. Muscularis Externa Layer

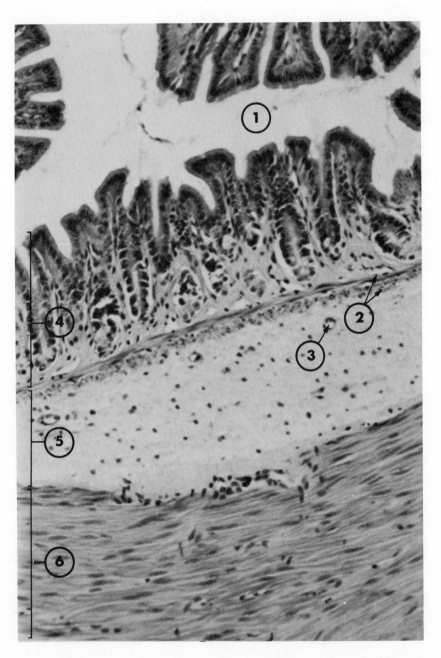

e. *Esophagus*

The mucosa is comprised of stratified squamous epithelium, lamina propria, and muscularis mucosae layers. The submucosa contains *esophageal glands* that appear similar to serous alveloi of the parotid gland. The remainder of this layer is connective tissue containing adipose cells and blood vessels. The muscularis externa consists of two layers—an inner one of circular fibers and an outer longitudinal muscle layer. The muscularis externa of the superior third of the esophagus contains striated muscle, the middle third both striated and smooth muscle fibers, and the inferior third smooth muscle only. The outer layer of the esophagus is primarily adventitia, except at the inferior end where it is serosa. *SKETCH and LABEL.*

f. *Stomach*

The stomach has several regions that differ slightly in their histology. The four basic layers are present in all regions and will be described in general. Observe your section first under low power and then under high power to examine the layers in detail.

The mucosal layer of the stomach consists of a surface of simple columnar epithelium supported by lamina propria (*see* Figure 107). If you are observing the fundic, or superior, portion of the stomach, notice that the epithelial layer dips into the mucosa, forming **gastric pits.** Farther down in the mucosal layer beneath the gastric pits are blue-staining cells, known as **chief** (or *zymogenic*) **cells** that contain **pepsinogen,** the precursor to the **enzyme pepsin.** There are also cells that stain red, or **parietal cells,** that secrete hydrochloric acid. Chief and parietal cells are arranged to form *fundic glands* (*see* Figures 108 and 109). Toward the muscularis mucosae you may see some fundic glands in cross section.

The gastric mucosa of the pyloric region of the stomach, which is more proximal to the small intestine, has a different appearance from the fundic region. The gastric pits of the pylorus are deeper, and *pyloric glands* are located between the gastric pits and muscularis mucosae. Cells comprising the pyloric glands stain the same as cells of the gastric pits but are slightly larger in volume. Pyloric gland cells secrete mucus and hydrolytic enzymes. In all areas of the stomach, there are darkly staining lymph nodules at intervals immediately superior to the muscularis mucosae. Lamina propria serves as the connective tissue both in the region of the gastric pits and in the region of the fundic and pyloric glands. The submucosa of the stomach contains blood vessels and adipose and connective tissue.

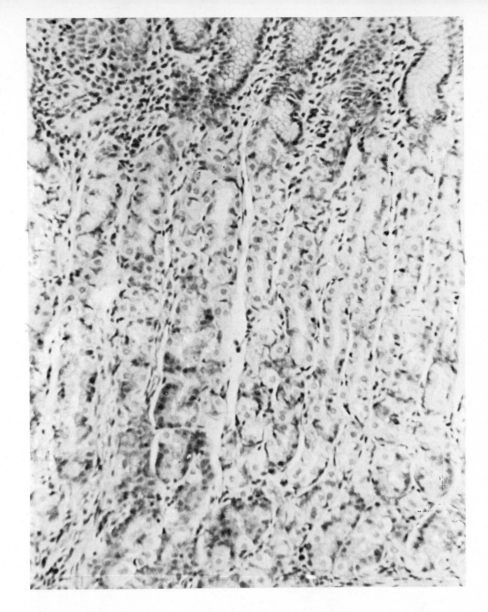

FIGURE 108 *Detail of fundic mucosa of stomach*

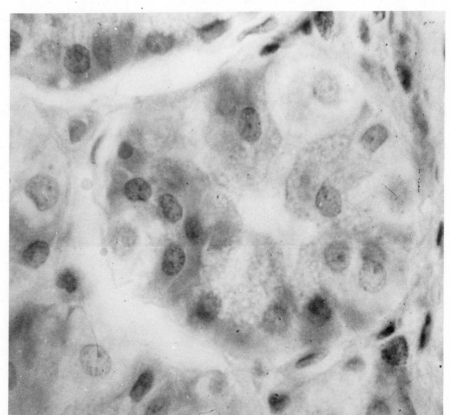

FIGURE 109 *Detail of fundic gland of stomach*

The muscularis externa region of the stomach consists of three layers. An inner layer of *oblique fibers* is present in addition to the middle and outer layers of circular and longitudinal fibers, respectively. The serosa is comprised of connective tissue. *SKETCH and LABEL* your slide of the stomach.

THE SMALL AND LARGE INTESTINES

There are several features common to both the small and large intestines. The mucosa forms deep pits known as *intestinal crypts of Lieberkühn*, or **intestinal glands**. The intestinal mucosa contains four cell types. The frequency with which these cell types appear depends upon the particular region of the intestine. Simple **columnar** epithelial cells line the inner surface of the intestinal mucosa. *Goblet cells* (*see* Figure 110 and **Color Plate 6**) secrete mucus and are thought to be formed through the transformation of

FIGURE 110 *Detail of goblet cells from a jejunum villus*

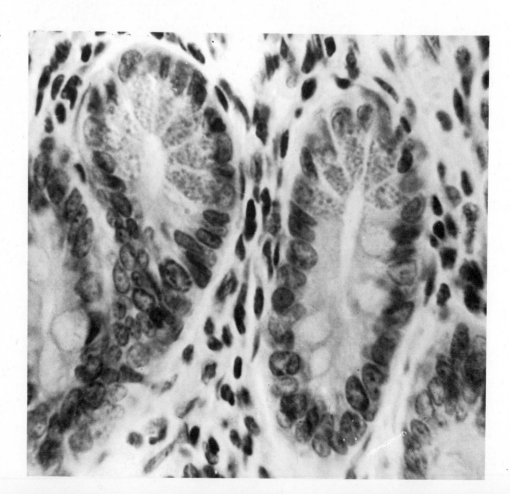

simple columnar cells. *Argentaffin cells* are usually found between other cells that line intestinal glands. *Paneth cells* are located at the base of intestinal glands. Argentaffin cells have a granular cytoplasm and are thought to secrete *serotonin*, a powerful stimulant of smooth muscle contraction. Paneth cells are pyramidal in shape, with a basophilic cytoplasm that stains darkly and granules that stain with acid dyes, usually red or orange, depending upon the particular stain used. Other common features of the intestinal tract mucosa are lamina propria, which is found underlying the mucosal epithelium and between intestinal glands; lymph nodules, which are more prominent in certain parts of the intestine than others; and a thin muscularis mucosae, which has an inner layer of circular and an outer layer of longitudinal smooth muscle fibers.

The submucosa of the intestinal tract consists of loose areolar tissue, adipose tissue, blood vessels, and nerves. There may be other structures present in the submucosa, depending on the particular intestinal region. The muscularis externa consists of an inner circular smooth muscle layer and an outer longitudinal layer. The outer layer of the intestinal tract is serosa.

g. *Small Intestine*

The small intestine is divided into three regions: the **duodenum,** which is approximately 10–12 inches long; the **jejunum,** which is about 8 feet long; and the **ileum,** which is about 12 feet long. The entire small intestine is characterized by the presence of **plicae circulares** (or *valves of Kerckring*), which are folds in the mucosa and submucosa, and by **villi** (*see* Figure 106), which are microscopic fingerlike projections of the mucosa overlying the plicae. Villi further increase the absorptive surface area of the small intestine (*see* Figure 111). Crypts of Lieberkühn can be seen inferior to the villi and superior to the muscularis mucosae.

Observe a section of duodenum. The primary histological characteristic of this region is the presence of *duodenal glands*, or Brunner's glands, that are found primarily in the submucosa (*see* Figure 112) and, to some extent, in the mucosa. The secretions of these glands are of a mucoid nature. Also, notice the four basic layers in this section: the mucosa, with mucosal goblet cells, argentaffin cells, and simple columnar cells, as well as the crypts of Lieberkühn, and muscularis mucosae; submucosa; muscularis externa; and serosa (*see* Figures 106 and 113). *SKETCH and LABEL.*

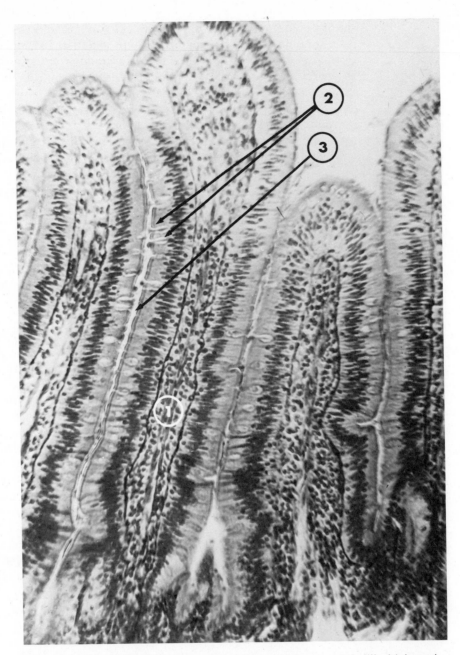

FIGURE 111 *Longitudinal section showing general structure of villi with lacteal detail visible*

1. Lacteal
2. Goblet Cells
3. Striated Border

FIGURE 112 *Detail of Brunner's glands of duo-denum*

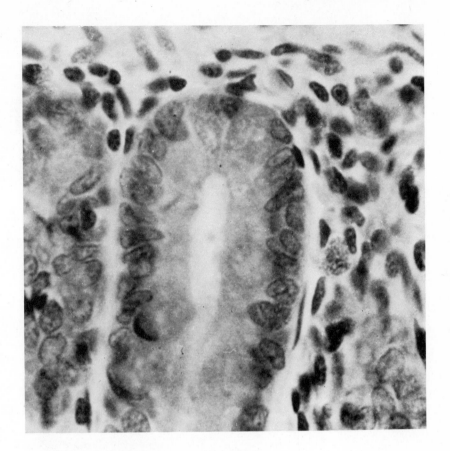

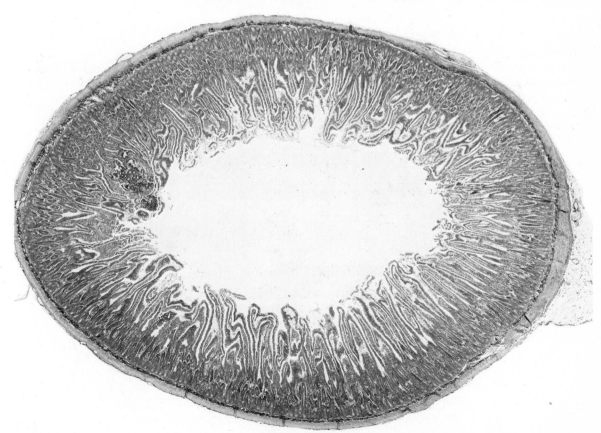

FIGURE 113 *Cross section of jejunum showing muscular wall, villi, and glands* (Courtesy Carolina Biological Supply Company)

FIGURE 114
*Microvilli, detail of
small intestine*
(Courtesy Carolina
Biological Supply
Company)

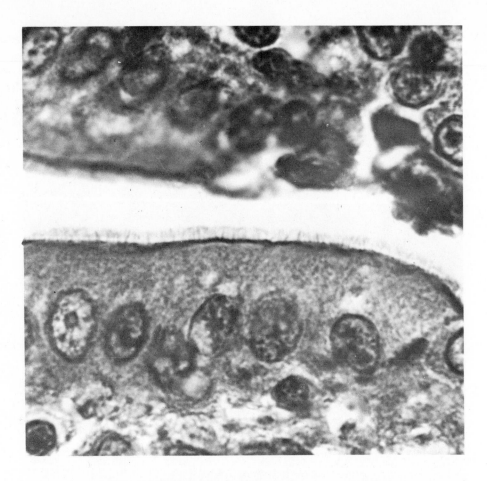

FIGURE 115 *Detail of ganglion cells
of Auerbach's plexus*
(Courtesy Carolina
Biological Supply
Company)

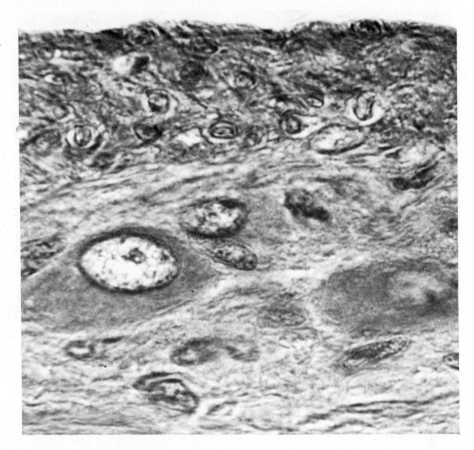

The sections of the jejunum and ileum are similar. Notice the presence of darkly staining lymphatic nodules in the submucosa. In the ileum, aggregations of these nodules are known as **Peyer's patches.** Also, notice that the villi become more elongated as one progresses from the duodenum to the jejunum to the ileum. Goblet, Paneth, and simple columnar cells should be visible on these sections (*see* Figures 114 and 115). *SKETCH and LABEL.*

h. *Large Intestine*

The large intestine lacks villi (*see* Figure 116). Intestinal glands, or crypts, in the mucosa are very prominent. Notice the numerous goblet cells (*see* Figure 117) in the lining of the intestinal glands in this section. *Solitary lymph nodes* can be seen in the lamina propria of the mucosa immediately adjacent to the muscularis mucosae. Also, observe the submucosa, muscularis externa, and serosa in this section. *SKETCH and LABEL.*

FIGURE 116 *General structure of colon*

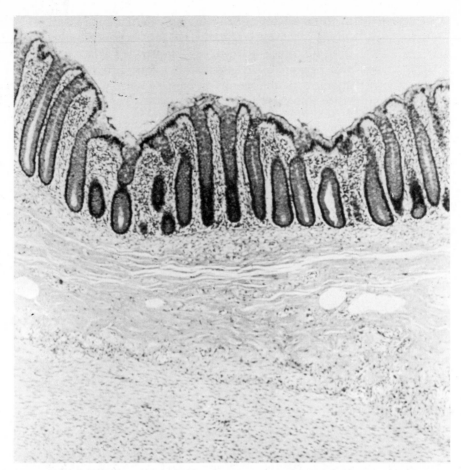

FIGURE 117
Colon, mucosa layer

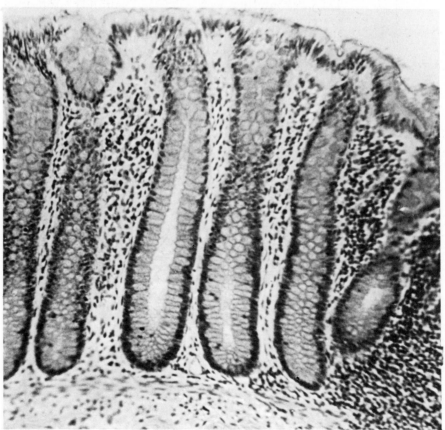

i. *Appendix*

Notice the poorly developed intestinal glands and numerous lymph nodules in this section. *SKETCH and LABEL.*

j. *Rectum*

The rectum is characterized by the presence of large longitudinal folds of mucosa and submucosa known as *rectal columns*. These columns are seen in transverse section on your slide. Notice the four basic layers here, including the crypts of Lieberkühn in the mucosa. *SKETCH and LABEL.*

k. *Anus*

The mucosa of the anus is stratified squamous epithelium. The lamina propria contains a plexus of large veins that when dilated are known as *hemorrhoids. SKETCH and LABEL.*

B. Digestive Chemistry

■ *Purpose*

The purpose of Unit VII-B is to enable you to understand the basic chemical principles of digestion and to introduce to you the nature and scope of enzymatic activity that is representative of metabolic reactions within cells.

■ *Objectives*

In order to complete these exercises, you must be able to do the following:

1. Summarize the digestion of proteins, carbohydrates, and fats.
2. Perform chemical tests for the identification of various amino acids, proteins, carbohydrates, and fats.
3. Summarize enzyme activity in terms of pH, temperature, and concentration ratios.

SUMMARY OF PROTEIN, CARBOHYDRATE, AND FAT DIGESTION

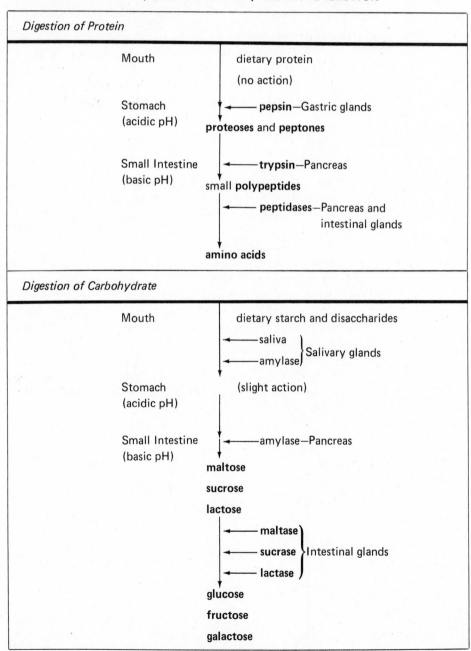

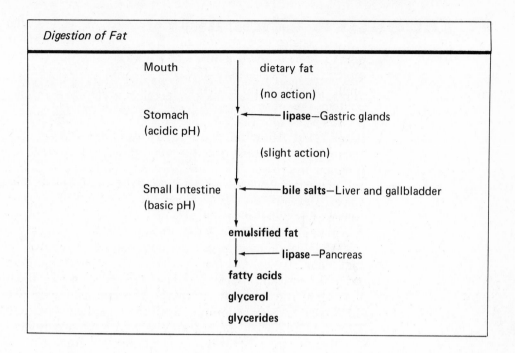

Digestion of Fat

Mouth   dietary fat

               (no action)

Stomach   ← **lipase**—Gastric glands
(acidic pH)

               (slight action)

Small Intestine ← **bile salts**—Liver and gallbladder
(basic pH)

emulsified fat

               ← **lipase**—Pancreas

fatty acids

glycerol

glycerides

EXERCISE 1

Thin Layer Chromatography of Amino Acids

Materials

1 mg/ml solutions of the amino acids proline, phenylalanine, cystine, and asparagine dissolved in an alcoholic acid solution—that is, in 0.5MHCl made up in ethanol; mix well; if amino acids do not dissolve, add a slight amount of water

Pasteur pipettes or Hamilton syringes

Concentrated ammonium hydroxide solution

60°C oven

Silica gel thin-layer chromatography sheets

Developing tank

Migrating solvent: butyl alcohol acetic acid, water—80:20:20 volume/volume

Ninhydrin solution: 0.3% ninhydrin in butyl alcohol containing 3% glacial acetic acid; or commercially prepared ninhydrin spray

Procedure

Score a silica gel sheet with a pencil or pasteur pipette in the following manner:

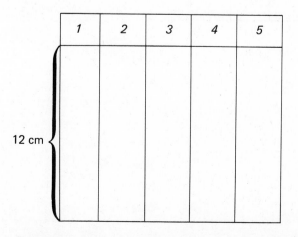

Spot 1 microliter (μl) (about two drops) of amino acid solution with a Pasteur pipette or syringe about 2 cm from the bottom of the sheet in columns 1 to 4, applying a different amino acid in each column. Be sure to keep a record of which amino acid you spotted in each column. In the fifth column, spot an amino acid taken from a vial labeled "unknown" that actually contains one of the four amino acids you are using.

Keep the spot in each column as small as possible, blowing gently to facilitate drying. Allow the spots to dry 15 minutes. To neutralize the acid, hold each of the five spots over the open mouth of a bottle of concentrated ammonium hydroxide for a few seconds. Without touching the surface of the gel, place the sheet into a developing tank containing the migrating solvent. Cover. Develop the chromatogram for approximately 3 hours or until the solvent has traveled a distance of 12 cm (10 cm above the spots). Place a maximum of two sheets in each tank.

When the migration is complete, carefully remove the sheet from the tank, mark the solvent front (distance the solvent traveled up the sheet), and set aside until dry. Spray the developed sheet with 0.3% ninhydrin or dip it in a ninhydrin solution. Place the chromatograms on a tray and heat in a 60°C oven for several minutes until the separated zones appear clearly visible. Do not place the chromatogram directly on the oven shelf.

Most amino acids will appear purple, but some will be brown or yellow. Measure the distance traveled by each amino acid and by your "unknown." Calculate an R_f (ratio of fronts) for each column:

$$R_f = \frac{\text{distance traveled by spot (in centimeters)}}{\text{distance traveled by solvent front (in centimeters)}}$$

The R_f value is always a value between 0 (spot did not move) and 1 (spot moved with the solvent front). Each amino acid has a different R_f value.

Record the appearance of spots on your plate in this space:

1	2	3	4	5

R_f value of each spot   Name of amino acid

1. . . . .   1. . . . .

2. . . . .   2. . . . .

3. . . . .   3. . . . .

4. . . . .   4. . . . .

5. . . . .   5. . . . .

Identify your unknown in column 5 by comparing its R_f value with the R_f's of the known amino acids in a chart provided by your instructor.

EXERCISE 2

Tests and Properties of Proteins

Proteins are compounds comprised of amino acids joined together with peptide bonds:

All amino acids and proteins contain carbon, hydrogen, oxygen, and nitrogen. A few proteins also contain other elements.

Materials

1% albumin solution or egg white	0.1% freshly prepared ninhydrin	Ring stands
10% NaOH	Undiluted egg white	5% trypsin solution
1% $CuSO_4$	Distilled water	Bunsen burners
Millon's reagent	Test tubes	Test tube holders
0.1% alcohol solution of glycine or alanine	Beakers	Test tube racks
		Medicine droppers

Procedure

a. *Biuret Reaction*

The **Biuret reaction** is specific for compounds containing two or more peptide bonds. To 3.0 cc of 1% albumin solution or egg white, add 3.0 cc of 10% NaOH and one drop of 1% $CuSO_4$. Mix. Add additional drops of $CuSO_4$ until a violet color is obtained, indicating a positive reaction.

b. *Millon Reaction*

The **Millon reaction** is specific for the amino acid tyrosine, which contains a benzene ring to which is attached a hydroxyl (–OH) group. Add a few drops of Millon's reagent to 5 cc of 1% albumin solution. Boil carefully. A brick-red color indicates the presence of tyrosine, a common amino acid in most proteins.

c. *Ninhydrin Reaction*

The **Ninhydrin reaction** is a test for alpha amino acids. To 5 cc of 0.1% alcohol solution of glycine or alanine, add 0.5 cc of 0.1% ninhydrin (freshly prepared). Heat. Note color formation. Repeat, using 3 cc of 1% albumin solution instead of an amino acid. Heat to boiling the albumin to which ninhydrin was added and cool. *Color?* . . . .

d. *Heat Coagulation*

Place 5 cc of undiluted egg white in a test tube and place in a water bath. Boil. *Describe your results. Is this reaction reversible upon cooling and agitation?*

e. *Digestion of Protein*

Place a small piece of cooked egg white from Exercise 2(d) into each of two test tubes. To Tube 1, add 10 ml distilled water (buffered to pH 8). To Tube 2, add 5% trypsin or pancreatin solution. Allow to stand in a warm place for several hours. Observe. *DESCRIBE your results:*

RECORD the results of Exercises 2(a)–2(e) in the table in Question 33 of the Discussion following this unit.

EXERCISE 3

Tests and Properties of Carbohydrates

All **carbohydrates** contain the elements carbon, hydrogen, and oxygen, the last two of which are present in the same ratio as in water (2:1). Carbohydrates may be classified as mono-, di-, or polysaccharides. The **mono-** and **disaccharides** resemble each other considerably, but polysaccharides bear little resemblance to the other two classes.

Materials

4% glucose solution	Barfoed's reagent	Beakers
5% alcoholic solution of	4% fructose solution	Test tubes
alpha naphthol	Seliwanow's reagent	Medicine droppers
Concentrated H_2SO_4	1M HCl	Ring stands
Benedict's solution	1M NaOH	Test tube holders
1% glucose solution	Thin starch paste	Test tube racks
1% sucrose solution	Lugol's iodine	Goggles

Procedure

a. *The Molisch Reaction*

The **Molisch reaction** is a general test for carbohydrates. In a test tube, place 5 cc of 4% glucose solution. Add two drops of 5% alcoholic solution of alpha naphthol. Mix. Place 5 cc concentrated sulfuric acid (*CAREFUL*) in a second tube. With the tube containing glucose in your left hand, pour the sulfuric acid (H_2SO_4) slowly and gently down the side of the tube containing glucose. The object is to get a layer of H_2SO_4 under the glucose solution. If a purple or pink color is formed at the boundary, the substance is a carbohydrate. *RESULT?*

b. *Benedict's Test*

Benedict's test is a test for reducing sugars, which include glucose, fructose, galactose, mannose, maltose, and lactose. To 5 cc of Benedict's solution, add four to five drops of 1% glucose solution. Boil in a water bath for 2 minutes. Cool slowly. A red, green, or yellow precipitate (ppt) indicates a positive reaction. Repeat, using sucrose. *RESULTS?*

c. *Barfoed's Test*

Barfoed's test is a test for monosaccharides. To 5 cc of Barfoed's reagent, add five to six drops of 4% glucose solution. Boil in a water bath for 1 minute. Set aside for 15–20 minutes and watch for a red ppt, which indicates a positive reaction. *RESULTS?*

d. *Seliwanow's Test*

Run this test in duplicate, using one tube with glucose and another tube with fructose. Add five drops of 4% glucose solution to one tube containing 5 cc Seliwanow's reagent and five drops of 4% fructose to another. Boil for 30 seconds. What does **Seliwanow's test** enable you to do?

e. *Inversion of Sucrose*

Add 5 cc of 1% sucrose solution to each of two test tubes. To the first tube, add 5 cc of Benedict's solution and bring to a boil. Remove from heat immediately. To the second tube, add two drops of dilute ($1M$) HCl and boil in a water bath for 1 minute. Neutralize with an equal amount and strength of NaOH. (Why?) Test this solution with 5 cc of Benedict's solution. How do you explain the different results for these two procedures?

f. *Test for Starch and Starch Hydrolysis*

Place 5 ml of thin starch paste into a test tube. Add one drop of Lugol's iodine solution. Is there a color change? . . . . Put 10 ml of the same starch solution into a test tube and add four or five drops of saliva. Set in a 37°–40°C water bath for ½ hour or more. Is there a color change when iodine is added? *EXPLAIN*.

RECORD the results of Exercises 3(a)–3(f) in the table in Question 33 of the Discussion following this unit.

EXERCISE 4

Tests and Properties of Fats

Fats are the esters (organic salts) formed by the union of a fatty acid and an alcohol. The three common fats in our foods and body are oleic, palmitic, and stearic, which are formed, respectively, from oleic, palmitic, and stearic acids united chemically with the alcohol glycerol (glycerin). One molecule of glycerol can unite with three fatty acids as follows:

$$H_3C-(CH_2)_{16}-\overset{O}{\overset{\|}{C}}-OH$$

$$H_3C-(CH_2)_{16}-\overset{O}{\overset{\|}{C}}-OH + \begin{matrix} HO-CH_2 \\ HO-CH \\ HO-CH_2 \end{matrix} \quad \underset{\longleftarrow}{\overset{-3H_2O}{\longrightarrow}} \quad$$

$$H_3C-(CH_2)_{16}-\overset{O}{\overset{\|}{C}}-OH$$

3 molecules of stearic acid   1 molecule of glycerol

$$H_3C-(CH_2)_{16}-\overset{O}{\overset{\|}{C}}-O-CH_2$$
$$H_3C-(CH_2)_{16}-\overset{O}{\overset{\|}{C}}-O-CH +3H_2O$$
$$H_3C-(CH_2)_{16}-\overset{O}{\overset{\|}{C}}-O-CH_2$$

tristearin (a triglyceride)

Materials

Fresh cottonseed oil	Sweet cream (fresh)
Distilled water	0.5% Na_2CO_3
Ethyl alcohol	Test tubes
Ether	Test tube racks
5% trypsin	Bunsen burners
Beakers	Thermometers
Benzene	Ring stands
Carbon tetrachloride	Blue litmus solution

Procedure

a. *Solubility of Fat*

Set up five test tubes as follows. **CAUTION: Make sure that there are no Bunsen flames burning in the area in which you are working.**

Tube No.	Solvent (5 ml)
1	water
2	ethyl alcohol
3	ether
4	benzene
5	carbon tetrachloride

Add five drops of fresh cottonseed oil to each tube and shake well. Observe the tubes immediately and again after 15 minutes. *RESULT?*

b. *Digestion of Emulsified Fat*

1. Add 10 ml sweet cream to a test tube.
2. Add 8 to 10 drops of 1% blue litmus solution (as indicator) to the cream to impart a blue color. Mix.

3. Pour 5 ml of this mixture into another test tube.
4. Add 2 ml of 5% trypsin to one tube and 2 ml 0.5% Na_2CO_3 to the second tube.
5. Place both tubes in a beaker of 40°C water. Observe after 1 hour. Did you observe a color change? *EXPLAIN*.

RECORD the results of Exercises 4(a) and 4(b) in the table in Question 33 of the Discussion following this unit.

EXERCISE 5

Enzymes: Chemical Activity in Living Systems

The purpose of this experiment is to introduce you to the nature and scope of biochemical activity that is indicative of metabolic reactions within all cells. The liver contains enzyme systems, one of which is catalase, the enzyme that will be studied in this exercise.

Materials
The following materials are sufficient for each group of four students:

(1) 5-cc disposable syringe
(1) Ring stand
(1) Buret clamp
(1) 50-ml buret
(1) 18-in. of aquarium tubing
(8) 50-ml Erlenmeyer flask
(1) 1 one-hole rubber stopper
(1) 3-in. glass tubing (¼ in. O.D.)
(1) Water trough
    Stock liver extract solution—20%, 50%, 70%, (500 ml each/30 students)
(1) Wax marking pencil
    Blender (for entire class)
    3% H_2O_2 (520 cc/30 students)
    Ringer's solution (1480 cc/30 students)
    Buffer solutions (pH 2, 5, 7, 8, and 12) of 100 ml quantity
    Liver (fresh calf) 2 lbs/30 students
(2) Wood splints
    Ice
(1) Finger bowl
(1) °C thermometer
    Paper towels

A. APPARATUS SETUP PROCEDURE

1. Several students should begin to assemble the apparatus as shown in Figure 118 while the other partners prepare the liver extract. Fill the water trough three-fourths full with tap water. Fill the 50-ml buret to the top with water. Check to see that the stopcock is closed. Place your thumb over the top of the opening of the buret; invert it and submerge in the water trough. Clamp the buret securely and place a paper towel between the clamp and buret in order to avoid breakage.

2. Cut glass tubing to approximately 3 in. and insert in a twisting motion into the rubber stopper until 2 in. are below the stopper.

3. Attach the aquarium tubing securely to the outside portion of the glass tubing and set this aside. The apparatus should now appear as shown in Figure 118.

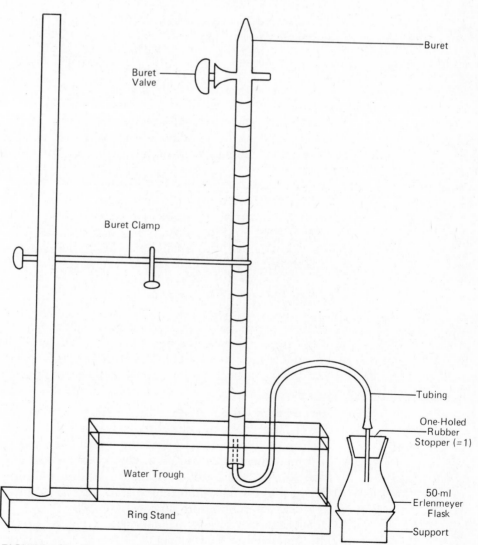

FIGURE 118 *Enzyme apparatus*

B. LIVER EXTRACT PREPARATION

While your partners prepare the apparatus, you may now help to prepare the liver extract. *SEVERAL STUDENTS MAY PREPARE EXTRACT FOR THE ENTIRE CLASS.*

1. Take ½ pound of fresh calves liver, dice it, and place in a blender. Pour 500 ml of Ringer's solution into the blender and blend to the consistency of a very thin malt. When completed, pour into a flask or beaker and label *Stock Liver Extract.* This stock solution will be the basis for determining the effects of temperature and substrate concentrations.

2. Solutions of liver extract must now be prepared with the following buffered pH solutions: 2, 5, 7, 8, and 12 (others may be substituted). Use ¼ pound of liver diced and chopped. Place in blender and add 330 ml of buffer of pH 2. Blend to the consistency of a thin malt. Pour into a flask and label *Liver Extract—pH 2.* Rinse out the blender container with distilled water and *REPEAT this procedure* for all the other pH solutions.

3. You have now prepared all of the liver extract stock solutions that will be utilized during the exercise. You will need to make dilutions of the stock liver extract solutions that you prepared in Step 1. Use the following dilutions and label:

Concentration Ratios	Add
20%	8 ml of Ringer's solution and 2 ml of stock liver extract
50%	5 ml of Ringer's solution and 5 ml of stock liver extract
70%	3 ml of Ringer's solution and 7 ml of stock liver extract

The enzyme solutions have now been prepared. These solutions will be utilized later in the experiment in order to demonstrate the effects of enzyme substrate concentration.

C. EXPERIMENTAL PROCEDURES

You have now completed the task of setting up the apparatus and producing the stock liver extract at the various concentrations and pH values. The only other variable that must be considered will be that of the effects of temperature on the stock liver extract solution. This variable will be adjusted at the time of use.

1. Place exactly 10 ml of stock liver extract solution into the 50-ml Erlenmeyer flask. Place the one-holed stopper with the glass fitting securely in the top of the flask and insert the opposite end of the aquarium tubing into the open end of the buret that is submerged in the water. You may have to shift the buret in order to position the tubing correctly. Water will go into the aquarium tubing. Check to make sure that all connections of tubing are tight before proceeding to Step 2.

2. Obtain a disposable syringe and draw 1 cc of hydrogen peroxide (H_2O_2) into the barrel of the syringe.

3. Insert the syringe needle into the aquarium tubing rapidly at the junction of the tubing and the glass fitting, injecting the H_2O_2 into the Erlenmeyer

flask. Rapidly push the plunger down the syringe and quickly remove from the tubing.

4. Now shake the Erlenmeyer flask with a slow and constant rotary motion.
5. In order to determine the rate of this reaction, watch the amount of water in milliliters that is displaced into the water trough. Observe the bubbles of gas over a period of 3 minutes. Begin this 3-minute timing the instant you introduce the hydrogen peroxide (H_2O_2).
6. Slowly turn the buret valve releasing the accumulated gas into an inverted test tube held over the tip of the buret. (Do not permit the water to enter the buret valve!) Quickly insert a glowing wood splint into the test tube and check the gas.
   a. *Hydrogen*—it will whistle loudly as it ignites and the test tube will turn warm to the touch.
   b. *Oxygen*—it will ignite the glowing splint into flame.
   c. *Carbon dioxide*—it will smother the glowing splint.

What kind of gas have you collected? . . . .
7. *REPEAT Steps 1-5*, since you are probably unfamiliar with this procedure and have introduced errors into the exercise. It is important that you perform the previously mentioned steps accurately throughout the remainder of the exercise. You have now seen the rate of reaction for 100% substrate enzyme concentration. This will be the basis for comparing the rest of the parameters that we will investigate, namely, temperature, pH, and varying enzyme concentrations.
8. *RECORD* the ml of H_2O displaced in 3 minutes under the *100% Enzyme Stock Solution (Control)* headings in Questions 26–28 of the Discussion.

D. TEMPERATURE INVESTIGATION

1. Use a clean, dry 50-ml Erlenmeyer flask. At the end of each exercise rinse out the flask with distilled water and dry it.
2. The 20°C room temperature is to be used first in investigating the amount of gas that will be produced by the enzyme stock solution. *RECORD* your results in the table in Question 26 of the Discussion at the end of this unit.
3. Pour 10 ml of stock enzyme solution into a clean 50-ml Erlenmeyer flask and place this flask in a finger bowl that has cold water and ice cubes in it. The water temperature should be 20°C. Place a one-holed stopper in the top of the flask with the aquarium tubing attached to the stopper.
4. Check to make sure that the opposite end of the aquarium tubing is in the open end of the buret and that the end is completely submerged in the water trough. Double check to make sure that all connections of tubing are tight.
5. Obtain the disposable syringe and draw 1 cc of hydrogen peroxide (H_2O_2) into the barrel of the syringe.
6. Insert the syringe needle into the aquarium tubing rapidly at the junction of the tubing and the glass fitting. Rapdily push the plunger down and quickly remove from the tubing.
7. *Immediately shake the Erlenmeyer flask with a slow and constant rotary motion.*
8. In order to determine the rate of this reaction, watch the amount of water in milliliters that is displaced into the water trough. Observe the bubbles of gas over a period of 3 minutes. Begin the 3-minute timing at the instant you introduce hydrogen peroxide (H_2O_2).
9. *REPEAT* the previous procedures for 37°C and 45°C temperatures. Pour water into the finger bowl. Adjust the hot and cold faucets to produce a

temperature that is 37°C. Place the Erlenmeyer flask in the finger bowl containing 37°C water. Wait 1 or 2 minutes and then *repeat* Steps 3–8. *Repeat* using 45°C water in the finger bowl. *RECORD* your results in the table in Question 26 of the Discussion.

10. Again, *RECORD* the amount of gas displacement in milliliters that is present within the buret. (*NOTE:* IF THE WATER LEVEL IS LOW IN THE BURET, THEN REFILL THE BURET SO THAT THERE IS NO AIR WITHIN IT.)

E. pH INVESTIGATION

Earlier, you prepared liver extract solutions with various buffers and you labeled these *Liver Extract—pH——*. The pH solutions of 2, 5, 7, 8, and 12 (substitution may be made by your instructor) will be the ones that we will investigate in terms of the effect pH has on enzyme substrates.

1. Use five clean, dry Erlenmeyer flasks and label them *2, 5, 7, 8,* and *12.*
2. Pour 10 ml of the liver extract—pH 2 into a labeled flask. Investigate each of the pH values.
3. Place the stopper and glass tubing arrangement into your Erlenmeyer flask labeled *pH 2* and check to make sure all tube fittings are tight and that there is sufficient water within the buret.
4. Obtain the disposable syringe and draw 1 cc of hydrogen peroxide (H_2O_2) into the barrel of the syringe.
5. Insert the syringe needle into the aquarium tubing at the junction of the tubing and the glass fitting and rapidly push the plunger down and then quickly remove the syringe.
6. *Immediately shake the Erlenmeyer flask while you observe for 3 minutes.*
7. Again, observe the rate of this reaction over a period of 3 minutes. Begin this 3-minute timing the instant you introduce the hydrogen peroxide with the syringe and shake the flask.
8. *RECORD* in the table in Question 28 of the Discussion at the end of this unit the amount of water displacement.
9. *REPEAT* the previous procedure for pH's 5, 7, 8, and 12.
10. *RECORD* all results in the table in Question 28 of the Discussion.

F. ENZYME CONCENTRATION INVESTIGATION

1. Earlier, you prepared from the stock liver extract solution concentrations that were labeled *20%, 50%,* and *70%.* You will use these to investigate the amount of enzyme necessary to produce the enzyme/substrate reaction.
2. Pour 10 ml of the 20% liver extract solution into a clean 50-ml Erlenmeyer flask. Then pour 10 ml of the 50% and 70% solutions into the other Erlenmeyer flasks. Label each flask.
3. Again, take the disposable syringe and insert 1 cc of hydrogen peroxide (H_2O_2) into the system.
4. Gently shake the flask for the 3-minute observation period.
5. Again, observe the rate of this reaction by the amount of water in milliliters that is being displaced by the gas from this reaction.
6. *RECORD* in the table in Question 27 of the Discussion at the end of this unit the amount of water in milliliters that was displaced by the gas from this reaction.
7. *REPEAT* the above procedures for the 50% and 70% concentration ratios and record in the table in Question 27 of the Discussion.

Digestive System

DISCUSSION

Word Relationships

Indicate which of the following terms in each group is *unrelated* to the others:

_____ 1. a. mucosa   c. muscularis
       b. submucosa   d. rugae

_____ 2. a. insulin and glucagon   c. usually is 6–9 inches in length
       b. is roughly fish shaped   d. accumulates bile

_____ 3. a. is the largest body gland   c. assists in maintaining blood sugar
       b. has two main lobes   homeostasis
                              d. the duct of Wirsung extends the entire
                                 length of the gland

_____ 4. a. mastication   c. peristalsis
       b. deglutition   d. hydrolysis

_____ 5. a. completes digestion of foods   c. 20 feet long
       b. duodenum   d. serves as a reservoir

_____ 6. a. dentin   c. pulp
       b. periodontal membrane   d. haustra

_____ 7. a. ninhydrin reaction   c. Molisch reaction
       b. Biuret test   d. Millon reaction

_____ 8. a. transverse colon   c. trypsin
       b. haustra   d. water reabsorption

Multiple Choice

_____ 9. The organ that produces bile is the:
       a. gallbladder   c. stomach
       b. pancreas   d. liver

_____ 10. The stomach:
       a. secretes gastric juice   c. provides for limited absorption of alco-
       b. mechanically churns food into   hol and water
            chyme   d. all of the above

_____ 11. The structure that connects the pharynx to the stomach is the:
       a. trachea   c. esophagus
       b. larynx   d. epiglottis

_____ 12. The hard covering over the crown of a tooth is known as:
       a. enamel   c. dentin
       b. cementum   d. pulp cavity

13. The waves of muscular contraction that push food down through the esophagus, stomach, and rest of the digestive tube are known as:
   a. periosteum   c. peristalsis
   b. pericardium   d. peritoneum

14. The most important accessory digestive organ that secretes digestive enzymes to digest carbohydrates, proteins, and lipids is:
   a. the spleen   c. the small intestine
   b. the pancreas   d. the liver

15. An example of a monosaccharide is:
   a. glucose   c. maltose
   b. glycogen   d. cellulose

16. Which of the following is the most important factor in determining the rate at which a chemical reaction takes place in cells?
   a. pH   c. enzyme/substrate ratio
   b. temperature   d. all of the above are important

17. Which of the following is *not true* of biological enzymes?
   a. slows all chemical reactions   c. may be used over and over
   b. not used up in a reaction   d. requires only a small amount

18. Which of these lines the wall of the abdominal cavity?
   a. visceral peritoneum   c. parietal peritoneum
   b. mesentery   d. adventitia

19. In which portion of the colon would feces have the most liquid consistency?
   a. ascending   c. descending
   b. transverse   d. rectum

20. Maltase is to maltose as peptidase is to:
   a. polypeptide   c. sucrose
   b. trypsin   d. proteose

21. In which of these solvents would a fat be insoluble?
   a. water   c. benzene
   b. ethyl alcohol   d. carbon tetrachloride

22. Which of these chemical elements is found in proteins, but not in lipids or carbohydrates?
   a. carbon   c. nitrogen
   b. hydrogen   d. oxygen

23. The final breakdown products of carbohydrate digestion are primarily:
   a. amino acids   c. enzymes
   b. monosaccharides   d. glycerol and fatty acids

24. Digestion of a piece of bread begins in the:
   a. stomach   c. duodenum
   b. ileum   d. mouth

25. In which portion of the alimentary canal does the greatest absorption of digested food take place?
   a. esophagus   c. small intestine
   b. stomach   d. large intestine

_____ 26. Record the results of temperature investigation in this table:

100% Enzyme Stock Solution (Control)	ml H_2O Displaced in 3 Mintues
20°C	
37°C	
45°C	

_____ 27. Record the results of your investigation of enzyme concentrations necessary to produce a reaction in this table:

Enzyme Solution Concentrations	ml H_2O Displaced in 3 Mintues
20%	
50%	
70%	
100% Enzyme Stock Solution (Control)	

_____ 28. Record the results of your investigation of the effect of pH on enzyme substrates strates in this table:

pH Values	ml H_2O Displaced in 3 Minutes
2	
5	
7	
8	
12	
100% Enzyme Stock Solution (Control)	

_____ 29. Summarize your results of the protein, fat, and carbohydrate exercises in this table:

Test	Result	Conclusion
1. Biuret reaction		
2. Millon reaction		
3. Ninhydrin reaction		
4. Heat coagulation of protein		
5. Digestion of protein		
6. Molisch reaction		
7. Benedict's test		
8. Barfoed's test		
9. Seliwanow's test		
10. Inversion of sucrose		
11. Starch test		
12. Starch hydrolysis test		
13. Solubility of fat		
14. Digestion of emulsified fat		

DISCUSSION QUESTIONS (Optional)

1. Why is it important to have increased surface area in the small intestine?

2. Compare the anatomy of the colon in the cat and human being with respect to the presence or absence of:

   a. appendix

   b. cecum

   c. hepatic flexure

   d. sigmoid colon

3. Which digestive glands secrete the following?

   a. amylase ....

   b. pepsin ....

   c. trypsin ....

   d. peptidase ....

   e. lactase ....

   f. lipase ....

4. Graph the results you obtained, using separate ordinates for temperature, pH, and concentration results. For the x-axis, use the parameter that you investigated; and for the y-axis, use the rate of enzyme reaction that you determined.

5. After graphing, indicate the optimum activity in terms of pH, temperature, and enzyme concentration. Indicate the minimum enzyme activity in terms of pH, temperature, and enzyme concentration.

6. Briefly explain why there is an optimum point for enzyme activity within the investigated parameters. Is there any proportionality (inverse or direct) related to increasing or decreasing the parameter investigated?

7. During a viral infection, afflicted individuals may have temperatures 104°F over an extended period of time. What happens to the enzyme systems within human cells? How are they affected by this increase in body temperature?

UNIT VIII
Circulatory System

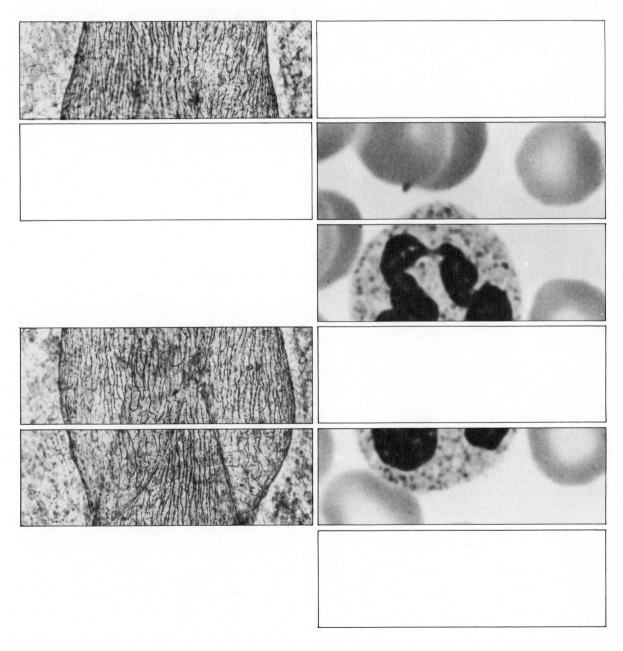

A. Anatomy of Blood, Blood Vessels, and Heart

■ *Purpose*

The purpose of Unit VIII-A is to enable you to identify the blood vessels, anatomical features of the heart, characteristics of blood, and patterns of systemic and pulmonary circulation.

■ *Objectives*

To complete Unit VIII-A, you must be able to do the following:

1. Observe the differences in the microscopic structure of an artery, a vein, a capillary, and a lymph vessel.
2. Microscopically identify the components of normal and abnormal human blood and frog blood.
3. Become familiar with the location and function of the major blood vessels in the cat with reference to the human system.
4. Identify the structures of the cat and sheep hearts.
5. State the functions of arteries, veins, capillaries, lymph vessels, blood, and the heart.

■ *Materials*

Compound microscope

Prepared microscope slides of artery, vein, capillary, lymph vessels, normal human blood smear (Wright's stain), sickle-cell anemia blood smear, frog blood, and leukemia blood smear

EXERCISE 1

Microscopic Examination of Artery, Vein, Capillary, and Lymph Vessels

The circulatory system is designed to transport blood to and from the capillaries. An exchange of vital respiratory gases and other metabolic substances occurs between the capillaries and the cells.

Arteries carry blood from the ventricles of the heart to the capillaries in progressively smaller vessels called *arterioles* and *metarterioles*. Arteries vary in cross section from large vessels, such as the aorta, to smaller tubes known as arterioles. The arterial wall is composed of three distinct layers: (1) the **tunica intima**, or innermost layer, consists of a thin layer of endothelium resting on an *internal elastic membrane*; (2) the **tunica media**, or middle layer, is composed of a layer of thick smooth muscle fibers and elastic connective tissue; and (3) the **tunica adventitia**, or outermost layer, is made up of loose collagenous fibers and adipose tissue. Larger blood vessels contain minute vessels termed **vasa vasorum** in the tunica adventitia that nourish the walls of larger arteries and veins. Because of the elastic fibers and thick walls of the arteries, these vessels do not readily collapse during periods of low blood pressure and are capable of expanding to accommodate high blood pressures.

Capillaries are thin-walled single-layered vessels. Through this endothelial layer, the exchange of oxygen, carbon dioxide, and other metabolic substances occurs. Squamous epithelium together with loose connective tissue comprises the endothelium of capillaries.

Blood flows into the veins after passing through a capillary network. Veins, like arteries, are composed of three layers; however, the layers are thinner. The **tunica media** is composed of several layers of smooth muscle; the **tunica adventitia** is a well-developed layer of collagenous fibers, which indicates that connective tissue predominates in veins. Veins, such as those in the extremities, contain **semilunar valves** in their walls that assist the return of blood to the heart against the force of gravity. The valves consist of semilunar flaps that open to permit the flow of blood to the heart and close to prevent the backflow of blood. Lymphatic vessels contain valves serving the same purpose.

Procedure

1. Examine a prepared slide of a cross section through an artery under low power. Identify the three layers of this vessel (*see* **Color Plate 15**).
2. To see the detail of the arterial wall, observe it under high power (*see* Figure 119). Observe the **endothelium** of the **tunica intima** layer. Note the wavelike **internal elastic membrane** that separates the tunica intima from the *tunica media*. Locate the tunica media layer, which contains

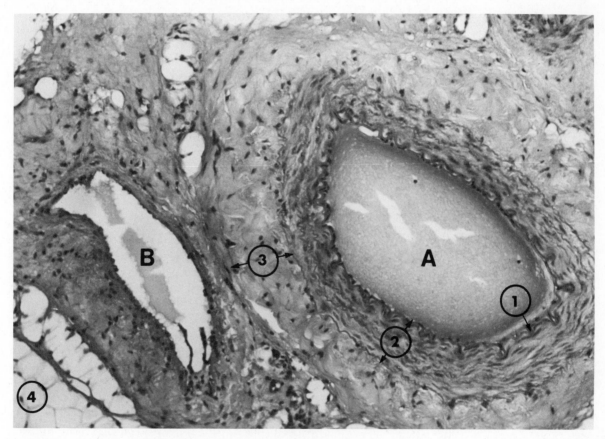

FIGURE 119 *Artery and vein of medium size stained to show layers*

A. **Artery, with Blood Cells in Lumen**
B. **Vein**
   1. Endothelium and Internal Elastic Membrane
   2. Tunica Media
   3. Tunica Adventitia
   4. Adipose Tissue

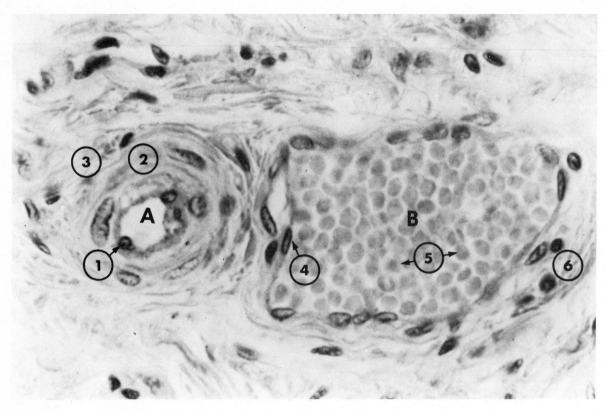

FIGURE 120 *Arteriole and vein, cross section*

A. Arteriole Lumen
1. Endothelial Cell
2. Tunica Media, Muscle
3. Tunica Adventitia

B. Venule Lumen with Blood Cells
4. Endothelial Cell
5. Blood Cells
6. Tunica Adventitia

thick smooth muscle and elastic connective tissue. Identify the *external elastic membrane* that separates the tunica media from the tunica adventitia. Try to locate the vase vasorum. *DRAW and LABEL* in detail a small section through the arterial wall in Box *A*, "Labeled Drawing of Artery (cross section)," following this exercise.

3. Examine a prepared slide of a cross section through a vein under low and high power. Identify the *tunica intima*, *tunica media*, and *tunica adventitia* layers of the vein (*see* **Color Plate 15**). Note the differences between the arterial and venous walls. *DRAW and LABEL* in detail a small section through the venous wall in Box *B* following this exercise.

4. Under high power, examine a prepared slide of a capillary network. Note the single layer of endothelial cells of the capillary wall that are separated by a connective tissue membrane (*see* Figure 120). *DRAW* in Box *C* provided following this exercise.

5. Under low power, examine a prepared slide of a longitudinal section of a lymphatic vessel containing a valve (*see* Figure 121). Lymphatic vessels contain numerous valves that are more closely spaced than those found in veins. The valves have their free margin oriented in the center of the vessel and have a swollen beaded appearance. *DRAW* in Box *D* following this exercise.

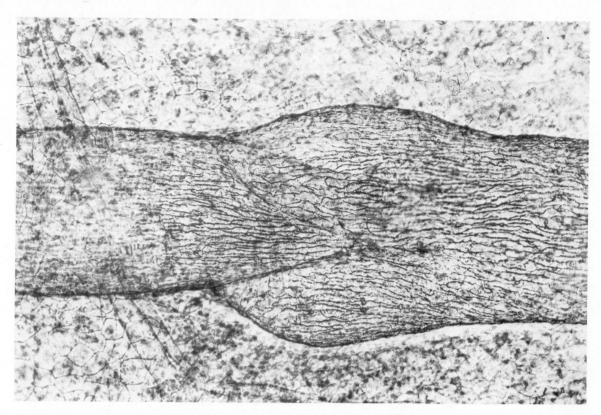

FIGURE 121   *Valve of lymphatic vessel* (Courtesy Carolina Biological Supply Company)

Labeled Drawings for Exercise 1

A.   *Labeled Drawing of Artery (cross section)*   **B.**   *Labeled Drawing of Vein (cross section)*

C. *Drawing of Capillary Network*   **D.** *Drawing of Lymphatic Valve*

EXERCISE 2

Microscopic Identification of Prepared Blood Slides

Blood is a specialized form of connective tissue and consists of formed and fluid elements. Blood is transported through vessels to receive oxygen from the lungs and collect metabolic products from the alimentary tract. Other substances are also contained in the blood so they can be transported in the bloodstream to all parts of the body.

Fresh whole blood, when centrifuged, separates into two distinct portions—fluid and formed—known as *elements*. The fluid portion is a clear straw-colored fluid called **plasma**. The plasma fraction of the blood accounts for more than 50% of blood volume. The formed elements consist of red blood cells, **erythrocytes**; white blood cells, **leukocytes**; and **platelets**, or *thrombocytes*.

The *erythrocytes* are highly specialized cells for the transportation of oxygen. Each cell is biconcave, and mature cells lack a nucleus. Erythrocytes are elastic and are capable of considerable distortion as they pass through capillary walls. In human males, there are 5 million red blood cells per cubic millimeter, and in females, 4.5–5 million. If a sample of blood is spread thinly on a microscope slide and stained properly with Wright's blood stain, the erythrocytes stain red. Erythrocytes have a tendency to adhere to one another along their concave surfaces, a phenomenon known as **Rouleaux formation**.

White blood cells, *leukocytes*, are cells with various shaped nuclei. Leukocytes average 5–9 thousand per cubic millimeter in normal blood. The count in children is higher and identifiable variations from the normal number occur in pathological conditions. Different types of leukocytes may be identified in a differential white blood cell count. Leukocytes are classified as *agranular* and *granular*. *Agranular leukocytes* have a homogeneous cytoplasm and large round nuclei. **Lymphocytes** and **monocytes** are agranular. *Granulocytes* can be distinguished by the presence of cytoplasmic granules and polymorphic nuclei. The granular leukocytes are of three types: **acidophils** (or **eosinophils**), **basophils**, and **neutrophils**.

Agranular Leukocytes Include (*see* **Color Plate 14**):

1. **Lymphocytes.** These leukocytes have a relatively large, round dark-staining nucleus surrounded by a narrow rim of clear blue cytoplasm. Lymphocytes may be classified as small, medium, or large. They constitute 25%–30% of the leukocytes of normal blood (*see* Figure 122).
2. **Monocytes.** These are larger than lymphocytes and have an oval, or kidney-bean-shaped, nucleus surrounded by a pale grayish blue cytoplasm when stained with Wright's stain. These cells constitute 3%–8% of the leukocytes of normal blood (*see* Figure 123).

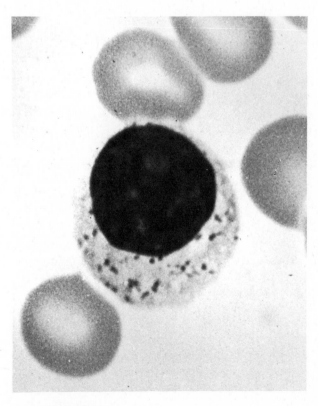

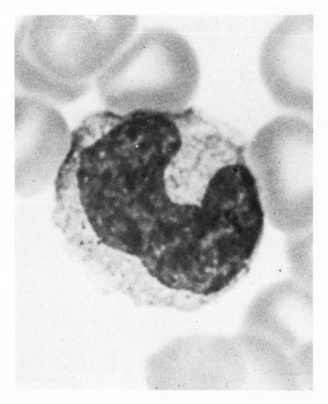

FIGURE 122 *Lymphocyte*   **FIGURE 123** *Monocyte*

Granular Leukocytes Include (*see* **Color Plate** 14):

1. **Neutrophils.** These are the most numerous leukocytes and constitute 65%–75% of the total white blood cell population. The neutrophil is a **polymorphonuclear** leukocyte—that is, the nucleus shows a variety of forms but usually consists of from three to five irregularly oval-shaped lobes connected by thin strands of nuclear material. This is sometimes referred to as the "drumstick" nucleus. The cytoplasm is filled with fine pale lilac-colored granules (*see* Figure 124).

2. **Acidophils (Eosinophils).** These cells are somewhat larger than neutrophils. The nucleus is usually irregularly shaped and partially constricted into two lobes. The cytoplasm is filled with coarse, round uniformly sized red-orange granules. They constitute about 2%–4% of the normal white blood cell count (*see* Figure 125).

3. **Basophils.** These cells are difficult to find in human blood since they constitute only 0.5%–1.0% of the total number of white blood cells. The nucleus is irregularly shaped and may appear as two lobes. The cytoplasm is filled with irregular blue-purple granules that usually obscure most of the bilobed nucleus (*see* Figure 126).

The function of leukocytes is to combat infection. White blood cells are one of the body's major defense mechanisms against bacteria, viruses, and other foreign substances that have entered the bloodstream. Neutrophils and monocytes are termed **phagocytes.** They have the unique ability to phagocytose, or engulf, foreign substances. As a result, most foreign invading substances are digested and inactivated by this process. At the site of inva-

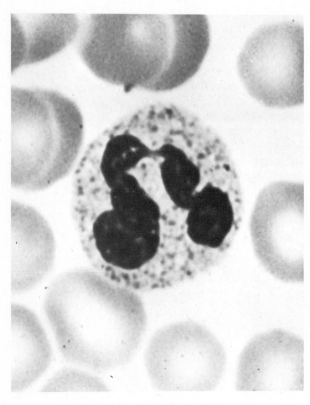

FIGURE 124 *Neutrophil*

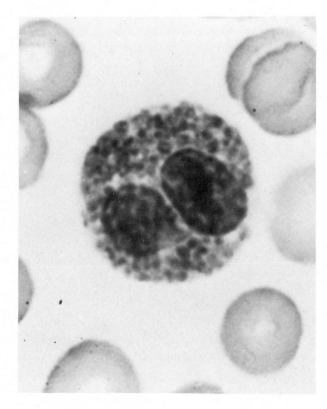

FIGURE 125 *Eosinophil*

sion, death of some surrounding tissue takes place. This **necrotic** (dead) mixture gives rise to **pus**. Lymphocytes are all-purpose cells that may be transformed into other white blood cell types. Lymphocytes may in effect become **antibodies**. Eosinophils (acidophils) seem to have the role of functioning in clot dissolution. Basophils secrete the anticoagulant *heparin*. Heparin is a normal constituent of blood and is necessary in order to prevent *intravascular clotting*.

Blood platelets, sometimes referred to as *thrombocytes*, are small disk-shaped structures that circulate freely in blood. The normal platelet count in adults varies between 200 thousand and 400 thousand per cubic millimeter of blood. Their essential role is **hemostasis**—the process by which platelets adhere to injured regions of blood vessels, accumulating to produce a blood clot known as a *thrombus*. Platelets are a factor in forming extrinsic *thromboplastin*, which is important in the clotting mechanism. Tissue thromboplastin (clotting factor III) is derived from extravascular sources.

Procedure

A. NORMAL HUMAN BLOOD

Examine, first under the high and dry objective and then under oil immersion, a prepared human blood smear stained with Wright's blood stain. Identify the following cell types: *erythrocytes* (which are anucleate), *monocytes, neutrophils, lymphocytes, basophils, acidophils,* and *platelets*. Basophils and acidophils may be difficult to identify in this preparation (*see* Figure 127).

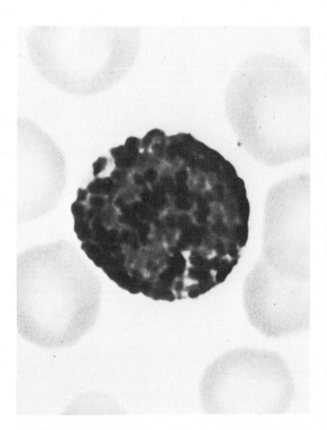

FIGURE 126 *Basophil*

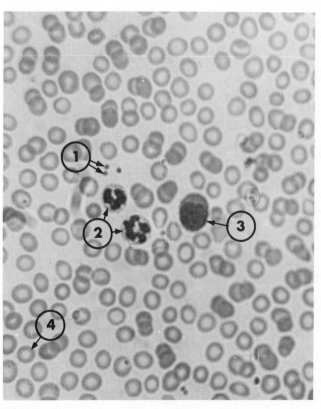

FIGURE 127 *Human peripheral blood smear*

| 1. Platelet | 3. Monocyte |
| 2. Neutrophil | 4. Erythrocyte |

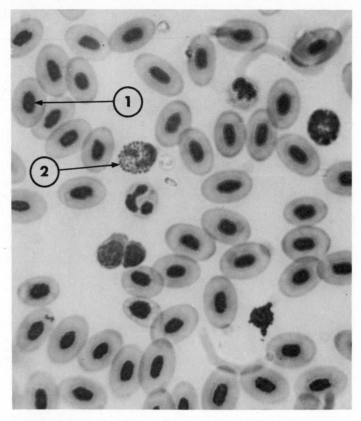

FIGURE 128 *Frog blood*

1. Nucleus of Red Blood Cell
2. Eosinophil

B. FROG BLOOD

Examine a prepared frog blood smear for *acidophils* and *basophils*, which appear similar to those in human cells. Notice that frog erythrocytes are large nucleated cells (*see* Figure 128).

C. SICKLE-CELL ANEMIA

Notice the sickle-shaped erythrocytes on this slide. These cells have a reduced capacity for carrying oxygen. This is an inherited condition in which the hemoglobin of the abnormally shaped cells contains the amino acid valine substituted for the amino acid glutamic acid that is present in normal hemoglobin.

D. LEUKEMIA

Leukemia is a malignant condition characterized by an abnormal increase in white blood cells and the presence of immature white blood cells in the peripheral circulation. If your slide is of *myelogenous* leukemia, there will be an increase in granular leukocytes. If you have a *lymphogenous* leukemia slide, you will see an increase in lymphocytes in addition to immature cells.

EXERCISE 3

Dissection of the Cat Circulatory System

Your specimen has been properly prepared for the dissection of the arterial and venous blood vessels by the injection of colored latex. This latex simulates the natural texture of the blood vessel. In order to facilitate the identification of blood vessels, the arteries have been injected red and the veins blue. Arteries are injected via the carotid and femoral arteries. Veins are also injected via the jugular vein in the neck and the femoral vein in the leg.

In order for you to identify blood vessels, all arteries and veins must be teased free from the adjacent supporting tissue for these structures to be clearly visible. Take great care to avoid severing these vessels while you expose them.

Retract the ribs laterally and, using dissecting pins, pin the superior portions of the ribs to the shoulder. Using your blunt metal probe, identify the lungs and trachea within the thoracic cavity before beginning dissection of the circulatory system.

Materials

Double-injected and preserved
  cat specimens
Demonstration specimens
Dissecting instruments

Procedure

DISSECTION OF ARTERIES WITHIN THE THORACIC CAVITY

(Figures 129 [**foldout preceding Unit I**] through 132)

1. If you have not already done so, remove the *thymus* gland—a tan to brown irregular granulated organ that functions in immunity—from the region anterior to the heart and trachea. The heart itself is covered with a protective membrane, the **pericardium**, that is not attached to the heart itself but to the ventral superior blood vessels. Determine where the pericardium is attached, then slit it open and carefully remove it. The ventral surface of the heart should be exposed.
2. The **aorta** is the largest artery emerging from the heart. It forms an **aortic arch** prior to descending along the dorsal body wall of the thoracic and abdominal cavities. Connective tissue should be carefully removed in order to expose this vessel.
3. The **pulmonary artery** extends from the right ventricle of the heart and carries oxygen-poor blood (deoxygenated) to each lung. (This vessel is probably injected with blue latex.) Follow this blood vessel and note that it branches into the right and left ~~pulmonary arteries~~ that enter each lung.
4. Locate the **coronary arteries**, which lie on the ventral surface of the heart. Note the groove—the **coronary sulcus**—that separates the right atrium from the right ventricle. A second groove may be found that separates the right ventricle from the left ventricle. This longitudinal groove is termed the *anterior longitudinal sulcus*.

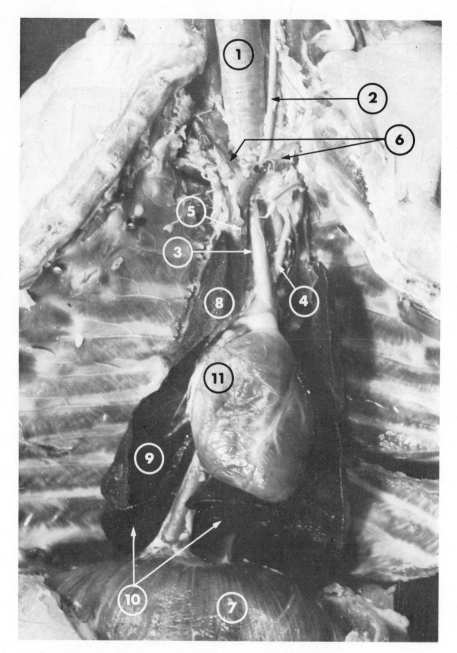

FIGURE 130 *Heart and lungs in situ, cat*

1. Trachea
2. Common Carotid Artery, Left
3. Brachiocephalic Artery
4. Subclavian Artery, Left
5. Superior Vena Cava
6. Right and Left Brachiocephalic Veins
7. Diaphragm
8. Right Superior Lobe of Lung
9. Right Medial Lobe of Lung
10. Right Inferior Lobe of Lung
11. Heart

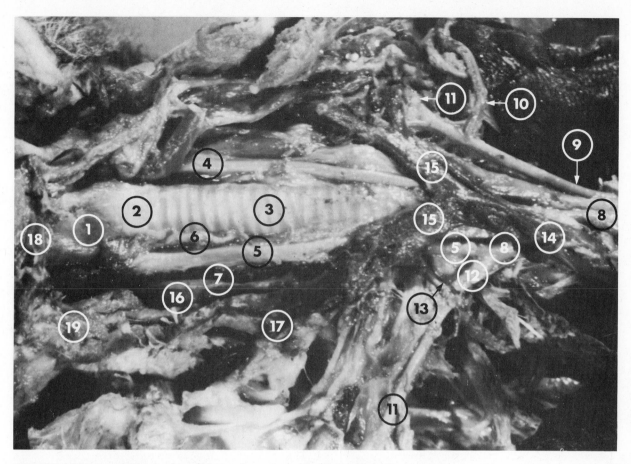

FIGURE 131 *Arteries and veins of the neck, ventral view* (From M. J. Timmons, *A Visual Guide to Dissection: Cat Anatomy Slides,* courtesy J. B. Lippincott Company)

1. Thyroid Cartilage of Larynx
2. Cricoid Cartilage of Larynx
3. Trachea
4. Common Carotid Artery, Left
5. Common Carotid Artery, Right
6. Vagus Nerve, Right Branch
7. Right Lymphatic Duct
8. Innominate (Brachiocephalic) Artery
9. Subclavian Artery, Left
10. Internal Mammary Artery
11. Axillary Artery
12. Subclavian Artery, Right
13. Thyrocervical Artery
14. Superior Vena Cava (Precava)
15. Right and Left Innominate (Brachiocephalic)
16. Internal Jugular Vein, Right
17. External Jugular Vein, Right
18. Transverse Jugular Vein
19. Lymph Node

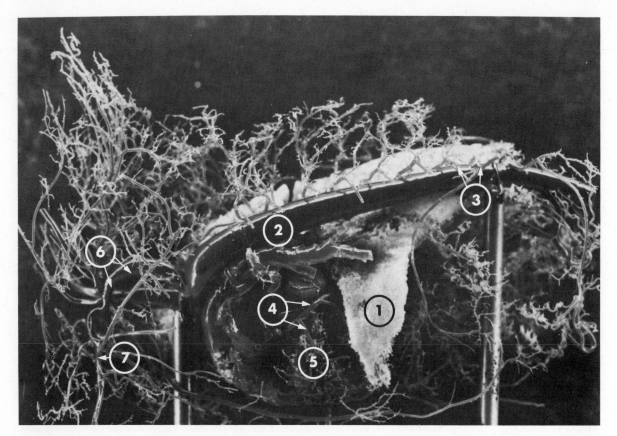

FIGURE 132 *Arterial corrosion cast of an adult cat, heart and lungs in situ*

1. Right Lung
2. Thoracic Aorta
3. Intervertebral Arteries
4. Coronary Arteries
5. Ventricle
6. Innominate Artery
7. Left Subclavian Artery

Divisions of the Brachiocephalic Artery

5. Locate the **brachiocephalic (innominate) artery**, which is the first and largest branch emerging from the aortic arch. In the cat, this blood vessel first divides into a **right subclavian artery**, a **right common carotid artery**, and a **left common carotid artery**. The second branch off the aortic arch is the **left subclavian artery**. In the human being, there are three branches off the aortic arch. The first branch is the brachiocephalic (innominate) artery, which divides into the right subclavian and right common carotid arteries. The second branch is the left common carotid artery. The third vessel branching off the aorta is the left subclavian artery.

6. Locate the left common carotid artery on your cat and you will observe two small arteries just lateral to the thyroid cartilage. The *superior thyroid artery* branches from the left common carotid artery and supplies the thyroid gland and hyoid muscle groups. The *inferior thyroid* artery is smaller and branches off the origin of the left common carotid artery.

7. The *occipital artery* branches immediately superior to the superior thyroid artery and continues anterior to supply the deep muscles of the neck and occipital region.

8. At the superior border of the larynx, the left common carotid artery divides into **external** and **internal carotid arteries**. The internal carotid artery is the smaller of the two and travels deep into the occipital region of the skull. The external carotid is much larger and branches many times to supply blood to the head.

9. If time permits, you may wish to locate the branches of the *right common carotid artery*, which are essentially the same as those of the left common carotid artery.

Divisions of the Right Subclavian Artery

10. Return to the **right subclavian artery** at its origin from the brachiocephalic. At the level of the first rib, locate the *vertebral artery* that emerges from the dorsal portion of the right subclavian artery. Follow this artery to the cervical vertebrae and notice that it branches into the dorsal musculature of the back. Note the *vagus* and *phrenic nerves* that pass next to or over the subclavian artery.

11. Again, at the level of the first rib, each subclavian artery gives off an *internal mammary artery* that supplies the ventral wall of the cat. If you have a pregnant cat, this blood vessel may be traced through the ventral thoracic wall to supply the mammary glands, which may have been removed when the cat was skinned.

12. The *costocervical artery* emerges from the subclavian artery at the level of the first rib. This vessel passes dorsally to supply the muscles of the scapular region. A short distance from the costocervical artery is the *thyrocervical trunk (axis)*. The thyrocervical branches further superiorly into the *ascending cervical*, which supplies the deep muscles of the neck, and the *transverse scapular artery*, which passes into the shoulder reigon.

13. As the right subclavian artery leaves the thoracic cavity, it is termed the **axillary artery**. As the axillary artery extends into the arm, it is termed the **brachial artery**. Below the elbow, the brachial artery becomes the **radial artery**.

Divisions of the Axillary Artery

14. Return to the *right axillary artery* and notice the slender *long thoracic artery*, that emerges from the ventral side of the axillary artery to supply the Pectoralis and Latissimus dorsi muscles.

15. The *subscapular artery* branches from the axillary artery at the proximal end of the brachial artery to supply the subscapular muscular region. The subscapular artery then branches laterally and posteriorly to give rise to the *thoracodorsal artery*, which supplies the Teres and Latissimus dorsi muscle groups.

Branches of the Aorta in the Thoracic Cavity

16. Return to the aortic arch area and note the descending thoracic aorta in the dorsal region below the aortic arch. In order to observe this, you will need to reflect the lungs and heart to one side. Use your probe to tease and separate the connective tissue that surrounds the aorta in this region.

17. Note the pairs of *intercostal arteries* (10 pairs) branching from the aorta, which supply the intercostal musculature.

18. The *esophageal arteries* have a variable origin but travel the length of the esophagus.

VEINS ASSOCIATED WITH THE HEART AND THE THORACIC CAVITY

You may find some variation in the way veins unite with one another in the cat. The veins in your specimen may not be as well injected as the arteries. Therefore, some of the veins may not be as easily identifiable as in the previously identified arteries of the thoracic region. Observe other specimens in the laboratory and any display specimens your instructor may have.

19. The **superior** (anterior in the cat) **vena cava** is the largest blood vessel entering the right atrium from the superior portion of the body, where it collects blood from the head and arms.
20. Use your probe to dissect out the **inferior** (*posterior* in the cat) **vena cava** that enters the right atrium from the posterior portion of the body.
21. The **pulmonary veins** can best be viewed by retracting the heart laterally and observing these vessels as they enter the left atrium. This group of veins carries blood from the lungs to the heart.

Divisions of the Superior Vena Cava

22. The **azygos vein** is the first branch of the superior vena cava. Retract the right lung and heart to expose it. Observe the azygos vein on the right side against the vertebral column. The branches of the azygos are the *intercostal* and *esophageal veins.*
23. The *internal mammary vein (sternal)* is the second branch entering the superior vena cava at the level of the third rib. It then passes through the thoracic body wall to supply the mammary glands.
24. The *costocervical vein* enters the *innominate vein (brachiocephalic)* at the anterior portion of the superior vena cava. The costocervical vein has one tributary, the *vertebral vein*, that drains the cervical vertebral area.
25. The right and left **innominate (brachiocephalic) veins** are the anterior divisions of the superior vena cava. These bood vessels drain blood from the head and arm regions.

Divisions of the Innominate (Brachiocephalic) Veins

26. The **external jugular vein** is the largest vein in the neck that drains from the cranial region. This vessel lies on the ventrolateral surface of the neck just under the platysma muscle. The **internal jugular vein** extends from the base of the skull to the inferior portion of the larynx, where it usually empties into the external jugular vein, although it may enter the innominate vein.
27. The **subclavian vein** is the short proximal segment of the main blood vessel that enters the forelegs. As soon as the subclavian vein passes out of the thoracic cavity, it is called the *axillary vein.* More distally, in the upper arm region, it is known as the *brachial vein*, which is formed by the junction of the radial and ulnar veins in the forearm. The *subscapular vein* drains the scapular region and enters the axillary vein.

Divisions of the External Jugular Vein

28. Remove excess fascia and musculature that may still be covering the external jugular vein. Observe the *transverse scapular vein* in the shoulder region just under the scapular region.
29. The *thoracic duct* of the lymphatic system enters the left external jugular, or subclavian, vein behind the peritoneum. It is usually difficult to identify the jugular veins, but sometimes they appear as elongated, beaded vessels.

At this point, it would be wise for you to review the arteries and veins in the thoracic cavity before dissecting the blood vessels of the abdominal cavity.

ARTERIES OF THE ABDOMEN

30. The **celiac artery** is the first major branch of the abdominal portion of the descending aorta. Move the viscera laterally and use your blunt probe to remove the peritoneum from the dorsal abdominal wall just below the diaphragm. The celiac artery divides into three main branches —the hepatic, left gastric, and gastrosplenic arteries.

31. The **hepatic artery** passes to the liver and lies parallel to the common hepatic duct and portal vein. The hepatic artery divides into a *gastroduodenal artery* that supplies the pyloric portion of the stomach, and a *pancreaticoduodenal artery* that supplies the duodenum and pancreas.

32. The *left gastric artery* supplies the lesser curvature and ventral surface of the stomach. The lesser omentum and greater omentum must be excised in order to see this vessel.

33. The **superior mesenteric artery** emerges from the ventral surface of the descending abdominal aorta about ½ inch posterior to the celiac artery. This large vessel supplies the small intestine and parts of the colon. The superior mesenteric artery branches into the *posterior pancreaticoduodenal artery* that supplies the pancreas and duodenum, and the *middle colic artery* that supplies the ileocecal region.

34. The **adrenolumbar arteries** arise about 1 inch posterior to the superior mesenteric artery on the lateral surface of the abdominal aorta. One branch supplies the suprarenal gland, while the other larger branch continues into the dorsal body musculature.

35. The **renal arteries** arise about ½ inch posterior to the adrenolumbar arteries on the lateral surface to supply each kidney. Normally, these arteries do not arise at the same level on both sides and, frequently, branching occurs before they enter the kidney.

36. The **genital** (*internal spermatic* or *ovarian*) arteries are a pair of slender arteries that arise posterior to the renal artery. In the male, trace the *internal spermatic artery* from its origin through the inguinal canal and down to the scrotum. In the female, trace this vessel along the dorsal body wall to the ovary and uterine horn. Usually, the ovarian artery is easier to locate than the internal spermatic artery.

37. The **inferior mesenteric artery** is a single vessel that arises from the ventral surface of the abdominal aorta to supply the descending colon and rectum. It has two branches—the *left colic artery* that supplies the descending colon, and the *superior hemorrhoidal artery* that supplies the proximal portion of the rectum.

38. The **iliolumbar arteries** are paired vessels arising about ½ inch posterior to the inferior mesenteric artery. These arteries arise just before the bifurcation of the abdominal aorta. These arteries supply the dorsal body wall of the lumbar region.

VEINS OF THE ABDOMEN

Branches of the Inferior Vena Cava (*see* Figure 129 (*Foldout*) and Figure 133)

39. You have identified the inferior vena cava, which enters the right atrium from the posterior portion of the body. This vessel proceeds through the diaphragm and lies adjacent to the thoracic portion of the aorta. You may observe many small branches, including the phrenic veins, that are embedded in the dorsal thoracic wall.

40. The **hepatic veins**, which may or may not be injected, are best viewed by scraping away some of the liver tissue. Use your blunt probe to expose these vessels in the right medial lobe of the liver.

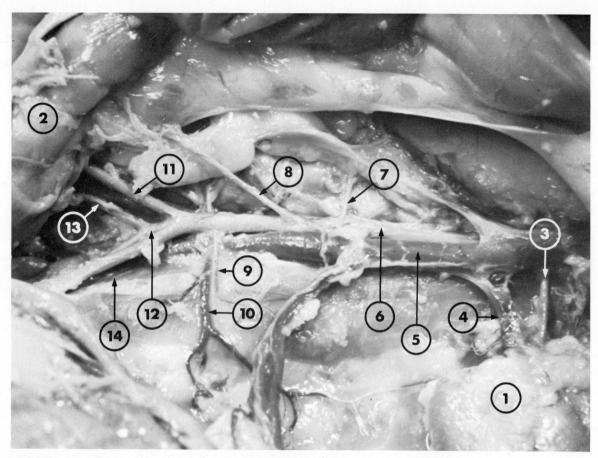

FIGURE 133 *Pelvic basin exposed to show blood vessels posterior to renal artery and vein*

1. Kidney, Left
2. Descending Colon
3. Renal Artery, Left
4. Renal Vein, Left
5. Inferior Vena Cava
6. Descending Abdominal Aorta
7. Ovarian Artery, Right
8. Posterior Mesenteric Artery
9. Iliolumbar Artery, Left
10. Iliolumbar Vein, Left
11. Right External Iliac Artery
12. Caudal Artery
13. Right Internal Iliac Artery
14. Left Common Iliac Vein

41. The *adrenolumbar veins* drain the suprarenal glands. The *right adreno-lumbar* vein drains into the inferior vena cava and the *left* drains into the left renal vein.

42. The **renal veins** drain blood from the kidneys. Normally, one vein enters the posterior vena cava more anteriorly than the other. Occasionally, you may note two renal veins on the right side.

43. The *genital (spermatic* or *ovarian) veins* are injected poorly and may not be easily observable. The *right spermatic vein* or *right ovarian vein* drains directly into the inferior vena cava, whereas the left empties into the left renal vein.

44. The **iliolumbar veins** lie adjacent to the iliolumbar arteries against the dorsal body wall. They serve to drain the back. The iliolumbar veins are

located just anterior (superior) to the bifurcation of the inferior vena cava into the hind legs.

45. The **hepatic portal system** is comprised of branches of the *hepatic portal vein* that drain the stomach, spleen, intestines, and colon. Unless your specimen is triple injected, these branches of the hepatic portal vein will be extremely difficult to observe. Consult Figure 129 and a display specimen for the interrelationship of the hepatic portal system to the systemic circulation.

At this point, it would be wise for you to review the arteries and veins in the abdominal cavity before dissecting the blood vessels of the hindleg.

ARTERIES OF THE HINDLEG

46. Locate the **dorsal abdominal aorta** and note that it gives off many branches into the hindlegs. The dorsal aorta continues as the small *caudal (sacral) artery* into the tail and divides into many branches in the pelvic basin.

47. The descending abdominal aorta bifurcates into the **external iliac arteries** which are large vessels that pass through the body wall and into the hindlegs. The external iliac artery extends a short distance in the thigh; than the **deep femoral artery** branches medially off the external iliac artery, which continues as the **femoral artery**. The **popliteal artery** branches off the femoral artery and supplies the ventral and lateral portions of the thigh.

48. The **internal iliac artery** is found immediately posterior (inferior) to the origin of the external iliac artery. This artery supplies the uterus and rectum. The following are tributaries of the internal iliac artery but are often not injected: the *umbilical artery*, *superior gluteal artery*, *hemorrhoidal artery*, and *inferior gluteal artery*. Consult Figure 133 for this detail.

VEINS OF THE HINDLEG

49. Identify the inferior vena cava in the pelvic basin and note that it bifurcates into a right and left **common iliac vein**. These veins are closely associated with the iliac arteries and should be found without much teasing of connective tissue. The common iliac vein continues as a large vein into the thigh—the **external iliac vein**. The external iliac continues as the **femoral veins** on the ventral and median surface of the leg and terminates as the *saphena magna vein*, or *greater saphenous vein*, on the posterior superficial surface of the calf. The *popliteal vein* may be found entering the femoral vein posterior to the knee, draining the posterior region of the thigh.

50. Return to the base of the common iliac veins and observe the *internal iliac veins* that branch about ¼ inch from the bifurcation of the vena cava into the common iliac veins. The tributaries of the internal iliac vein are deep in the pelvic basin and may be poorly injected. The anterior branch is the *hypogastric vein*, or *middle hemorrhoidal vein*, that runs from the urinary bladder. The posterior branch of the internal iliac vein is the *gluteal vein* that is adjacent to the pubic symphysis and rectum.

EXERCISE 4

Dissection of the Cat and Sheep Hearts

The cat and sheep hearts, like the heart of a human being, consist of four separate chambers—two superior **atria** and two inferior **ventricles**. Blood from all regions of the body empties into the right atrium by way of the **superior** (*anterior* in cat and sheep) **vena cava** and **inferior** (*posterior* in cat and sheep) **vena cava**. Blood then passes to the **right ventricle**. Between the right atrium and right ventricle is the **tricuspid valve**, which has three flaps, or cusps. From the right ventricle, blood is pumped through the **pulmonary artery** that branches and leads to the right and left lungs where the blood is oxygenated. The oxygenated blood returns from the lungs to the **left atrium** through the **pulmonary veins** and then passes into the **left ventricle**. Between the left atrium and left ventricle is the **bicuspid, or mitral, valve**. Attached to the cusps of the bicuspid and tricuspid valves are slender fibers—the **chordae tendineae**—that are attached to the muscular wall of the heart by **papillary muscles**. The *tricuspid* and *bicuspid valves* prevent the backward flow of blood from the ventricles into the atria during ventricular systole. The *chordae tendineae* and the *papillary muscles* serve to prevent the cusps of the valves from being forced open into the atria during the contraction phase of the heart cycle. During ventricular **systole**, blood from the left ventricle flows through the aortic arch. The **aortic semilunar** and **pulmonary semilunar valves** prevent the backflow of blood from the aorta and pulmonary artery into the left and right ventricles, respectively.

The thick muscular wall of the heart is the **myocardium**. The *antrioventricular* **septum** is the thick portion of the myocardium that separates the two atria and ventricles. More specifically, the muscular wall that divides the atria is the **interatrial septum**, and the wall that divides the ventricles is the **ventricular septum**.

Materials

Preserved and double-injected cat
Preserved sheep heart
Dissecting instruments and tray

Procedure

DISSECTION OF THE CAT HEART

1. If you have not done so in the preceding exercise, remove any excess fat, the thymus gland, and any connective tissue that may obstruct your view of the heart.
2. Identify the following external features of the cat heart (*see* Figure 134):
   a. **Pericardium.** The pericardium consists of an outer fibrous **parietal** pericardium (pericardial sac) that surrounds the heart. The inner serous layer is the **visceral pericardium (epicardium)** that lies immediately over the myocardium. The space between the inner and outer walls of the pericardial sac is the **pericardial cavity**. The pericardial sac will normally not be present on the sheep heart specimen.
   b. **Base.** This is the superior end of the heart where the blood vessels emerge.
   c. **Apex.** This is the inferior end of the heart that is pointed and just above the diaphragm.

FIGURE 134 *Detail of heart in situ with pericardium removed, ventral aspect*

d. **Atria.** The atria are two irregular thin-walled chambers at the base of the heart. The two atria are separated from the ventricles by a horizontal groove—the **coronary sulcus.** The scalloped border of each atrium is the **auricle.**

e. **Ventricles.** These are large muscular posterior chambers of the heart. The left ventricle is more muscularized and, therefore, firmer than the right ventricle. Externally, the left and right ventricles are separated by a faint groove, or depression, that extends obliquely across the ventral surface from the base to the apex.

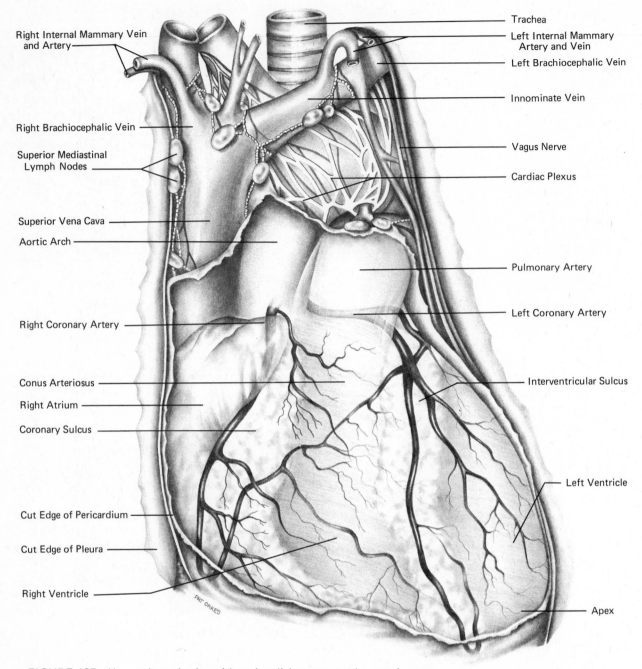

Right Internal Mammary Vein and Artery

Right Brachiocephalic Vein

Superior Mediastinal Lymph Nodes

Superior Vena Cava

Aortic Arch

Right Coronary Artery

Conus Arteriosus

Right Atrium

Coronary Sulcus

Cut Edge of Pericardium

Cut Edge of Pleura

Right Ventricle

Trachea

Left Internal Mammary Artery and Vein

Left Brachiocephalic Vein

Innominate Vein

Vagus Nerve

Cardiac Plexus

Pulmonary Artery

Left Coronary Artery

Interventricular Sulcus

Left Ventricle

Apex

FIGURE 135 *Human heart in situ with pericardial sac removed, ventral aspect*

f. **Anterior vena cava** (*superior* vena cava in the human). This vessel empties into the anterodorsal surface of the right atrium (*see* Figure 135).

g. **Posterior vena cava** (*inferior* vena cava in the human). This posterior vessel empties just inferior to the superior vena cava on the anterodorsal surface of the right atrium.

h. **Pulmonary artery.** This vessel emerges from the right ventricle and passes laterally toward the left, where it divides into the right and left pulmonary arteries. After this division, the vessel continues as the **ductus arteriosus** and connects with the aorta. This vessel bypasses the majority of the blood from the lungs to the systemic circulation. Normally, this vessel atrophies after birth; however, persistence of the open vessel represents a congential heart defect known as **patent ductus arteriosus.**

i. **Aortic arch and aorta.** The aortic arch emerges from the left ventricle at a superior angle. In the cat, two major branches emerge from the arch—the brachiocephalic (innominate) artery on the right side and the subclavian artery to the left.

j. **Pulmonary veins.** These are small vessels that empty into the left atrium on the anterodorsal surface.

k. **Coronary arteries.** These are blood vessels that supply the myocardium of the heart. These arteries originate at the base of the aortic arch. One coronary artery may be viewed deep between the pulmonary artery and the right atrium.

3. Identify the following internal features of the cat heart after you have removed the heart from the thoracic cavity by cutting the major blood vessels. Using your scalpel, make a frontal incision from apex to base, dividing the heart into dorsal and ventral halves. Wash and probe out any latex that may be present in the chambers (*see* Figures 134, 135, and 136).

a. **Right and left atria.** The inner surface of these chambers is comprised of endocardium. Note the openings in the dorsal wall of the left

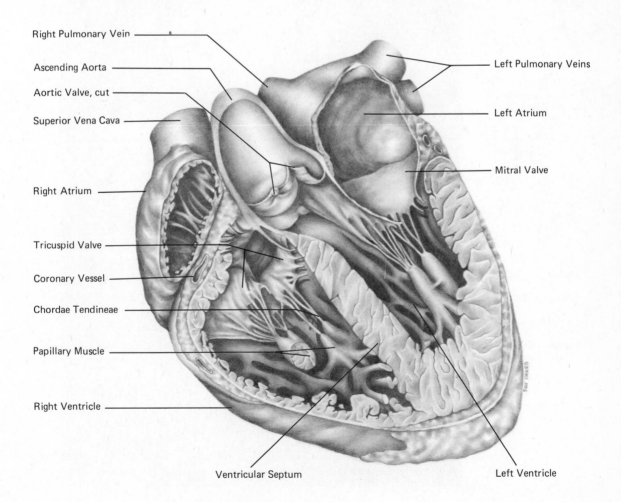

Right Pulmonary Vein

Ascending Aorta

Aortic Valve, cut

Superior Vena Cava

Right Atrium

Tricuspid Valve

Coronary Vessel

Chordae Tendineae

Papillary Muscle

Right Ventricle

Left Pulmonary Veins

Left Atrium

Mitral Valve

Ventricular Septum

Left Ventricle

FIGURE 136   *Human heart showing chamber detail, ventral aspect*

atrium for the pulmonary veins. In the fetal heart, an opening exists between the two atria. This opening is termed the **foramen ovale**. Normally, this closes after birth; however, failure to close results in a defect known as **atrial septal defect**.

b. **Right ventricle.** The inner surface of the ventricular wall is corrugated by muscular cords termed **trabeculae cornae**. Identify the tricuspid valve between the right atrium and right ventricle. Each of the three cusps is connected by thin fibers—the chordae tendineae—and by pillars of **papillary muscles** to the myocardial wall. Note the pulmonary semilunar valves at the base of the pulmonary artery.

c. **Left ventricle.** Identify the **bicuspid valve**, or mitral valve, between the left atrium and left ventricle. The components of this valve are attached in the same manner as those of the tricuspid valve. Note the aortic semilunar valves that open into the aortic arch.

FIGURE 137
Sheep heart in situ with pericardial sac removed, ventral aspect

1. _____
2. Pulmonary Arteries
3. Pulmonary Artery
4. Atrium, Left
5. _____
6. _____
7. Coronary Sulcus
8. Ventricle, Right
9. Atrium, Right
10. Superior Vena Cava
11. Aorta

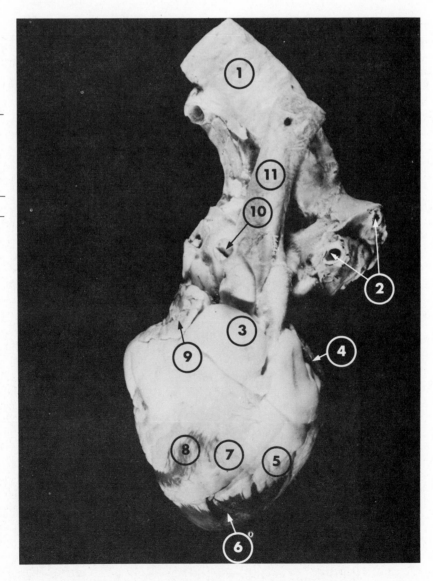

DISSECTION OF THE SHEEP HEART

1. Rinse the sheep heart with cold water in order to remove excess preservative or blood. Make sure the water flows through all of the blood vessels to irrigate any blood clots out of the chambers.
2. Identify the external features of the sheep heart from the description given previously under the external features of the cat heart (*see* Figures 137 and 138).
3. Identify the internal features of the sheep heart from the description given previously under the internal features of the cat heart (*see* Figures 139 and 140).

FIGURE 138
Sheep heart in situ with pericardial sac removed, dorsal aspect

1. _____
2. Pulmonary Veins
3. Inferior Vena Cava
4. Right Ventricle
5. _____

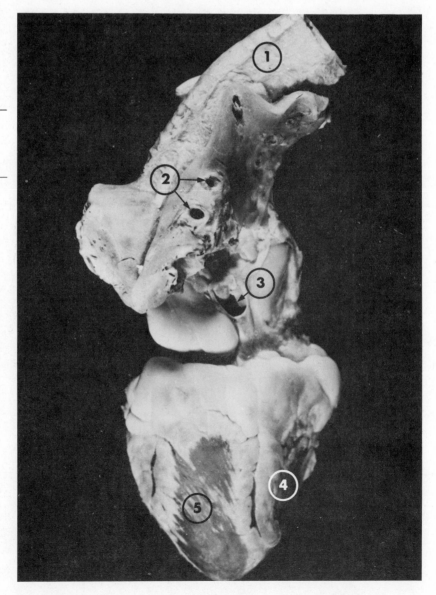

FIGURE 139

Sheep heart with ventral portion reflected to show chamber detail

1. _____

2. Aorta

3. Pulmonary Arteries

4. Semilunar Valve

5. Mitral Valve (Bicuspid)

6. Chordae Tendineae

7. _____

8. _____

9. Myocardium

10. Interventricular Septum

11. Pericardium

12. Tricuspid Valve

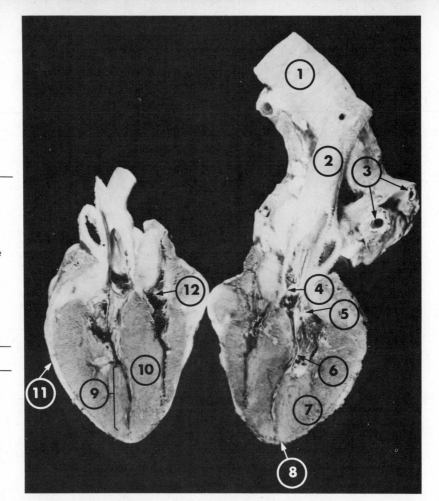

FIGURE 140

Sheep heart showing chamber and valve detail

1. Aorta

2. _____

3. Tricuspid Valve

4. _____

5. Interventricular Septum

6. Apex

7. Myocardium of Left Ventricle

8. Aortic Semilunar Valve

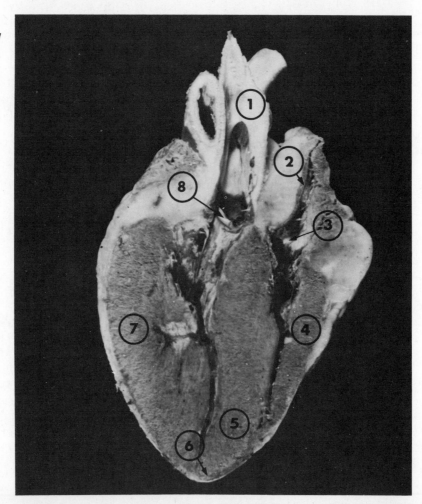

B. Blood Physiology

- *Purpose*

  The purpose of Unit VIII-B is to familiarize you with basic human blood physiology.

- *Objectives*

  To complete Unit VIII-B, you must be able to do the following:

  1. Discuss the significance of coagulation time and the mechanism of normal blood clotting.
  2. Demonstrate an understanding of and ability to determine ABO and Rh blood types.
  3. Demonstrate an understanding of and ability to determine coagulation times, hemoglobin, and hematocrit percentages.
  4. Demonstrate proficiency in blood pressure determination and an understanding of its medical significance.

EXERCISE 1

Coagulation Time (Clot Formation)

Coagulation time is the period of time required for a blood sample to form a clot. The clot consists of a mass of tangled strands of **fibrin** in which red and white blood cells are enmeshed. The four essential substances necessary for coagulation to occur are (1) *prothrombin*, which is normally present in the circulatory system and is produced by the liver; (2) *thromboplastin*, which is liberated from injured tissues and platelets; (3) *calcium* (in free ionic form); and (4) *fibrinogen*, a soluble protein present in the plasma. Vitamin K, which is found in green leafy foods and is produced by certain intestinal bacteria, is utilized by the liver in order to produce prothrombin. In the absence of vitamin K or when there is a reduced supply, the liver fails to produce the necessary prothrombin. As a result, the blood either fails to clot or clots very slowly. A clot that is formed within a blood vessel and remains attached at the site of formation is termed a *thrombus.* Any clot that has broken loose from its site of formation and is freely floating is termed an **embolus.**

Materials

Sterile cotton
Unheparinized capillary tubes
70% ethyl alcohol

Sterile lancets (Hemolets)
Clock with a second hand

Procedure

1. Finger puncture provides the best method of obtaining blood for the tests that follow in the blood studies series. The following finger puncture procedure is to be repeated for obtaining blood samples. Clean the fourth or fifth finger with cotton moistened with 70% ethyl alcohol. Allow the alcohol to evaporate. Remove the sterile Hemolet from the envelope by its blunt end. It is best to hold the Hemolet near its pointed tip and then puncture the finger with a quick jabbing motion. The blood should flow freely of its own accord from the wound. *Do not squeeze the tip of the finger* to obtain blood. If you must squeeze, start at the base of the finger

and move toward the tip. If an additional puncture is required, discard the used Hemolet and obtain a fresh sterile Hemolet for the subsequent puncture.

2. Obtain a drop of blood from the puncture, but do not use the first drop because it usually coagulates fast. After the second drop has formed, place the end of the unheparinized capillary tube in the drop and allow the tube to fill by capillary action. There should be no air bubbles in the column of blood. *REPEAT* the above procedure with another unheparinized capillary tube. Note the time when both tubes were filled and *RECORD* these times. *Times:*

3. At the *end of 2 minutes*, break off about ¼ inch of the capillary tubes and observe whether the blood has clotted, as evidenced by a thread of coagulated blood connecting the two pieces of tubing. *REPEAT* this procedure with both unheparinized capillary tubes at 30-second intervals until the blood has clotted. *RECORD* the times. The coagulation times of the two capillary tubes should be approximately the same. Normal coagulation times fall within the range of 2 to 4 minutes. What explanation can you give for this? Is there any difference in coagulation times between the two tubes?

4. After completion of the exercise, dispose of all equipment.

EXERCISE 2

ABO and Rh Determination

Red blood cells contain proteins called **antigens** (agglutinogens) that are located on the **erythrocyte** (red blood) cell surface. Common antigens found in the human system are A, B, AB (both A and B), O, and D (Rh). Research has identified more than 100 erythrocyte antigens, each an expression of an inherited gene. To the individual who possesses it, an antigen is beneficial, but this same antigen can be harmful to someone else who did not inherit that particular antigen. When an individual is exposed to *foreign antigens*, that individual may produce a chemical whose purpose is the destruction of the foreign antigen. This chemical substance, which is contained in the serum fraction of the blood, is called an **antibody**. Once produced, an antibody remains in the circulation for many years, ready to destroy foreign antigens if introduced into the blood. *Antibodies are specific;* they will destroy only an antigen that is identical to the one that stimulated their production and are, therefore, harmless to other antigens within the same individual.

In order to determine the ABO blood type, two common antibodies contained in *anti-A* and *anti-B* sera will be used. Research indicates that 41% of the United States Caucasian population possess only the A antigen, 9% have only the B antigen, 4% have both A and B antigens, and the remainder of the population, 46%, have neither A nor B antigen. Thus, individuals are said to belong to the blood groups A, B, AB, or O, respectively. If the blood type is A, there will be **agglutination** (clumping) with the anti-A serum. If type B is present, agglutination will take place with anti-B serum. If type AB is present, agglutination by both anti-A and anti-B sera will occur. If type O is present, neither anti-A nor anti-B sera will produce agglutination.

Human erythrocytes are also classified according to the **Rh factor** as either Rh positive or Rh negative, depending on whether the Rh antigen is present. This antigen is more specifically termed the **D-antigen** in the Fisher-Race nomenclature. In order to determine the presence or absence of the Rh antigen, a small sample of whole blood is tested with anti-D serum.

An individual is Rh positive (Rh+) if agglutination of erythrocytes takes place with the antiserum and Rh negative (Rh-) if no agglutination takes place. About 85% of the U.S. Caucasian population is Rh positive and 15% Rh negative.

Materials

70% alcohol and sterile cotton or wipes   Anti-A, anti-B, and anti-D sera
Sterile lancets (Hemolets)   Light warming box
Toothpicks   Microscope slides
Wax marking pencil

Procedure

1. Use a wax pencil to divide a clean microscope slide into two sections or use a prepared slide for blood typing.
2. Label one side of the slide A, the other side, B. Label a second slide D.
3. Puncture your finger and wipe away the first drop of blood. Place the second and third drops on the slide, one on side A, the other on side B. Place another drop of blood on the second slide.
4. Add one drop of anti-A serum to the drop of blood marked A and one drop of anti-B serum to the B side. Place one drop of anti-D serum on the other drop of blood, which is on the second slide.
5. Mix the blood and antisera with a toothpick. In order to avoid contamination, use a separate toothpick for each mixing.
6. Let stand for 2 minutes. Examine closely for agglutination *during this 2-minute* period. If doubtful, use low power of the microscope.
7. Place the microscope slide marked D on the warming box (40°C) and mix by rocking the slide back and forth for *2 minutes*. Examine for agglutination during this time period.
8. Determine your ABO and Rh types. *Record* your results in the tables provided. Mark agglutination as (+) and no agglutination as (-).

Type	Anti-A	Anti-B	Number of Individuals	Class Percentage	Theoretical Percentage
A					41
B					9
AB					4
O					46

D Antigen	Number of Individuals	Class Percentage	Theoretical Percentage
Rh +			85
Rh -			15

EXERCISE 3

Hemoglobin Estimation by the Tallquist Method

The amount of *hemoglobin* present in red blood cells is a good indicator of the oxygen-carrying capacity of the blood. A simple method of measuring hemoglobin is to compare a small piece of Tallquist paper that has been saturated with a sample of blood with a Tallquist color chart.

Materials

Tallquist scale and test paper
70% ethyl alcohol
Sterile Lancets (Hemolets) and cotton

Procedure

1. Remove one square of paper from the Tallquist booklet.
2. Puncture your finger as previously described. Place the second drop of blood in the center of the paper. Wait *15 seconds* and compare your sample with the Tallquist scale.
3. *RECORD* the scale estimate in grams of hemoglobin per 100 cc of blood. Note differences in male and female comparisons and any anemic conditions that may be indicated.

EXERCISE 4

Hematocrit, or Packed Cell Volume

The percentage of erythrocytes found in a set volume of blood is known as the *hematocrit*, or packed cell volume (PCV). The PCV is a reliable indicator of the percentage of red blood cells in a given sample. Centrifugation of the blood sample results in the bottom portion of the tube being packed with red blood cells. The upper portion will contain the plasma fraction of the blood. A hematocrit reader is used to determine the percentage of packed red blood cells in the sample. The normal hematocrit in males is 40%–54%, with 47% as the average. In females, it is 37%–47%, with 42% the average.

Materials

Clay-Adams microhematocrit reader (or substitute)
70% alcohol and sterile cotton
Heparinized (0.5 mm) capillary tubes

Clay-Adams Seal-Ease
Microhematocrit centrifuge
Sterile lancets (Hemolets)

Procedure

1. Puncture your finger as previously described. After the blood flows freely, wipe away the first drop of blood and place the red end of the capillary tube on the wound area and allow the blood to flow downward into the capillary tube. *Fill the tube only with blood* (no bubbles).

2. Place your finger over the opposite end of the capillary tube and place the red end of the tube into the Seal-Ease.
3. Place the sample tube into the microhematocrit centrifuge *with an empty capillary tube opposite the sample tube for balance;* otherwise, the sample in the capillary tube will spill into the head of the centrifuge.
4. After loading the centrifuge, tighten the inside and outside covers. Set the timer for 4 minutes.
5. After the 4-minute spin period, remove the sample and place it in the microhematocrit reader. Determine the percentage of blood volume from the scale.
6. *RECORD* your results.

EXERCISE 5

Pulse and Blood Pressure Determinations

The beating heart makes certain characteristic sounds that can be heard through an instrument called a *stethoscope.* The stethoscope amplifies well-defined sounds, described as lubb-dubb. The first sound (lubb) occurs during ventricular **systole**—during the contraction of the ventricles—and is caused by the closing of the atrioventricular valves. The second sound (dubb) is produced during ventricular **diastole** (relaxation) and is the sound produced by the closing of the semilunar valves. The sounds heard between these two may indicate the condition of the valves of the heart. Ventricular systole produces a wave of pressure, the **pulse**, in the arteries. The pulse can normally be felt where an artery is near the surface of the skin and over the surface of a bone. Pulse rates may vary in individuals because of changes in temperature, stressful conditions, the time of day, and other factors.

Materials

Stethoscope
Sphygmomanometer
SphygmoStat or other electronic blood pressure monitor

Procedure

a. *Taking the Radial Pulse*

Use your index and middle fingers to palpate the pulse. *Do not use the thumb* because a clear pulse is present in the thumb itself. Place your partner's forearm with the palm up on the laboratory table. Reach over the wrist, not under it. The pulse can be felt behind your partner's thumb between the large tendon and bony prominence found on the lateral aspect of the wrist. Too much pressure will obliterate the pulse. Practice until you can achieve an accurate pulse rate by counting the beats per 15 seconds and multiplying by 4 to obtain the beats per minute. The normal adult range is 72–80 beats per minute. Concentrate fully on the pulse and screen out other noises.

b.   *Determination of the Carotid Pulse*

Place your fingers on either side of your partner's larynx (Adam's apple). Gently press medially and toward the back. Apply pressure gradually until you feel the pulse. Relocate your fingers slightly upward or downward until the pulse is clearly felt with at least two of the fingers. Practice several times to achieve an accurate pulse rate.

c.   *Determination of the Arterial Blood Pressure by Ausculation*

Extend your partner's forearm across a table with the palm up. The upper arm should be approximately at heart level. Either arm may be used, but the left is preferable. Use the same arm for all exercises involving blood pressure determinations.

Blood Pressure Values for Men and Women

		Systolic			Diastolic		
	Age	Normal Range	Mean	Hypertension Lower Limit*	Normal Range	Mean	Hypertension Lower Limit*
Men							
	16	105-135	118	145	60-86	73	90
	17	105-135	121	145	60-86	74	90
	18	105-135	120	145	60-86	74	90
	19	105-140	122	150	60-88	75	95
	20-24	105-140	123	150	62-88	76	95
	25-29	108-140	125	150	65-90	78	96
	30-34	110-145	126	155	68-92	79	98
	35-39	110-145	127	160	68-92	80	100
	40-44	110-150	129	165	70-94	81	100
	45-49	110-155	130	170	70-96	82	104
	50-54	115-160	135	175	70-98	83	106
	55-59	115-165	138	180	70-98	84	108
	60-64	115-170	142	190	70-100	85	110
Women							
	16	100-130	116	140	60-85	72	90
	17	100-130	116	140	60-85	72	90
	18	100-130	116	140	60-85	72	90
	19	100-130	115	140	60-85	71	90
	20-24	100-130	116	140	60-85	72	90
	25-29	102-130	117	140	60-86	74	92
	30-34	102-135	120	145	60-88	75	95
	35-39	105-140	124	150	65-90	78	98
	40-44	105-150	127	165	65-92	80	100
	45-49	105-155	131	175	65-96	82	105
	50-54	110-165	137	180	70-100	84	108
	55-59	110-170	139	185	70-100	84	108
	60-64	115-175	144	190	70-100	85	110

Adapted with permission from the *Journal of the American Medical Association* **1950,** *143,* 1465. Copyright 1950, American Medical Association.

*The lower limit for hypertension has more recently been established at 140/90.

Check the **sphygmomanometer** cuff for markings indicating the area to be applied against the brachial artery. Some cuffs have markings for both the right or left arm and some have no such markings. In any case, apply the center of the inflatable area of the cuff to the inside of the upper arm and wrap the long end of the cuff snugly about the limb. If it is not snug, inflate the sphygmomanometer until it does not slip down under the influence of gravity. Place the gauge, or mercury column, where it can be easily read. Put the earpieces of the stethoscope in your ears. Place the bell, or diaphragm, of the stethoscope over the blood vessels visible under the skin covering the inside of the elbow.

Inflate the cuff to 180 mm mercury (please, no higher; work quickly). Slightly open the valve at the bulb and listen carefully with the stethoscope as the pressure in the cuff decreases. The point at which the *pressure sounds*, or sounds of Korotkoff, are first heard is the *systolic* pressure. The point at which the pressure sounds assume a muffled quality represents the *diastolic* pressure. If you cannot detect the change to the muffled sound, then consider the point at which all pressure sounds disappear to be the diastolic pressure. *However, you must consistently use one method or the other.* Consult the table, "Blood Pressure Values for Men and Women." *REPEAT* three times and determine pressure. Wait 15–30 seconds before repeating each determination.

EXERCISE 6

Electronic Blood Pressure Measurement: Indirect Method Using the SphygmoStat Electronic Blood Pressure Monitor

In addition to height and weight, there are four medical measurements, or *vital signs*, used to evaluate a person's general health—*blood pressure*, *pulse rate*, *temperature*, and *respiratory rate*. Of these, blood pressure is very important and is subject to the greatest variation. Variations in blood pressure readings in an individual may be due to true variations in arterial blood pressure caused by factors such as emotional status, physical activity, physical position, and room temperature. Variations in blood pressure readings may also be due to errors in measurement, either because of faulty instrumentation or observer errors. It is the purpose of this exercise to introduce you to the principles and method of operating an electronic blood pressure device that will eliminate many instrumentation and observer errors.

The ausculatory method for indirect measurement of blood pressure has long been accepted as standard, even though use of a stethoscope to hear the sounds of Korotkoff is known to introduce a wide range of possible errors. Such errors occur with both mercury columns and aneroid manometers. The principal source of error lies in the nature of the sounds and the inability to hear and interpret them correctly. In Exercise 5, you may have had difficulty hearing and interpreting the Korotkoff sounds. New advances in electronic technology now make blood pressure measurement totally objective and accurate, eliminating errors caused by misinterpretation of sounds heard.

The SphygmoStat Electronic Blood Pressure Monitor and other similar instruments have overcome the disadvantages of the stethoscope in indirect blood pressure measurement. Solid-state circuitry allows rapid, accurate measurement of blood pressure without using a stethoscope. A special microphone in the cuff detects the Korotkoff sounds as the cuff is deflated. These signals are then processed electronically, causing a red light to flash on the

FIGURE 141 *SphygmoStat, a type of electronic blood pressure monitoring device* (Courtesy Technical Resources, Inc.)

instrument panel. (The SphygmoStat is shown in Figure 141.) The internal portion of the instrument panel contains an aneroid manometer and a battery-powered electrical system. A microphone converts the Korotkoff sounds into an electrical signal that is then fed into an amplifier containing sound filters that remove all external noises. A red light flashes in response to the sounds, the first and last flashes being systolic and diastolic pressures, respectively.

Materials

Electronic blood pressure apparatus, such as SphygmoStat, produced by Technical Resources, Inc., Waltham, Mass.

Procedure

1. The cuff should be securely wrapped around the upper arm of your partner with the microphone (circle on cuff) over the *brachial artery*, and not over the antecubital fossa as is done with a stethoscope. The correct position of the microphone is about 2 inches above the elbow on the flat medial portion of the arm, where the biceps appears to flatten out. (*Note:* When using the cuff on a heavy arm, it may be necessary to shift the position of the microphone. In order to place the microphone as close as possible to the artery, move it more laterally on the arm and toward the antecubital fossa. This assures a closer acoustical coupling between the artery and the microphone.)
2. Plug the cuff into the instrument case.

3. Turn on the instrument and close the air valve by rotating the silver cap-screw on the hand bulb in a clockwise direction.
4. Inflate the cuff by pumping the hand bulb. The light may flash during manipulation of the cuff. This is caused by the noise of air entering the arm-cuff region. The flashing should stop once the cuff pressure reads at a pressure greater than the systolic pressure. *STOP INFLATING.*
5. Observe the pressure dial on the instrument. Begin to deflate the cuff slowly by gradually opening the air valve by turning the cap counter-clockwise. Deflate at the rate of 2–3 mm Hg/pulse. **The point on the pressure dial where the first regular flash of light occurs is the systolic pressure.** The light will continue to flash synchronously with the pulse until the diastolic pressure is reached. **THE LAST FLASH IS THE DIASTOLIC PRESSURE.** *RECORD* the systolic and diastolic pressures.
6. Practice this procedure several times until you have mastered the operation of the instrument.
7. Once you have mastered the operation of the instrument, take two additional blood pressure readings, and *RECORD* the average systolic and diastolic pressures in the table provided.

Average Systolic and Diastolic Pressures

	Trial 1		Trial 2		Average	
	Partner's	*Yours*	*Partner's*	*Yours*	*Partner's*	*Yours*
Systolic Pressure						
Diastolic Pressure						

8. Compare the readings obtained by the electronic blood pressure apparatus versus the mercury sphygmomanometer of Exercise 5. *RECORD* your findings in the table provided.

Comparison of Blood Pressure Readings Obtained Auscultatorily and Electronically

	Mercury Sphygmomanometer		Electronic Blood Pressure	
	Partner's	*Yours*	*Partner's*	*Yours*
Average Systolic Pressure				
Average Diastolic Pressure				

9. List and describe three instrumentation errors that may occur when using the SphygmoStat:

a.

b.

c.

10. List and describe three observer errors in reading of blood pressures using the mercury sphygmomanometer:

a.

b.

c.

EXERCISE 7

Electronic Pulse Rate Measurement

The most widely used method for pulse rate measurement involves feeling the pulse along a major artery with the fingertips and timing with a watch. This is the method you employed in Exercise 5. There are obvious disadvantages to this method, including an increased chance for human error, inability for constant monitoring, and the need for complete observer attention during measurement.

Electronic pulse rate instruments can provide automatic and continuous measurement of the pulse rate through either an optical or a bioelectric signal pickup. The purpose of this exercise is to demonstrate the accuracy of pulse rate measurements using an electronic pulse rate monitor. The pulse rate instrument described in this exercise is the Cardio Tach Series 4600.

The Cardio Tach Series 4600 pulse rate monitor measures the peripheral pulsation by means of a photoelectric finger sensor (*see* Figure 142). Each time the heart beats, capillaries in the fingertips fill with blood, causing a slight change in the optical properties of the fingertips. A sensitive photoelectric cell within the finger sensor detects this optical change at each pulse when the light reflected to the cell changes intensity. The resultant photoelectrical pulse is amplified and processed by electronic circuits powered by rechargeable batteries.

The time interval between two consecutive pulses is measured electronically and updated with each pulse. This time interval is then converted into frequency and displayed in a digital readout as the number of beats per

FIGURE 142
Cardio Tach Series 4600 apparatus shown monitoring pulse rate (Courtesy Gulf + Western Applied Science Laboratories)

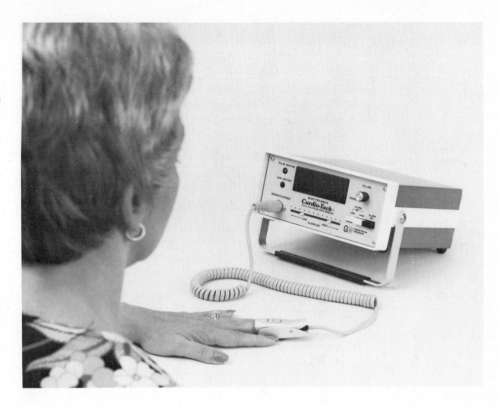

minute. The conversion from time interval to frequency and the display of pulse rate is accomplished before the arrival of the next pulse, thereby permitting instantaneous and continuous measurement of the pulse rate. In addition, an audible tone and flashing light indicate each pulse. The display of pulse rate is either for instantaneous values or for a four-pulse average and is selected by a switch on the rear of the instrument.

Materials
Cardio Tach Series 4600 Pulse Rate Monitor or similar electronic monitor

Procedure
1. Plug the optical finger sensor into the SENSOR/CHARGER socket.
2. Switch on power by turning the VOLUME control in a clockwise direction.
3. Gently push one of the three buttons under ALARM so that none of the three buttons is in the depressed position.
4. Attach the sensor to the end of the middle or index finger as shown in Figure 142. The finger must be inserted all the way against the stop within the sensor clip.
5. Once the sensor is attached, an audible beep will be heard for each heart beat. The red PULSE RHYTHM light will flash concurrently with the sound at each beat. (*Note:* Proper operation of the instrument may be impaired by any factor that affects blood circulation in the finger, such as bending the finger, poor peripheral circulation, excessive pressure on parts of the hand, or cold fingers. These factors may make pulse measurement difficult and unreliable. If the finger is not in the sensor clip, there may be one or two extra beats. After these beats, a double zero (00) will appear on the numerical display, indicating the absence of any pulse signal output.)

6. When the INST-AVG switch on the rear of the instrument is set at the INST mode, the readings after the second beat reflect the true instantaneous readings of the pulse rate. When the switch is set at the AVG position, the reading is the average over every four consecutive instantaneous beats. If the pulse rate falls outside of the operating range (40–199 BPM) of the unit, a *Flashing* 00 will appear on the screen.

7. Turn the unit off when not in use.

8. Practice the operating procedures for the instrument. Take two consecutive resting pulse readings of your partner and yourself, then *RECORD* your results in the table provided.

9. *RECORD* the pulse rates of your partner and yourself in the table provided after the following activities: running in place (1 minute) and hyperventilation (1 minute).

Pulse Rates

Resting		Running in Place		Hyperventilation	
Partner's	*Yours*	*Partner's*	*Yours*	*Partner's*	*Yours*

10. Compare your resting pulse rate from Exercise 5a to the rate recorded in this exercise. Are both values reliable and accurate? *Explain.*

C. Cardiac Muscle Physiology

■ *Purpose*

The purpose of Unit VIII-C is to observe the action of the heart and to demonstrate some of the properties of cardiac muscle when stimulated by various agents.

■ *Objectives*

To complete Unit VIII-C, you must be able to do the following:

1. Observe the normal cardiac cycle of the frog heart by recording a tracing of the frog's heartbeat.
2. Determine the effect of temperature change on the rate of heartbeat and the significance of that effect.
3. Observe the refractory period, extrasystole, and compensatory pause in the cardiac cycle.
4. Observe the effect of vagal nerve inhibition on the cardiac cycle.
5. Determine the effect of adrenalin on the cardiac cycle and its influence on the cardiovascular system.

The frog heart is similar in action to the human heart, even though it is a three-chambered heart. The heartbeat is influenced by internal and external factors. Special conducting tissues of the heart are located in small masses of tissue called the *sinoatrial* (SA) and *atrioventricular nodes* (AV). These specialized centers establish a particular rhythm that may be accelerated or slowed by specific nerve fibers of the central nervous system. The medulla oblongata and other chemical and physical factors also influence the heart rate. In these exercises, you will study the nature of the heartbeat and some factors that influence its rate.

EXERCISE 1

Observation of the Normal Cardiac Cycle of the Frog

Materials

Dissecting instruments	Frog board or dissecting pan
Dissecting pins	Ringer's solution
Mechanical or electronic recording device and related apparatus	Living frog(s)

Procedure

1. Your instructor will explain the use of the recording equipment that is to be used in this exercise.
2. Place a double-pithed frog (*see* Unit VI-B, "The Physiology of Muscle Contraction," for pithing procedure), ventral surface up, on a frog board or dissecting pan. Pin the feet and lower jaw to the pan.
3. Use your scissors to make a midventral incision through the skin from the pubic symphysis to the midline of the mandible. Cut laterally on each side at the pectoral and pelvic girdles and pin back the flaps of skin. Cut superiorly through the ventral musculature from the pubic symphysis. Carefully cut through the sternum area. Pin back the cut edges of the pectoral girdle in order to expose the heart, lungs, and major blood vessels.

Identify all of the external structures of the heart before proceeding to record the cardiac cycle.

4. Connect the mechanical or electronic recording device according to your instructor's directions.

5. *RECORD* the tracing of the frog's cardiac cycle for 5 minutes. You may moisten the heart with several drops of Ringer's solution during this 5-minute period. The average rate for a frog is 50–70 beats per minute.

6. Determine the rate of contraction of the atria and the ventricle for 15 seconds by counting the number of beats occurring during 15 seconds. Multiply each figure by 4 to determine the beats per minute. Observe the sequence of contraction of the chambers of the frog heart during this 5-minute recording period.

7. Refer to Figure A for a tracing of the *normal cardiac cycle* of the frog. Attach a labeled recording of the normal cardiac cycle of the frog's heart. (*Note:* Keep your frog preparation moist and save for the next exercise, which *must* follow immediately.)

FIGURE A *Normal frog cardiac cycle*

a–b   Atrial Systole

b–c   Atrial Diastole
        (Complete at *f*)

c–e   Ventricular Systole

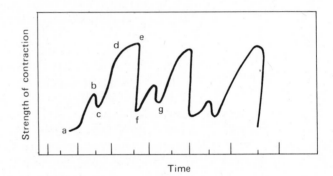

EXERCISE 2

Observation of the Effect of Temperature on the Heart Rate

The heart rate can be influenced by various factors such as activity, age, body weight, and body temperature. The average heart rate for a frog is between 50 and 70 beats per minute. In this exercise, you will note how temperature change can influence the rate of heartbeat. All biological reactions increase with temperature within physiological limits (2°C–40°C). An increase within limits in temperature speeds up the rate of biological reactions. Biologists often use the change of rate of a reaction after a 10°C temperature change (Q_{10}) as a measurement of thermal energy:

$$Q_{10} = \frac{\text{Rate at a higher temperature}}{\text{Rate at a 10°C lower temperature}}$$

Materials

Double-pithed frog—same frog as used in previous exercise

8°C, 18°C, and 28°C water baths

Thermometer

Ringer's solution

Mechanical or electronic recording device

Wax pan

Ice cubes made with Ringer's solution

Procedure

1. Use the same frog preparation as in the previous experiment.
2. Immerse the frog in the 8°C water bath. This should be done slowly by adjusting the temperature of the water bath through the addition of Ringer ice and by heating. After the frog has been immersed at 8°C for 15 minutes, remove the frog and record the cardiac cycle with a mechanical or electronic recording device.
3. *REPEAT* the immersion in water of 18°C and 28°C. Calculate the Q_{10} and record your results. *RECORD* the cardiac cycle and note changes in the rate and amplitude of contraction.
4. Compute the Q_{10} for the 8°C, 18°C, and 28°C temperatures.

Temperature	Q_{10}
8°C	
18°C	
28°C	

5. Use the same frog for the next exercise. Moisten with Ringer's solution.

EXERCISE 3

Observation of Refractory Period, Extrasystole, and Compensatory Pause in the Cardiac Cycle

The refractory period, during which additional stimuli have no effect, is longer for cardiac muscle than for skeletal muscle. This extended refractory period is an important factor in setting the rhythmicity and in preventing tetany of the heart. If an extra stimulus is applied to the ventricle during systole or diastole, the ventricle will respond to this stimulus by giving an extra stroke, or contraction (extrasystole), after which the ventricle may pause and "skip a beat" (compensatory pause). What takes place is that the refractory period to stimuli from the atria causes a longer delay that results in a compensatory pause of the ventricle.

Materials

Double-pithed frog
Mechanical or electronic recording device
Electronic stimulator and event-time marker
Dissecting instruments
Wax pan or frog board

Procedure

1. Use the same frog preparation as in the previous exercise. Keep moist with Ringer's solution.

2. Position the mechanical or electronic recording apparatus. Record a tracing of 8–10 normal beats of the heart.

3. Using an electrode holder, position the electrodes so that the wire tips are in contact with the heart just slightly below the atrial-ventricular groove. Use a single stimulus of 40 volts, stimulate the ventricle *for 2 seconds*, and then break the circuit. *Wait 10 seconds* before applying the next stimulus. Stimulate the ventricle during the systolic and diastolic phases of the cardiac cycle. Continue stimulating until the tracing shows three extra systoles. Note any compensatory pauses that occur between stimuli (*see* Figure B).

FIGURE B *Refractory period tracing*

a   Refractory Period
b–c   Extrasystole
d   Compensatory Pause

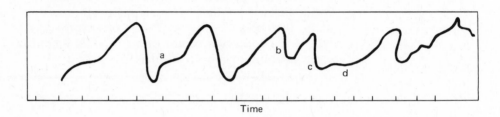

Time

4. Observe the following during the recording period:
   a. The point at which the ventricle exhibits absolute refractoriness:

   . . . .

   b. The length of the refractory period: . . . .

   c. The duration of a compensatory pause: . . . .
5. Save the frog for Exercise 4. *Keep moist with Ringer's solution.*

EXERCISE 4

Observation of the Effect of Vagal Nerve Inhibition and Adrenalin Acceleration

The inherent rhythm of cardiac muscle may be modified by the central nervous system as well as by thermal effects. There are two branches of the autonomic nervous system that innervate the heart: the *vagus nerve* by way of the parasympathetic branch and the *nerve fibers* from the sympathetic chain of **ganglia** in the cervical region. When the vagus nerve is stimulated, the cardiac rate may decrease or cease beating for a few moments. This is referred to a vagal inhibition. The secretion of the vagus nerve is **acetylcholine**, which momentarily slows the heart rate; however, this inhibition is only temporary because acetylcholine is destroyed within 0.05 seconds by the enzyme cholinesterase. The sympathetic fibers secrete **norepinephrine**, which accelerates the heart rate. This effect is also short in duration because enzymatic breakdown also occurs. **Adrenalin (epinephrine)** released from the adrenal medulla during physical or emotional stress has an action similar to that of norepinephrine.

Materials

Double-pithed frog
Dissecting instruments
Wax pan or frog board
Epinephrine (1:10,000)
26-gauge hypodermic needle

Mechanical or electronic recording device
Electronic stimulator and event-time marker

Procedure

1. Use the same pithed frog preparation as in the previous exercise. If the condition of the frog is poor, obtain a fresh specimen, pith it, and make a similar preparation.
2. Locate and isolate the vagus nerve, which runs along the external jugular vein. After identifying the nerve, loosely tie a loop of thread around it. A dissecting microscope may be used to assist in isolating this nerve.
3. Using an electrode holder, position the electrodes so the wire tips are in contact with only the nerve.
4. Record a normal cardiac cycle for 15 seconds. Stimulate the vagus nerve with multiple stimuli of 200 volts for a period of 5 seconds. After obtaining a normal cardiac cycle, repeat with multiple stimuli of 10-, 15-, and 20-second durations. From your tracing, determine and *LABEL* where the following occur: point of vagal inhibition and termination of vagal inhibition (refer to Figure C).

FIGURE C *Vagal inhibition*

a Point of Vagal Stimulation
b Vagal Inhibition
c Recovery

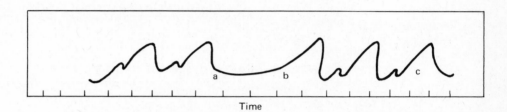

Time

5. Record a normal cardiac tracing for 15 seconds before injection of epinephrine. Inject 0.2 cc of 1:10,000 epinephrine intraperitoneally or intramuscularly. Record for 10 minutes and note the height of the ventricular systole. Determine the effects on the heart rate and note any other visible bodily changes (*see* Figure D).

FIGURE D *Epinephrine stimulation*

a Normal Cardiac Systole and Diastole
b Point of Epinephrine Stimulation

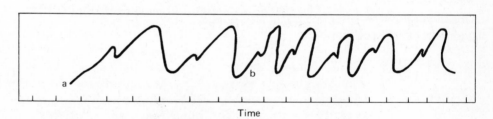

Time

EXERCISE 5

Human Heart Rate Measurement Using Electrocardiograph (ECG) Signal Amplification (Optional or Demonstration)

A more sophisticated mode of measuring heart rate involves the sensing of bioelectric potentials produced by the heart. The electrocardiographic (ECG) signal occurring during each heartbeat is detected by three skin electrodes and is then amplified. Figure 143 shows the necessary assemblies for this mode of operation: the basic Cardio Tach instrument, ECG cable with grabbers for connection to electrodes, and three reusable skin electrodes. The principle of operation for the ECG pickup is identical to that of the optical finger sensor in Exercise 7, Part B, of this unit. The time interval between consecutive R waves (or S waves) is measured and converted into frequency. The readout is either instantaneous or a four-beat average.

The purpose of this exercise is to demonstrate the accuracy of heart rate measurement using the Cardio Tach electric monitor. Your instructor will provide you with operating instructions for any other electronic monitor you use.

Materials

Cardio Tach with ECG module
ECG cable with grabber connections
Reusable or disposable skin electrodes

Procedure

1. Switch on power by turning the POWER-OFF/VOLUME control clockwise.
2. Plug the ECG cable into the SENSOR/CHARGER socket.
3. Attach each of the three color-coded grabber electrodes to three button electrodes on your partner as follows:

Black Grabber Electrode—designated as the indifferent electrode that can be attached anywhere on the body.

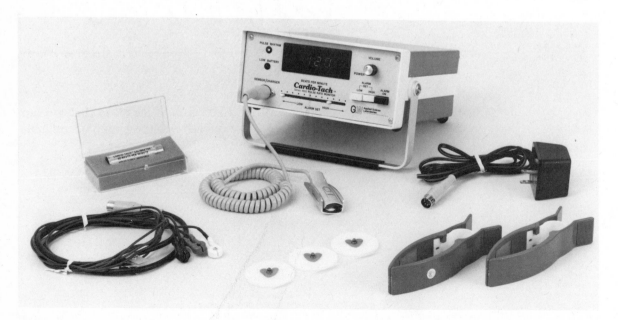

FIGURE 143 *Cardio Tach Series 4600 apparatus with ECG module* (Courtesy Gulf + Western Applied Science Laboratories)

White Grabber Electrode—designated as the upper electrode that is attached to a spot above the heart.

Green Grabber Electrode—designated as the lower electrode that is attached to a spot below the heart.

4. If wrist clamp electrodes are used, attach the *white grabber* to the right clavicle area, the *green grabber* to the left wrist, and the *black grabber* to the right wrist or right ankle.

5. The remaining operating instructions for the Cardio Tach are the same as for determining pulse rate. If you are not familiar with the use of this instrument, refer back to Exercise 7, Part B, in this unit for operating instructions.

6. Occasionally, the presence of an irregular heart rhythm may be displayed. This is the result of interference by less prominent ECG waves. Should this occur, interchanging the *white* and the *green grabber* leads often will eliminate such difficulties.

7. *RECORD the rate for your partner and yourself* in the table provided while performing the following activities in sequence: sitting at rest, running in place for 2 minutes, and holding your breath for 1 minute. (*Note:* Smoking a cigarette within one minute would have the same effect on heart rate as holding your breath.)

Heart Rates

Sitting at Rest		Running in Place		Holding Your Breath for One Minute	
Partner's	Yours	Partner's	Yours	Partner's	Yours

Note: Wrist electrodes must be firmly attached while running in place. Electrodes may be taped to the wrists in order to reduce electrode movement.

8. Compare your resting pulse rate from Exercise 5 of Part B to that of your resting heart rate as recorded in this exercise. Are these numbers reliable and accurate? *Explain.*

9. What were the effects of exercise and holding your breath on heart rate? List and describe the five factors that may have caused this change in heart rate:

a.

b.

c.

d.

e.

UNIT VIII

Circulatory System

DISCUSSION

Multiple Choice

_____ 1. Hematocrit may be defined as:
- a. a measure of the volume of RBC's in a given volume of blood
- b. an estimate that can be used to determine the blood volume of a person
- c. a measure of blood clotting time
- d. a hormone excreted in the kidneys in response to low numbers of RBC's

_____ 2. A clot that freely floats in blood vessels is known as a(an):
- a. thrombus
- b. infarction
- c. embolus
- d. thromboblast

_____ 3. Which of the following layers is normally thicker in arteries than in veins?
- a. tunica media
- b. tunica intima
- c. tunica adventitia
- d. serosa

_____ 4. The first blood vessel to branch from the aortic arch is the:
- a. innominate artery
- b. carotid artery
- c. subclavian artery
- d. coronary artery

_____ 5. The period of time in which the ventricles of the heart are filling with blood while the myocardium is relaxed is termed:
- a. a heart attack
- b. a stroke
- c. the diastolic phase
- d. the systolic phase

_____ 6. A person who is typed as B positive would have which type of antigen and antibodies?
- a. B antigen, D antigen, anti-D antibodies
- b. B antigen, D antigen, anti-A antibodies
- c. A antigen, anti-B antibodies
- d. A antigen, anti-D antibodies, anti-B antibodies

_____ 7. The QRS wave of the electrocardiogram represents:
- a. atrial depolarization
- b. ventricular depolarization and relaxation
- c. atrial relaxation
- d. ventricular depolarization

_____ 8. An implanted electronic pacemaker serves to replace the normal functioning of the:
- a. vagus nerve
- b. glossopharyngeal nerve
- c. SA node
- d. AV node

9. Why is it advantageous for red blood cells to be anucleate?
   a. oxygen-carrying capacity is increased
   b. it prevents excessive numbers of erythrocytes from being formed through mitotic division
   c. it makes it easier to distinguish them from leukocytes
   d. it enables them to pass through capillaries more easily

10. In the presence of infection, the leukocyte count:
    a. increases
    b. decreases
    c. remains the same

11. Which of these vessels would have the thickest wall?
    a. inferior vena cava
    b. internal carotid artery
    c. renal artery
    d. aorta

12. The distal colon receives its blood supply from the:
    a. superior mesenteric artery
    b. celiac artery
    c. inferior mesenteric artery
    d. iliolumbar artery

13. Blood goes from the heart to the lungs through the:
    a. aorta
    b. ductus arteriosus
    c. pulmonary artery
    d. pulmonary veins

14. Under which of these circumstances would counting the pulse for 15 seconds and multiplying by 4 be inaccurate?
    a. if the pulse is rapid
    b. if the pulse is slow
    c. if there is background noise
    d. if the pulse is irregular

15. A decrease in vagal impulses to the heart results in:
    a. increased heart rate
    b. decreased heart rate
    c. no effect on heart rate
    d. death

16. What is the significance of the diastolic blood pressure you obtained in Exercise 5, Part B, of this unit?
    a. it is the pressure of blood against the wall of the aorta when the ventricles contract
    b. it is the pressure of blood against the wall of the aorta when the ventricles relax
    c. it is the pressure of blood against the wall of the brachial artery when the ventricles relax
    d. it is the pressure of blood against the wall of the left ventricle when the ventricles contract

17. Which of these is not included in the "formed" elements of blood?
    a. erythrocytes
    b. monocytes
    c. plasma
    d. platelets

18. The mitral valve is found:
    a. between the left atrium and left ventricle
    b. between the right atrium and right ventricle
    c. between the two atria
    d. in the pulmonary artery

19. If one were to centrifuge fresh whole blood, the clear liquid resulting would be:
    a. red and white blood cells
    b. prothrombin
    c. serum
    d. plasma

_____ 20. Heart valves controlling the flow of blood from the atria to the ventricles are generally termed:
a. myocardial
b. semilunar
c. sinoatrial
d. cuspid

_____ 21. Veins from the intestines en route to the heart must first pass through the:
a. inferior vena cava
b. superior vena cava
c. portal vein
d. common iliac vein

_____ 22. Mature, anucleate blood cells are:
a. normoblasts
b. granulocytes
c. agranulocytes
d. erythrocytes

_____ 23. Large arteries carrying blood to the head are:
a. mesenteric
b. jugular
c. brachial
d. carotid

_____ 24. The saphenous vein drains blood from the:
a. leg
b. arm
c. brain
d. sacral portion of the vertebral column

_____ 25. In the human, the left common carotid artery branches off the:
a. left subclavian artery
b. aortic arch
c. brachiocephalic artery
d. innominate artery

DISCUSSION QUESTIONS (Optional)

1. Using a flow chart, describe the scheme of human blood clotting.

2. Where is hemoglobin produced and what is its purpose? Can you suggest reasons why hemo-globin values differ in males and females?

3. Identify the place of formation of each of the following and the function each performs in the coagulation of blood:

   a. prothrombin

   b. fibrinogen

   c. thromboplastin

4. Trace the path of a drop of blood from the rectum to the heart and back to the rectum, naming all the blood vessels and structures of the heart through which the blood will pass.

5. Match the following statements with the following structures of the heart:

aorta	pericardium	pulmonary artery	tricuspid valve
bicuspid valve	inferior vena cava	pulmonary vein	aortic semilunar valve
chordae tendineae	left atrium	right atrium	
coronary arteries	left ventricle	right ventricle	
endocardium	myocardium	septum	

   a. Chamber that receives blood from lungs: . . . .

   b. The valve between the left atrium and the left ventricle: . . . .

   c. Muscle tissue that composes the heart: . . . .

   d. Tissue that lines the heart: . . . .

e. Covering of serous tissue that surrounds the heart: ....

f. Heart's own blood vessels: ....

g. Left ventricle exit valve: ....

h. Chamber that forces blood into systemic circulation: ....

i. Ventricle partition: ....

6. Match the following statements with the following terms:

anemia	fibrin	lymphocytes	serum
antibodies	lymph	plasma	monocyte
erythrocytes	leukocytosis	platelets	

a. Liquid portion of the blood: ....

b. Phagocytic cells: ....

c. Cells that transport oxygen: ....

d. Noncellular irregular elements essential to clotting: ....

e. Protective immunity: ....

f. Low hemoglobin level: ....

g. Too many white blood cells: ....

Respiratory System

■ *Purpose*

The purpose of Unit IX is to familiarize you with the anatomy of the human and cat respiratory organs and to allow you to understand the use of the spirometer and pneumograph in measuring standard respiratory volumes.

■ *Objectives*

To complete Unit IX, you must be able to do the following:

1. Identify major organs of the human and cat respiratory systems.
2. Identify the respiratory organs and related circulatory structures of sheep pluck.
3. Demonstrate knowledge of basic lung and trachea histology.
4. Determine standard respiratory air volumes using a spirometer.
5. Use a pneumograph to determine respiratory variations.

■ *Materials*

Preserved cat
Sheep pluck
Physiological recording equipment
Spirometers and replaceable
   mouthpieces

Dissecting pan
Dissecting pins
Prepared slides of lung and tracheal
   tissue
Pneumograph and recording attach-
   ments

■ *Procedure*

EXERCISE 1

Anatomy of Human and Cat Respiratory Organs

The function of the respiratory system is to transport oxygen from the external atmosphere to the alveoli of the lungs. The blood then transports oxygen from the alveoli to the cells for metabolic use. The respiratory system also eliminates excessive carbon dioxide from the body. Anatomically, the lungs function as the center of gas exchange, but also included as accessory structures are the *nasal cavity* and surrounding **paranasal sinuses**; frontal, sphenoid, ethmoid, and maxillary bones; and the *pharynx, larynx, trachea, bronchi*, and divisions of the bronchi. (*See* Figures 144 and 145 for the structures of the human respiratory system). Notice the number of *lobes* in the human right and left lungs. Also, note the C-shaped cartilaginous rings of the trachea that prevent collapse during inspiration (*see* Figures 146 and 147).

If you have not already done so, continue the midventral incision you made when studying the digestive and circulatory systems to expose the pharynx, or throat, and larynx, which is a cartilaginous, elongated boxlike structure at the superior end of the trachea. The **trachea**, or "windpipe," can be recognized by its cartilaginous rings.

Expose the respiratory structures in your cat by retracting the cut muscle and ribs laterally. Secure the ribs and muscles to your dissecting pan with dissecting pins. Gently grip the larynx and free it from its surrounding muscle by cutting. Pull the larynx and trachea anteriorly and ventrally from the thoracic cavity and identify the major cartilages comprising the larynx. The superior **epiglottis** is quite large, the anterior **thyroid cartilage** is elongated, and the inferior **cricoid cartilage** is thick and compressed laterally. Follow the trachea inferiorly and observe its division into **right and left stem bronchi**.

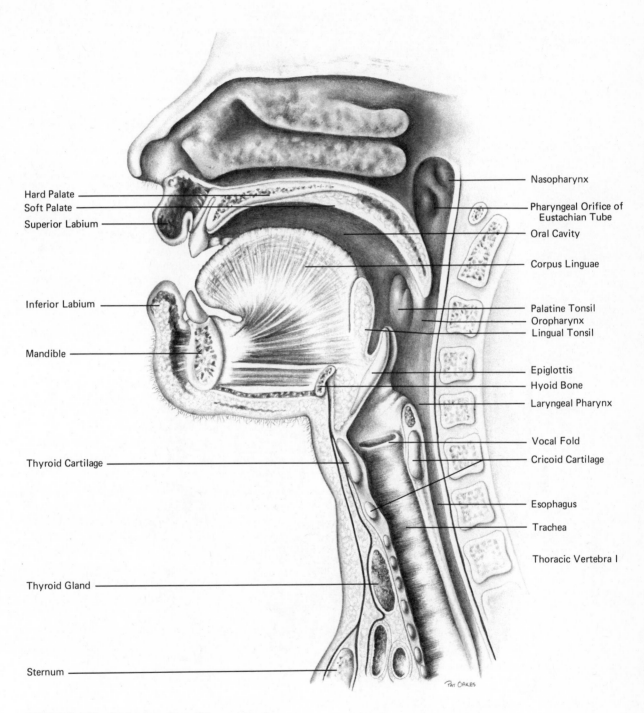

Hard Palate

Soft Palate

Superior Labium

Inferior Labium

Mandible

Thyroid Cartilage

Thyroid Gland

Sternum

Nasopharynx

Pharyngeal Orifice of Eustachian Tube

Oral Cavity

Corpus Linguae

Palatine Tonsil

Oropharynx

Lingual Tonsil

Epiglottis

Hyoid Bone

Laryngeal Pharynx

Vocal Fold

Cricoid Cartilage

Esophagus

Trachea

Thoracic Vertebra I

FIGURE 144   *Sagittal section of human head*

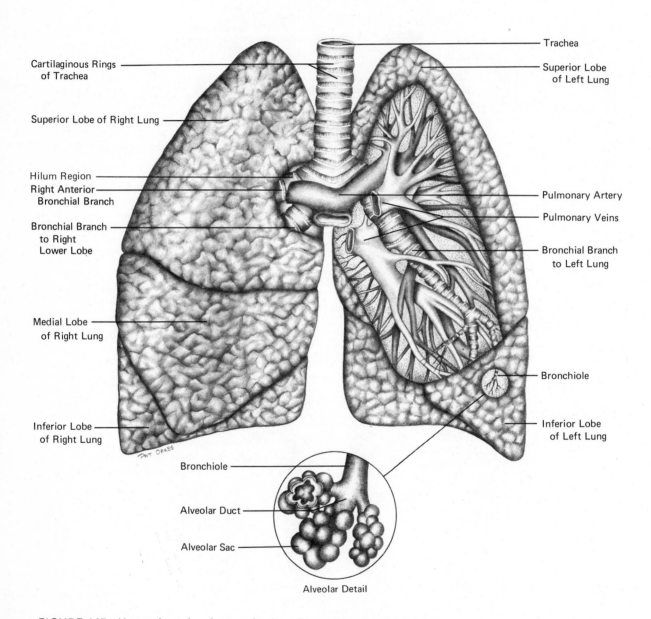

Trachea

Cartilaginous Rings
of Trachea

Superior Lobe
of Left Lung

Superior Lobe of Right Lung

Hilum Region
Right Anterior
Bronchial Branch

Pulmonary Artery

Pulmonary Veins

Bronchial Branch
to Right
Lower Lobe

Bronchial Branch
to Left Lung

Medial Lobe
of Right Lung

Bronchiole

Inferior Lobe
of Right Lung

Inferior Lobe
of Left Lung

Bronchiole

Alveolar Duct

Alveolar Sac

Alveolar Detail

FIGURE 145 *Human lung showing termination of bronchiole into alveolus*

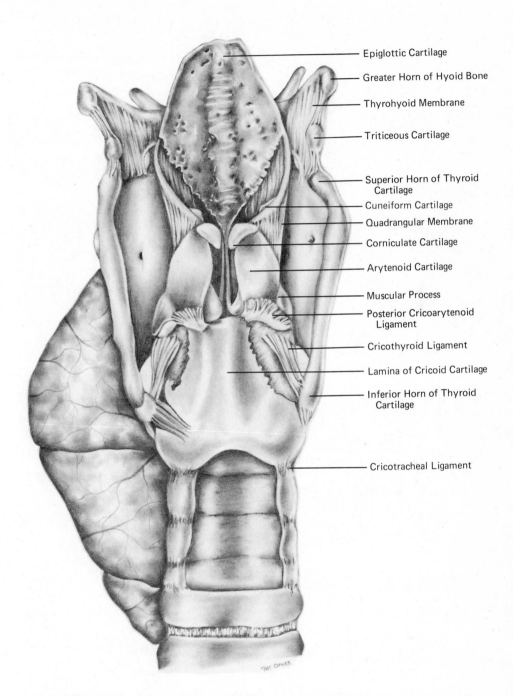

Epiglottic Cartilage

Greater Horn of Hyoid Bone

Thyrohyoid Membrane

Triticeous Cartilage

Superior Horn of Thyroid Cartilage

Cuneiform Cartilage

Quadrangular Membrane

Corniculate Cartilage

Arytenoid Cartilage

Muscular Process

Posterior Cricoarytenoid Ligament

Cricothyroid Ligament

Lamina of Cricoid Cartilage

Inferior Horn of Thyroid Cartilage

Cricotracheal Ligament

FIGURE 146 *Human larynx viewed from the dorsal and superior aspects* (Redrawn from B. J. Anson and C. B. McVay: *Surgical Anatomy,* Philadelphia: W. B. Saunders Co., 1971)

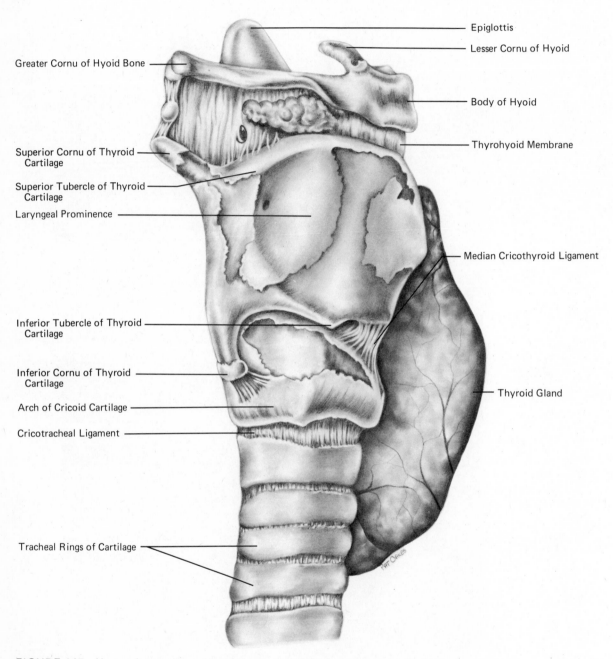

Epiglottis

Lesser Cornu of Hyoid

Greater Cornu of Hyoid Bone

Body of Hyoid

Superior Cornu of Thyroid
Cartilage

Thyrohyoid Membrane

Superior Tubercle of Thyroid
Cartilage

Laryngeal Prominence

Median Cricothyroid Ligament

Inferior Tubercle of Thyroid
Cartilage

Inferior Cornu of Thyroid
Cartilage

Thyroid Gland

Arch of Cricoid Cartilage

Cricotracheal Ligament

Tracheal Rings of Cartilage

FIGURE 147 *Human larynx viewed from the lateral and superior aspects* (Redrawn from
B. J. Anson and C. B. McVay: *Surgical Anatomy,* Philadelphia: W. B. Saunders Co., 1971)

In the right lung, the right stem bronchus sends branches to the right superior, right medial, and right posterior lobes. In the left lung, the left stem bronchus sends branches to the left superior, left medial, and left posterior lobes (*see* Figure 148).

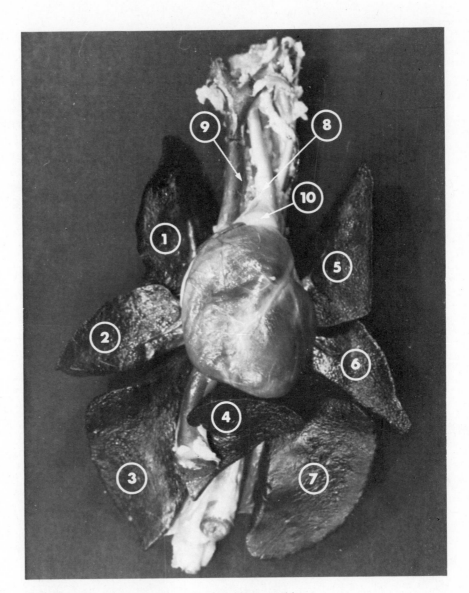

FIGURE 148 *Ventral aspect of isolated heart and lungs*

1. Right Superior Lobe of Lung
2. Right Medial Lobe of Lung
3. Right Posterior Lobe of Lung
4. Right Posterior Lobe of Lung
5. Left Superior Lobe of Lung
6. Left Medial Lobe of Lung
7. Left Posterior Lobe of Lung
8. Brachiocephalic Artery
9. Superior Vena Cava
10. Aortic Arch

EXERCISE 2

Sheep Pluck

Your instructor will have a sheep pluck available for organ identification. Included will be the lungs, heart, and pericardium, as well as the aorta, the pulmonary artery, and pulmonary veins. Parts of the trachea and esophagus may also be intact. Observe the structures, noting similarities and differences among the sheep, cat, and human structures.

EXERCISE 3

Examination of Prepared Slides of the Lung and Trachea

The **alveolus**, a saclike structure composed of *simple squamous epithelium*, is the main structural and functional unit of the lung. It is in the alveolus that gas exchange occurs. The structures associated with the alveoli (*see* Figures 149 and 150) are the bronchioles, which divide into the **alveolar ducts** and **alveolar sacs**. Obtain a prepared slide of the lung, locate and identify the previously named structures. Also obtain a prepared slide of the trachea (cross section) and observe the tissue types present. Note the ciliated pseudostratified epithelium, hyaline cartilage, and trachealis muscle. *SKETCH and LABEL* the structures that you observe.

Lung:

Trachea:

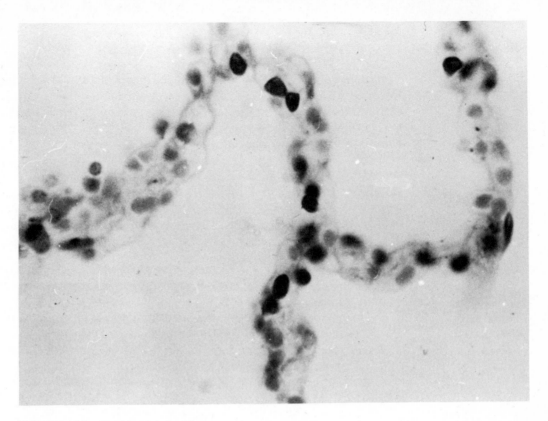

FIGURE 149   *Alveolar detail of lung*

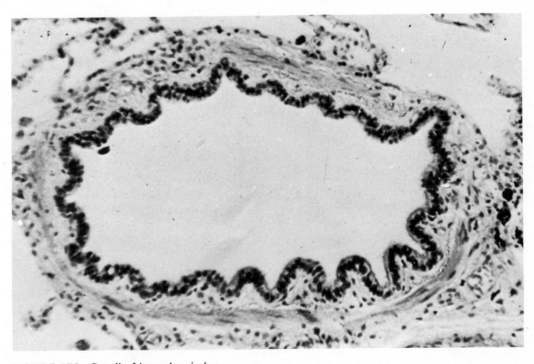

FIGURE 150   *Detail of bronchus in lung*

EXERCISE 4

Spirometric Measurement of Standard Respiratory Volumes

Variations in the size, age, and sex of an individual account for variations in respiratory volumes. Normal breathing usually moves about 500 ml of air in and out of the lungs and is referred to as *tidal volume* (TV). Even after a normal inspiration, an individual can forcibly inhale approximately 3000 ml of air into the lungs. This forced inspiration is called the *inspiratory reserve volume* (IRV). Also, after a normal expiration, an individual can forcibly exhale approximately 1100 ml of air—the *expiratory reserve volume* (ERV). The *vital capacity* is the sum total of the inspiratory reserve, tidal, and expiratory reserve volumes, or approximately 4600 ml of air; thus, the vital capacity is the total exchangeable air of the lungs. In addition, approximately 1200 ml of air remain in the conducting tubules. This air is not exchangeable, and is referred to as **residual volume** (RV).

Air volumes can be measured by means of a spirometer. Set the needle of the spirometer at zero (0) at the beginning of each use. Remember, each student should use a clean replaceable mouthpiece. Each of you should determine your *tidal volume*, *expiratory reserve volume*, *vital capacity*, and *inspiratory reserve volume* (vital capacity minus expiratory reserve and tidal volumes). See the following table to determine normal ranges for vital capacity. Enter your individual volumes in Question 1 of the Optional Discussion at the end of this unit.

NORMAL VITAL CAPACITY (IN CUBIC CENTIMETERS) OF ADULT MALES AND FEMALES*

Height in Inches		Age in Years					
		20	30	40	50	60	70
Males	60	3885	3665	3445	3225	3005	2785
	62	4154	3925	3705	3485	3265	3045
	64	4410	4190	3970	3750	3530	3310
	66	4675	4455	4235	4015	3795	3575
	68	4940	4720	4500	4280	4060	3840
	70	5206	4986	4766	4546	4326	4106
	72	5471	5251	5031	4811	4591	4371
	74	5736	5516	5516	5076	4856	4636
Females	58	2989	2809	2629	2449	2269	2089
	60	3198	3018	2838	2658	2478	2298
	62	3403	3223	3043	2863	2683	2503
	64	3612	3432	3252	3072	2892	2710
	66	3822	3642	3462	3282	3102	2922
	68	4031	3851	3671	3491	3311	3131
	70	4270	4090	3910	3730	3550	3370
	72	4449	4269	4089	3909	3729	3549

Adapted with permission from Propper Manufacturing Co., Inc., New York.

*Variations must be at least 20% below predicted normal to be considered subnormal. Variations can also exist depending upon size and body structure.

EXERCISE 5

Use of the Pneumograph in Determination of Respiratory Variations

Human respiratory variations are determined by use of a *pneumograph*, an instrument that records variations in breathing patterns. Your instructor will advise you how to set up the instrument so you can perform various exercises and read the results. You should be actively involved in these exercises and should not be giving attention to recordings until completion. It is best to work with a partner. One partner can operate the pneumograph and identify the activities the other person performs. The pneumograph tubing should be attached firmly but not tightly around the thoracic cavity. Space for chest expansion must be available.

1. Breathe normally for approximately 2 minutes in a sitting position. Inspiration should be recorded by a downward deflection of the stylus. Upon inspiration, the pneumograph tubing lengthens, thus increasing the volume but decreasing the pressure within the chest cavity, allowing atmospheric air to enter.

2. Vary your activities in the following manner, and have your partner record the results:

   a. talking
   b. standing
   c. talking and standing
   d. coughing
   e. contracting biceps
   f. deep breathing
   g. shallow breathing
   h. running in place
   i. concentrating
   j. sitting and extending legs forward
   k. drinking water
   l. holding your breath

3. Breathe normally for 2 minutes. *RECORD.* Hold your breath as long as you can. *RECORD. RECORD* the recovery period. CO_2 accumulation causes the recovery period. Recovery usually results when alveolar air

   reaches 7%–10% CO_2. How long was your recovery period? . . . . Average recovery is slightly longer than 1 minute. An individual usually breathes 14–18 inspirations and expirations per minute. Now hyperventilate for 45 seconds and then hold your breath. Can you hold your breath

   for a longer or shorter period of time now . . . . *Explain.*

4. Attach a labeled recording of your respiratory cycles to Discussion Question 5.

Respiratory System

DISCUSSION

Picture Identification

Use this diagram for Questions 1–4:

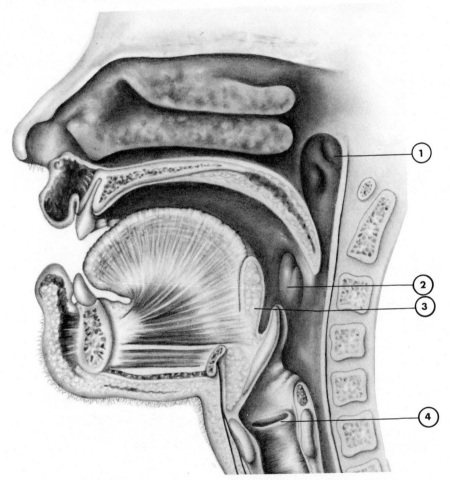

_____ 1. Chamber 1 is the:
    a. pharyngeal orifice
    b. oral cavity

    c. nasopharynx
    d. oropharynx

_____ 2. Structure 2 is known as the:
    a. lingual tonsil
    b. palatine tonsil

    c. oropharynx
    d. laryngeal pharynx

_____ 3. Structure 3 is the:
    a. corpus linguae
    b. epiglottis

    c. lingual tonsil
    d. hyoid bone

_____ 4. Structure 4 is (are) the:
    a. laryngeal pharynx
    b. epiglottis

    c. thyroid cartilage
    d. vocal folds

Multiple Choice

_____ 5. The anatomical structure that prevents food from entering the trachea when you swallow is the:
a. larynx
b. esophagus
c. epiglottis
d. glottis

_____ 6. The volume of air that moves into the lungs with each inspiration is termed:
a. vital capacity
b. tidal volume
c. residual volume
d. inspiratory reserve volume

_____ 7. The serous protective saclike membrane that encases the lungs is known as the:
a. pericardium
b. peritoneum
c. pleural cavity
d. pleura

_____ 8. The anatomical site where the pulmonary artery, vein, and bronchi enter and exit the lung is the:
a. apex
b. base
c. mediastinum
d. hilum

_____ 9. Which of the following tissue types is _not_ found in the respiratory system?
a. pseudostratified ciliated epithelium
b. goblet cells
c. osteoblasts
d. lamina propria with mucous glands

_____ 10. If an individual's tidal volume is 500 cc, expiratory reserve volume is 1200 cc, and vital capacity is 4500 cc, what is the individual's inspiratory reserve volume?
a. 1700 cc
b. 2800 cc
c. 6200 cc
d. cannot be calculated from the information given

_____ 11. Oxygen and carbon dioxide cross the alveolar epithelium by means of:
a. osmosis
b. active transport
c. diffusion
d. filtration

_____ 12. What is the advantage of the trachea's lining of ciliated epithelium?
a. to trap dust particles before they enter the lungs
b. to prevent food from becoming lodged in the trachea
c. to aid in the movement of oxygen and carbon dioxide molecules in and out of the lungs
d. to protect the tracheal epithelium against irritation

_____ 13. Which laryngeal cartilage forms the "Adam's apple"?
a. corniculate
b. cricoid
c. thyroid
d. epiglottis

_____ 14. Vital capacity in the adult tends to . . . . with age.
a. increase
b. decrease
c. remain the same

_____ 15. Vital capacity would be . . . . in a patient with pneumonia.
a. increased
b. decreased
c. remain the same

_____ 16. Under which of these conditions would breathing be most rapid?
a. while asleep
b. during normal quiet breathing
c. immediately after holding one's breath
d. immediately following hyperventilation

_____ 17. Air remaining in the lungs even after a forced expiration is:
a. tidal volume
b. inspiratory reserve volume
c. expiratory reserve volume
d. residual volume

_____ 18. The nasal septum consists of which of the following bones?
a. mandibular and zygomatic
b. occipital and parietal
c. frontal and temporal
d. ethmoid and vomer

_____ 19. Structurally, the larynx is primarily comprised of:
a. cartilage
b. bone
c. membranes
d. stratified epithelium

_____ 20. The lower end of the trachea divides into:
a. alveoli
b. bronchioles
c. bronchi
d. pleural cavities

_____ 21. The spirometer cannot directly measure:
a. vital capacity
b. residual volume
c. tidal volume
d. expiratory reserve volume

_____ 22. Which of the following tissues is most likely to be observed in a cross section of the trachea?
a. trachealis muscle
b. osseous tissue
c. adipose tissue
d. ciliated pseudostratified epithelium

_____ 23. Normal tidal volume is approximately how many milliliters of air?
a. 500 ml
b. 1000 ml
c. 700 ml
d. 750 ml

_____ 24. An alveolus is composed of which tissue type?
a. simple squamous epithelium
b. ciliated columnar epithelium
c. cuboidal epithelium
d. hyaline cartilage

_____ 25. Which of the following terminates in an alveolar duct?
a. bronchiole
b. bronchus
c. alveolar capillary
d. pleura

DISCUSSION QUESTIONS (Optional)

1. a. My tidal volume is . . . .

   b. My inspiratory reserve volume is . . . .

   c. My expiratory reserve volume is . . . .

   d. My vital capacity is . . . .

   e. My approximate residual volume is . . . .

   f. My total lung capacity is approximately . . . .

2. How do the cat and human lungs differ with respect to number and names of lobes?

3. Why is it advantageous that the epiglottis is cartilaginous?

4. What is the advantage of the alveolar sacs being simple squamous epithelium?

5. Label, interpret, and attach your pneumograph recordings here.

Nervous System

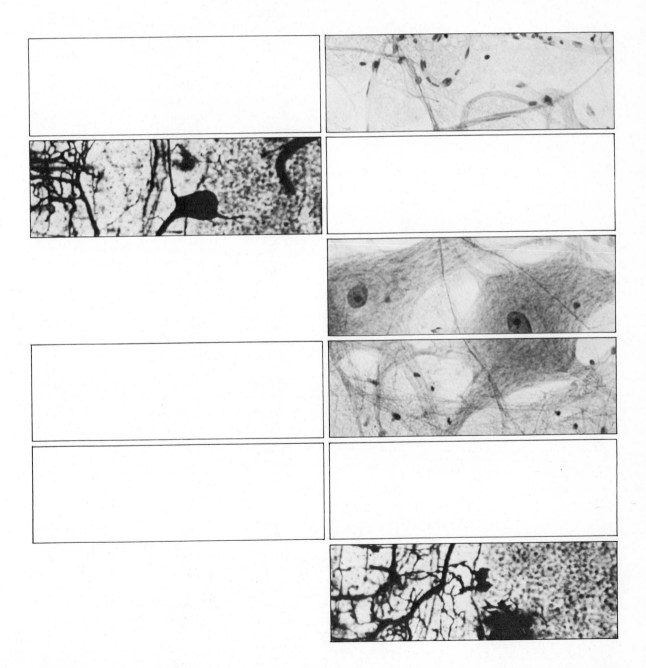

A. Nervous System Anatomy

■ *Purpose*

The purpose of Unit X-A is to familiarize you with the histological and gross anatomical features of the nervous system.

■ *Objectives*

In order to complete Unit X-A, you must be able to do the following:

1. Using a microscope, identify the characteristics of a neuron.
2. Using a microscope, identify the features of the spinal cord.
3. Using a microscope, differentiate between cerebral and cerebellar tissue.
4. Identify the meninges.
5. Identify the major gross anatomical characteristics of the brain.
6. Identify selected cranial and spinal nerves.

■ *Materials*

Prepared slides of motor neurons (nerve smear)

Preserved sheep or beef brains with meninges attached (if available)

Slides of cross sections of spinal cord, cerebrum, and cerebellum

Ox spinal cord (preserved)

Preserved cats

■ *Procedure*

Neurons, the basic structural and functional units of the nervous system, are characterized in a number of ways: (1) by the direction of impulse flow, (2) by the number of processes, and (3) by the absence or presence of myelinated sheaths.

EXERCISE 1

Identification of Neuron Types and Cytological Characteristics of Neurons

(*see* Color Plate 25)

Examine a prepared slide of a *nerve smear*. This section probably was prepared from the spinal cord. Carefully focus the slide; then, using high power or oil immersion, locate large stellate *cell bodies* exhibiting two or more extensions, or *processes*. These cells are **neurons**. Neurons with only two processes are *bipolar*. One process—the **axon**—carries impulses away from the cell body, and the other—the **dendrite**—carries impulses toward the cell body. *Multipolar* neurons have many processes including one axon and several dendrites. Select a distinct neuron and identify the basic structures (*see* Figure 151) described in the following text.

The *perikaryon*, or cell body, from which the processes extend contains a large nucleus, or *karyon*, that contains a rather large nucleolus in addition to other nuclear materials. The cytoplasm or neuroplasm contains *Nissl bodies* and *neurofibrils* in addition to the usual cytoplasmic inclusions. Nissl bodies are granules of RNA and are darkly stained. Neurofibrils are tabular structures, their function has not been identified.

Observe an axon extending from a cell body. Some axons are surrounded by a segmented protective sheath—the *myelin sheath*. Peripheral to the

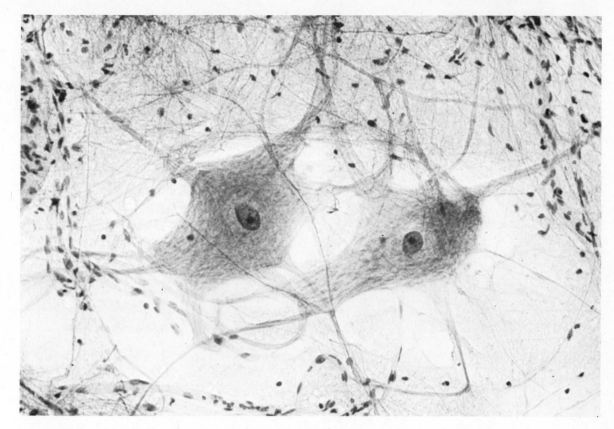

FIGURE 151 *Motor nerve cells of the spinal cord* (Courtesy Carolina Biological Supply Company)

myelin sheath is an additional layer, the *neurilemma*, that plays a role in regeneration of axons. The constrictions between the segments of myelin are termed *nodes of Ranvier*, which are surrounded by the neurilemma. Sometimes a smaller axon branches off the main axon and is termed an *axon collateral.*

EXERCISE 2

Microscopic Examination of the Spinal Cord

Obtain a prepared slide of a cross section of an ox or human spinal cord. Observe the slide under a dissecting microscope. (*see* Figures 152 and 153.) Notice the central gray matter, which is H-shaped, and the peripheral white matter tracts. Identify the *ventral* and *dorsal medial sulci.* These grooves are on the ventral and dorsal surfaces in the center of the spinal cord. The *dorsal gray columns*, or *dorsal horns*, are the dorsal extensions of the gray matter of the "H." The *ventral gray columns*, or *ventral horns*, are the ventral extensions of the "H." Fibers of **afferent** *sensory nerves* enter at the dorsal gray columns, and the fibers of **efferent** *motor nerves* extend from the ventral gray columns. The *dorsal median septum* is the smooth white area on either side of the dorsal median sulcus. The *funiculi* are divisions of white matter surrounding the gray dorsally, laterally, and ventrally.

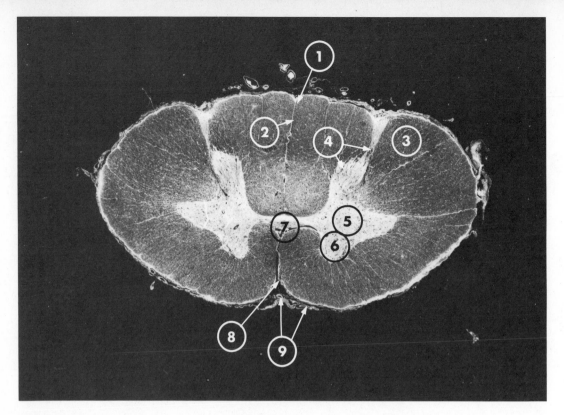

FIGURE 152 *Human spinal cord, transverse section at the cervical region*

1. Posterior Median Sulcus
2. Posterior Median Septum
3. White Matter
4. Posterior Gray Horn

5. Motor Neurons
6. Anterior Gray Horn
7. Central Canal
8. Anterior Median Sulcus

9. Anterior Spinal Artery and Vein with Pia Mater

FIGURE 153 *Spinal cord showing fiber tracts*

1. Dorsal Median Sulcus
2. Dorsal Root Ganglion (Sensory)
3. Ventral Root (Motor)
4. White Matter
5. Ventral Median Fissure
6. Ventral Horn
7. Dorsal Horn (Gray Column)
8. Central Canal

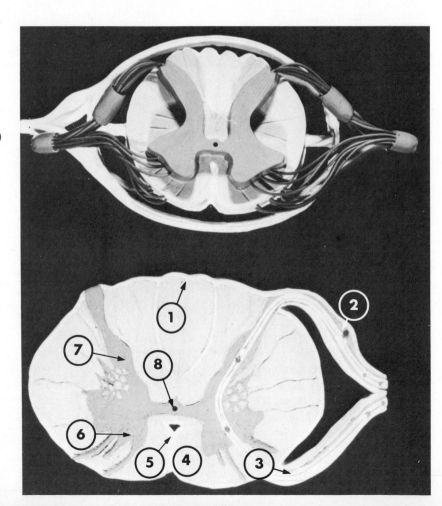

EXERCISE 3

Microscopic Examination of Cross Sections of the Cerebrum and Cerebellum

(*see* **Color Plates 26–29**)

In the cerebrum, notice the gray matter of the *cortex* and the white matter of the *corpus callosum*. The obvious *pyramidal cells* constitute most of the gray matter. Long threadlike fibers in the central motor regions of the cerebrum originate from pyramidal cells (*see* Figures 154–156).

FIGURE 154 *Human cerebral cortex*

1. Pyramidal Cells of Cortex
2. White Matter

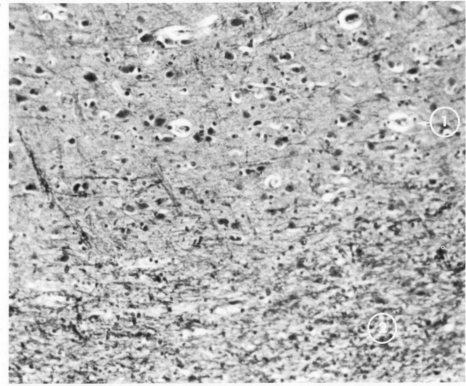

FIGURE 155 *Pyramidal cells as seen in the cerebral cortex*

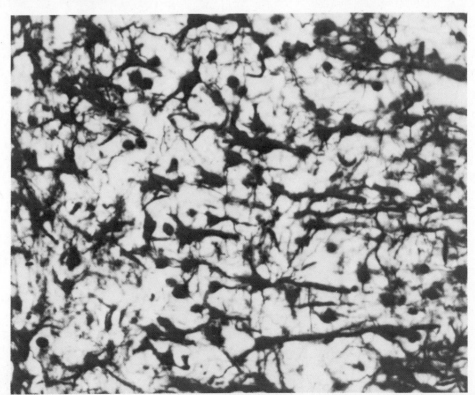

FIGURE 156 *Detail of pyramidal cells in cerebrum*

FIGURE 157 *Purkinje cells in cerebellum*

1. Purkinje Cell

Examine a cross section of cerebellum. Note the gray matter toward the periphery and the white matter of the **arbor vitae**. Also, notice the large **Purkinje cells** at the interface of the cerebellar cortex and medulla (*see* Figure 157).

EXERCISE 4
Examination of Meninges

The **meninges** are protective coverings around the brain and spinal cord. Examine a preserved brain. The outer thick whitish layer is the *dura mater*. Gently make a horizontal incision and then a vertical incision through this layer and reflect it back. Beneath the dura mater, you will see a whitish-to-transparent covering with a spider-web pattern. This is the *arachnoid layer*. Gently reflect this layer laterally and observe the innermost layer, which is adherent to the sulcus and fissures of the brain itself. This third innermost layer is the *pia mater*. Observe the spinal cord specimen and locate the meninges if they are present.

Between the meninges are spaces called the *subarachnoid* and the *subdural spaces*. The subarachnoid contains *cerebrospinal fluid*. Blood sinuses of the skull are in the dura mater and the dried blood may be obvious on your specimen. Fine vascularization may be obvious in the *pia mater*.

EXERCISE 5
Examination of a Preserved Brain

The brain of the cat is small; therefore, it is not suitable for dissection. Using a preserved sheep or beef brain, identify the **cerebrum**, **cerebellum**, **medulla oblongata**, **longitudinal cerebral fissure**, and **paired cerebral hemispheres** on the *dorsal* surface. You may be able to identify the **olfactory bulbs**, which extend anteriorly if they remain intact. Ventrally, identify anteriorly to posteriorly, the *olfactory bulbs* and *tracts*, the *optic nerves*, the **optic chiasma**, the **infundibulum**, the **hypophysis** (pituitary gland) and stalk, **mammillary body, pons, pyramidal tract, hemispheres** of the cerebellum, medulla oblongata, and spinal cord. (*See* Figures 158–163.)

On the *external lateral view* of the brain (Figure 160), locate the **frontal, parietal, temporal,** and **occipital lobes** of the cerebrum. Sulci are deep depressions in the cerebral cortex. Locate the *central sulcus* (fissure of Rolando) that runs superiorly to inferiorly in the central region. Anterior to the central sulcus and running in the same direction is the *precentral gyrus* (which is less deep), and posterior to the central sulcus is the *postcentral gyrus*. Immediately posterior to the frontal lobe on the lateral surface of the brain is the *lateral sulcus*.

Make a midsagittal incision through the longitudinal fissure of the cerebrum and continue the incision posteriorly through the cerebellum and spinal cord. Observe the following structures in *median view* (Figure 161): the **corpus callosum, fornix, thalamus, hypothalamus, septum pellucidum infundibulum,** and **pituitary gland**. The **third** and **fourth ventricles** lie between the cerebrum and cerebellum and between the cerebellum and spinal cord, respectively. In the cerebellum locate the **arbor vitae,** and in the spinal cord note the *reticular formation*. Observe the pons, medulla, and spinal cord toward the posterior inferior portion of the brain. In the central region, observe the *intermediate mass, pineal body, mammillary body,* and **midbrain**. Various *cranial nerves* may partially project from the **brainstem** (Figure 159). Examine a human brain model or specimen and locate the above structures. Use Figures 164 and 165 to help in your identification.

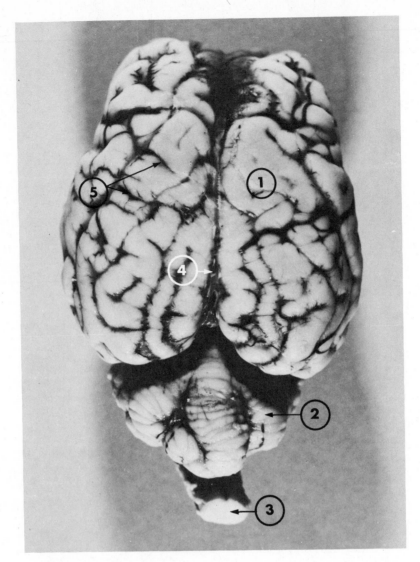

FIGURE 158 *Sheep brain, superior aspect*

1. Postcentral Gyrus of Cerebrum

2. Cerebellar Hemisphere

3. _____

4. _____

5. Sulci

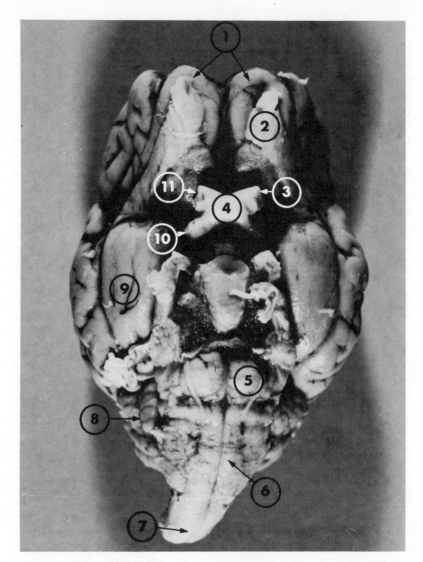

FIGURE 159 *Sheep brain, inferior aspect, showing gross detail*

1. _____
2. Olfactory Bulb
3. Left Optic Nerve
4. Optic Chiasma
5. _____
6. Medulia Oblongata
7. _____
8. Cerebellum
9. _____
10. Optic Tract
11. Right Optic Nerve

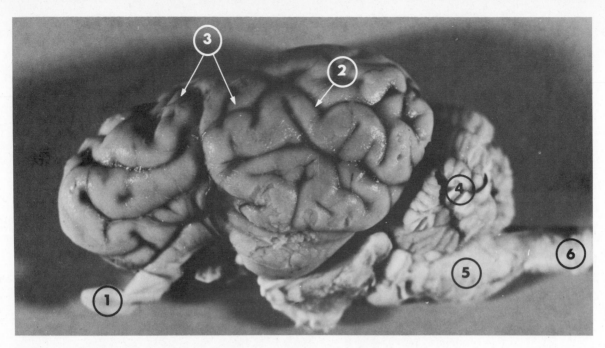

FIGURE 160 *Sheep brain, lateral aspect, showing gross structures*

1. Olfactory Bulb

2. Sulcus of Parietal Lobe

3. Percentral and Postcentral Gyri of Cerebral Hemisphere

4. _____

5. _____

6. _____

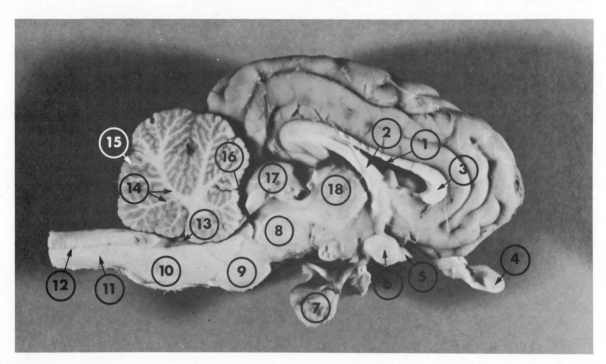

FIGURE 161 *Sheep brain, midsagittal section, showing internal structures*

1. Lateral Ventricle

2. Fornix

3. Corpus Callosum

4. Olfactory Bulb

5. Optic Nerve

6. _____

7. _____

8. Midbrain

9. Pons

10. _____

11. _____

12. Central Canal of Spinal Cord

13. Fourth Ventricle

14. Medullary Body of Cerebellum (Arbor Vitae)

15. _____

16. Third Ventricle

17. Corpora Quadrigemina

18. Thalamus

FIGURE 162 *Sheep brain, coronal aspect, showing gray and white matter detail*

1. Gray Matter of Cerebral Cortex
2. Corpus Callosum
3. Lateral Ventricle
4. Medullary Body of Cerebrum (White Matter)

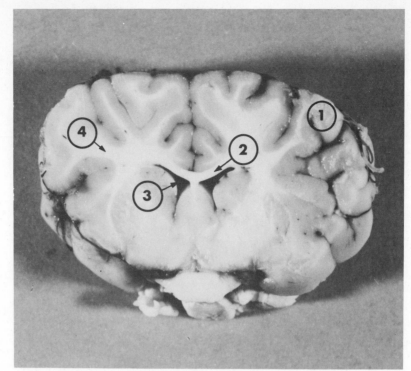

FIGURE 163 *Human brain, superior aspect, showing hemisphere detail*

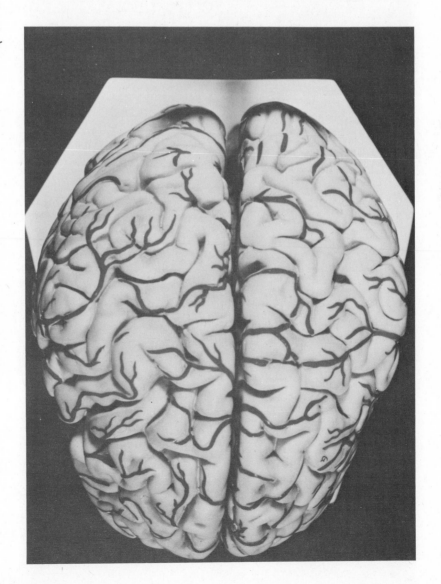

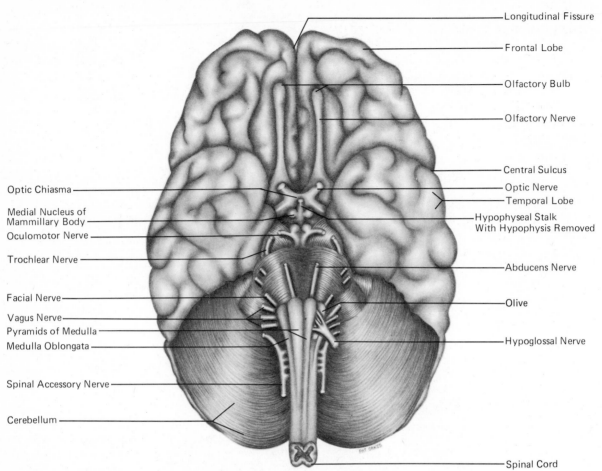

Longitudinal Fissure

Frontal Lobe

Olfactory Bulb

Olfactory Nerve

Central Sulcus

Optic Chiasma

Optic Nerve

Temporal Lobe

Medial Nucleus of Mammillary Body

Hypophyseal Stalk With Hypophysis Removed

Oculomotor Nerve

Trochlear Nerve

Abducens Nerve

Facial Nerve

Olive

Vagus Nerve
Pyramids of Medulla
Medulla Oblongata

Hypoglossal Nerve

Spinal Accessory Nerve

Cerebellum

Spinal Cord

FIGURE 164 *Human brain, inferior aspect*

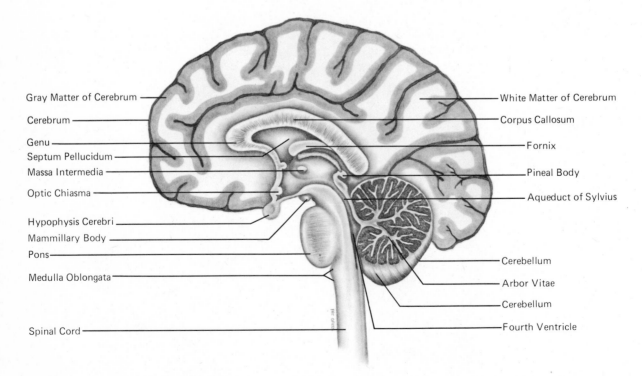

Gray Matter of Cerebrum

White Matter of Cerebrum

Cerebrum

Corpus Callosum

Genu

Fornix

Septum Pellucidum

Massa Intermedia

Pineal Body

Optic Chiasma

Aqueduct of Sylvius

Hypophysis Cerebri

Mammillary Body

Pons

Cerebellum

Medulla Oblongata

Arbor Vitae

Cerebellum

Spinal Cord

Fourth Ventricle

FIGURE 165 *Human brain, midsagittal region*

EXERCISE 6

Examination of Cranial Nerves

On the preserved brain with intact cranial nerves (*see* Figure 159), locate the 12 pairs of cranial nerves:

    I. Olfactory
    II. Optic
    III. Oculomotor
    IV. Trochlear
    V. Trigeminal
    VI. Abducens
    VII. Facial
    VIII. Acoustic (Vestibulocochlear)
    IX. Glossopharyngeal
    X. Vagus
    XI. Spinal accessory (Accessory)
    XII. Hypoglossal

EXERCISE 7

Examination of Spinal Nerves

On the preserved ox spinal cord or demonstration preparation, note the emerging nerves (*see* Figures 166 through 169). In the human being, there are 31 pairs: 8 cervical, 12 thoracic, 5 lumbar, 5 sacral, and 1 coccygeal.

To expose the **brachial plexus** in your cat, carefully remove the muscles, blood vessels, and connective tissue from the shoulder axillary region and chest on one side of the body. The brachial plexus, which consists of interconnected white, tough nerves emerging from the last three cervical and first thoracic vertebrae should now be visible. By removing the muscles, blood vessels, and connective tissue from the lumbosacral region, the **lumbosacral plexus** can be seen. It is comprised of branches from the last three lumbar and first sacral nerves.

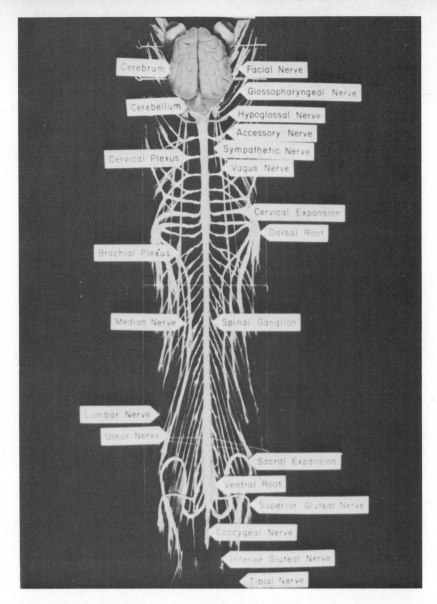

FIGURE 166 *An isolated complete cat nervous system showing all major nerves, dorsal aspect* (from M. J. Timmons, *A Visual Guide to Dissection: Cat Anatomy Slides,* courtesy J. B. Lippincott Company)

Cerebrum

Facial Nerve

Cerebellum

Glossopharyngeal Nerve

Hypoglossal Nerve

Accessory Nerve

Sympathetic Nerve

Cervical Plexus

Vagus Nerve

Cervical Expansion

Dorsal Root

Brachial Plexus

Median Nerve

Spinal Ganglion

Lumbar Nerve

Ulnar Nerve

Sacral Expansion

Ventral Root

Superior Gluteal Nerve

Coccygeal Nerve

Inferior Gluteal Nerve

Tibial Nerve

FIGURE 167 *Cat nervous system at the level of the frontal lobe to the cervical plexus showing cranial and spinal nerves, dorsal aspect* (From M. J. Timmons, *A Visual Guide to Dissection: Cat Anatomy Slides,* courtesy J. B. Lippincott Company)

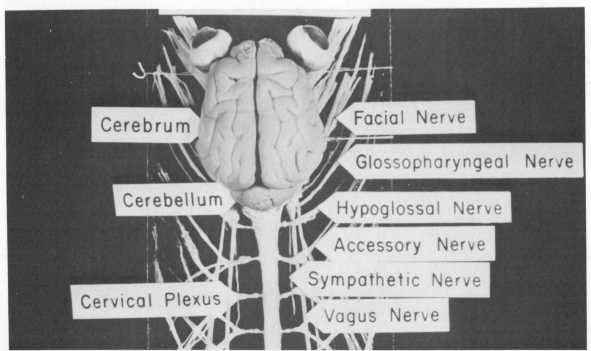

Cerebrum

Facial Nerve

Glossopharyngeal Nerve

Cerebellum

Hypoglossal Nerve

Accessory Nerve

Sympathetic Nerve

Cervical Plexus

Vagus Nerve

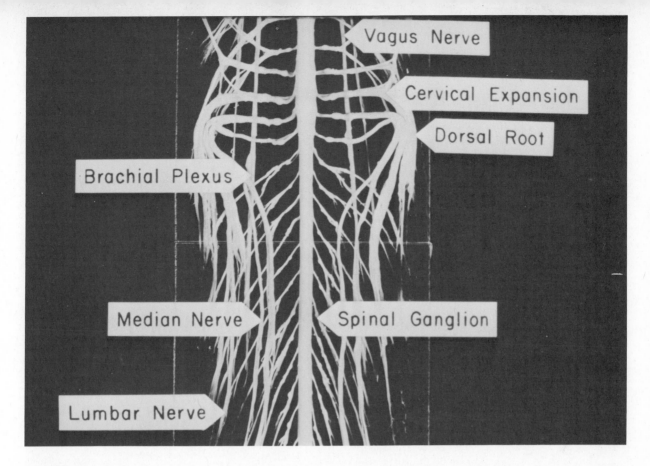

FIGURE 168 *Cat nervous system at the level of the cervical plexus and posterior to the dorsal root showing nerve detail, dorsal aspect* (From M. J. Timmons, *A Visual Guide to Dissection: Cat Anatomy Slides,* courtesy J. B. Lippincott Company)

FIGURE 169 *Cat nervous system posterior to the dorsal root showing nerve detail, dorsal aspect* (From M. J. Timmons, *A Visual Guide to Dissection: Cat Anatomy Slides,* courtesy J. B. Lippincott Company)

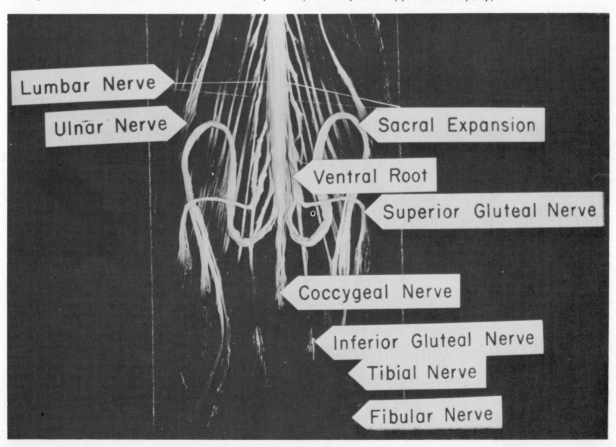

B. Nervous System Physiology

- *Purpose*

  The purpose of Unit X-B is to familiarize you with some of the physiological aspects of the nervous system.

- *Objectives*

  To complete Unit X-B, you must be able to do the following:

  1. Identify selected characterictics of nerve impulse transmission in the frog.
  2. Identify selected spinal **reflexes** in the frog.
  3. Identify and demonstrate some selected reflexes in the human being.
  4. Perform tests for cranial nerve and cerebellar function.

- *Materials*

Live frogs	Dissecting instruments
Source of heat	Facial tissues
1% HCl solution	5% acetic acid
Ringer's solution	Flashlight
Rubber reflex hammer	Penlight (optional)
Pans of water	Coffee, tobacco, or spices
Electric stimulator	Cotton
Kymograph or physiological recording aparatus	Cotton-tipped applicator
10% NaCl solution	Stopwatch
10% sugar solution	Tuning fork (512 cycles per second)
Tongue depressors	Swivel chair

- *Procedure*

  Nerves conduct impulses upon stimulation. For a nerve to be excitable and to change from a resting state, a difference in **ionic** concentration must exist between the external and internal surfaces of the cell membrane. Conductivity results from a continuous change in ionic concentration from the original point of stimulation along the nerve.

EXERCISE 1

Cranial and Spinal Nervous Responses of the Frog

In order to demonstrate nerve physiology in the frog successfully, the frog must be freed from pain by *pithing*. Pithing, in contrast to **anesthetizing** the frog, eliminates side effects and produces more valid responses. The frog will be alive, the muscles will contract, and respiration will continue. Two methods of pithing can be utilized, each evoking different responses. *Single pithing* involves destruction of the brain only, but the spinal cord remains intact and functional. In *double pithing*, both the brain and spinal cord are destroyed (refer to Unit VI-B, "The Physiology of Muscle Contraction," for pithing instructions).

Perform the following exercises and complete the chart. Work with an unpithed frog first; then repeat with the single-pithed frog and the double-pithed frog. *RECORD* your results in the table provided.

1. Pinch the toes of the hind leg. Repeat in 5 mintues.
2. Gently touch the cornea (corneal reflex).
3. Apply 5% acetic acid to the gastrocnemius muscle of one leg.
4. Introduce a light source (flashlight) to the eyes and then darken area over the eyes with your hand (pupillary reflex).
5. Place frog in a pan of water and see if it swims.
6. Place frog on its dorsal surface to see if it turns over to its ventral surface.

Exercises	Frog Responses			Conclusions
	Unpithed	Single-Pithed	Double-Pithed	
1. Pinching toes	1.	1.	1.	1.
2. Corneal reflex	2.	2.	2.	2.
3. Acetic acid	3.	3.	3.	3.
4. Pupillary reflex	4.	4.	4.	4.
5. Swimming	5.	5.	5.	5.
6. Righting self	6.	6.	6.	6.

Selected Characteristics of Nerve Impulse Transmission

A nerve-muscle preparation such as the one you used in Unit VI-B will facilitate your understanding of various characteristics of nerve physiology. In order to make a nerve-muscle preparation, follow the procedure given in Exercise 2 of Unit VI-B. Set up the physiological apparatus or **kymograph** according to your instructor's direction and do the following two exercises.

Materials (additional)

1% HCl solution
0.5% HCl solution
Hot water
Cold water
Electric shock apparatus
1% NaCl solution
NaCl crystals
Forceps (for pinching nerve)
Frog Ringer's solution
Novocain
Cotton

Procedure

a. *Effects of the Application of Various Stimuli*

Apply each of the stimuli listed in the "Variations in Stimuli" table to the end of a cut sciatic nerve. Cut off the destroyed nerve ending and rinse with frog Ringer's solution after each application. *RECORD* your data in the table.

b. *Conductivity*

Stimulate a freshly cut nerve ending with a mild electric shock and observe the responding muscle contraction. Keeping the nerve moist with Ringer's solution, soak a small amount of cotton in novocain and apply it at a point approximately 1 cm from its cut end. Allow the cotton to remain on the nerve for 60 seconds. Set up the nerve-muscle preparation as before. Stimulate the nerve with a mild electric shock in a region anterior to the anesthetized area. *RECORD* your results. *REPEAT* stimulation at the point where novocain was applied. *RECORD* your results. *REPEAT* stimulation at a point posterior to novocain application. Interpret your results here and record them in Discussion Question 6 at the end of this unit.

VARIATIONS IN STIMULI

Stimulus	Type of Stimulus: chemical, osmotic, electrical, mechanical	Response
1. 1% HCl	1.	1.
2. 0.5% HCl	2.	2.
3. Hot water	3.	3.
4. Cold water	4.	4.
5. Mild electric shock	5.	5.
6. Strong electric shock	6.	6.
7. 1% NaCl	7.	7.
8. NaCl crystals	8.	8.
9. Pinched nerve (with forceps)	9.	9.

EXERCISE 3

Reflexes in the Human Being

a. *The Swallowing Reflex*

Swallow the saliva in your mouth and immediately swallow again. Explain your result and compare it to the rapid succession of swallowing demonstrated by rapidly drinking a glass of water.

Try to stop yourself from swallowing. Explain in terms of a reflex action.

b. *The Patellar Reflex*

Sit so that your legs hang down freely from the knee. Have another student strike the patellar ligament (just below the knee) with a rubber reflex hammer. This may require a little patience, and it is best to divert the subject's attention. Notice that the leg is extended by the contraction of the quadriceps muscle group. Repeat on another student. Is the reflex obtained just as readily and is it equally extensive for all students? *RECORD* your data. (*See* Figure 170.)

c. *Photo-Pupil Reflex*

Close your eyes for 2 minutes. While facing a bright light, open them and let another student examine your pupils immediately. Describe the observed response. What is the purpose of this reflex action?

d. *The Accommodation Reflex*

In a moderate light, look at a distant object (20 ft or more removed) and have another student examine your pupils. Now look at a pencil held about 10 in. from your face (without changing the illumination) and have the student note your pupils. What is the purpose of this reflex?

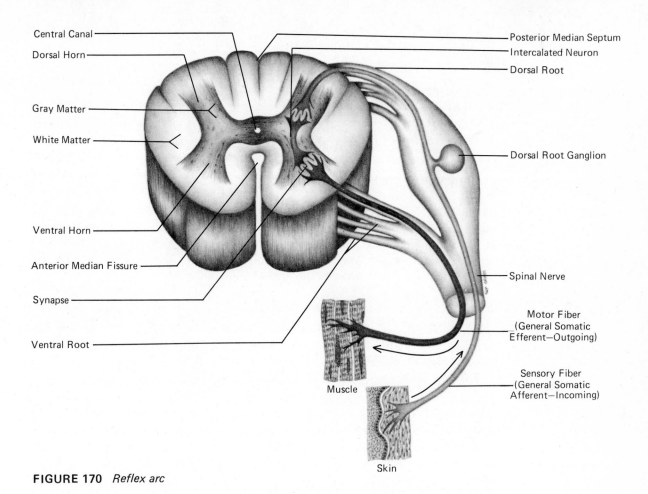

Central Canal

Dorsal Horn

Gray Matter

White Matter

Ventral Horn

Anterior Median Fissure

Synapse

Ventral Root

Posterior Median Septum

Intercalated Neuron

Dorsal Root

Dorsal Root Ganglion

Spinal Nerve

Motor Fiber
(General Somatic
Efferent—Outgoing)

Sensory Fiber
(General Somatic
Afferent—Incoming)

Muscle

Skin

FIGURE 170 *Reflex arc*

e. *Convergence Reflex*

Look at a pencil held 36 inches away. Have another student note the position of your eyeballs. Have your partner slowly bring the pencil closer until it nearly touches your nose. What change is observed in your eyeballs? This is convergence. What is the purpose of this reflex?

f. *The Achilles or Ankle Jerk*

Kneel on a chair; let your feet hang freely over the edge of the chair. Bend your foot to increase the tension of the gastrocnemius muscle. Have your partner tap the tendon of Achilles with a rubber reflex hammer. What reflex results?

g. *Corneal Reflex*

Touch the cornea of your eye with a piece of facial tissue. What is the result? What is the purpose of this reflex?

EXERCISE 4

Tests for Cranial Nerve Function in the Human Being

A superficial assessment of cranial nerve function can be made by performing relatively simple procedures. Working in pairs, test each other's cranial nerve function in the following manner. *RECORD* your results in the space provided as you perform each test and then fill in the table in Question 7 of the Optional Discussion at the end of this unit.

a. *Olfactory Nerve (Cranial Nerve I)*

Ask the subject to identify the odor of coffee, tobacco, or spices. The test may not be valid if the subject has a cold. Additional tests of olfactory discrimination will be performed in Unit XI, Exercise 3.
RESULT:

b. *Optic Nerve (Cranial Nerve II)*

Ask the subject to read a portion of a printed page with each eye, wearing glasses if necessary. Additional vision tests will be performed in Unit XI, Exercise 1.
RESULT:

c. *Oculomotor Nerve (Cranial Nerve III)*

Ask the subject to follow your finger, a pencil, or a penlight with both eyes, keeping his or her head still as you slowly move the object up, then down.
RESULT:

d. *Trochlear Nerve (IV) and Abducens Nerve (VI)*

Have the subject follow your finger, a pencil, or a penlight with both eyes, keeping his or her head still, as you slowly move the object laterally in each direction. As well as innervating eye movements through the extrinsic rectus ocular muscles, cranial nerves III, IV, and VI also innervate the upper eyelid and provide parasympathetic stimulation to the pupils. Therefore, you should observe your subject for signs of **ptosis** (drooping of one or both eyelids), and, with the room lights darkened, for pupillary reaction to light, if you have not already done so in Exercise 3c. To test for reaction to light, better results will be obtained if you bring the penlight (or flashlight) in from the side, rather than from the front of the subject. Observe the pupil. *RESULT:*

e. *Trigeminal Nerve (Cranial Nerve V)*

To test the motor responses of this nerve, ask the subject to clench his teeth. With the examiner providing resistance by holding his hand under the subject's chin, ask the subject to open his mouth. He should be able to do both.
RESULT:

    To test the sensory responses of this nerve, have the subject close his eyes. Test for light touch by whisking a piece of dry cotton over the mandibular, maxillary, and ophthalmic areas of the face. Wet the cotton with cold water and determine whether the subject is able to discriminate temperature in these same areas.

    Gently touch the subject's cornea with a piece of dry cotton or facial tissue, if you have not done so in Exercise 3g.
RESULT:

f. *Facial Nerve (Cranial Nerve VII)*

To test the motor response of this nerve, ask the subject to wrinkle his forehead, raise his eyebrows, puff his cheeks, and smile showing his teeth. Look for any asymmetry.
RESULT:

To test the sensory response of this nerve, touch a cotton-tipped applicator stick that has been dipped into a 10% NaCl solution to the tip of the subject's tongue. Repeat with 10% sugar solution on the anterior surface of the tongue. Additional exercises involving taste will be done in Unit XI, Exercise 4.
RESULT:

g. *Acoustic Nerve (Cranial Nerve VIII)*

To test the cochlear portion of this nerve, determine the subject's ability to hear a ticking stopwatch and repeat a whispered sentence. The Weber and Rinne tests (see Unit XI, Exercise 2), which use a tuning fork, may also be done at this time.
RESULT:

To test the vestibular portion of this nerve, have the subject sit in a swivel chair or stool, turn him 10 turns in approximately 20 seconds, then stop the chair suddenly. Observe the subject's eyeballs for rapid movement or quivering, known as **nystagmus**, which is a normal finding when one is dizzy. Nystagmus is not normal under nonexperimental conditions and may indicate a disorder of the inner ear or of cranial nerves III, IV, or VI.

In general, if the subject is able to keep his balance while walking, the vestibular branch of cranial nerve VIII is all right.
RESULT:

h. *Glossopharyngeal Nerve (IV) and Vagus Nerve (X)*

These nerves may be tested together. If you wish (and your subject is willing), test the gag reflex by touching the subject's uvula with a cotton-tipped applicator.
RESULT:

The motor portion of these nerves may be tested by: (1) asking the subject to swallow some water, (2) holding his tongue with a tongue depressor and asking him to say "ah" (the uvula should move), and (3) noticing any hoarseness when he speaks.
RESULT:

i. *Spinal Accessory Nerve (Cranial Nerve XI)*

This procedure tests the strength of the trapezius and sternocleidomastoid muscles, which are innervated by this nerve. To check the strength of the trapezius, place your hands on the subject's shoulders and determine whether he can raise them against resistance. To test the strength of the sternocleidomastoid muscle, place your hands on each side of the subject's head and ask him to turn his head to each side against resistance.
RESULT:

j. *Hypoglossal Nerve (Cranial Nerve XII)*

Ask the subject to stick out his tongue. The tongue should protrude straight, with no deviation.
RESULT:

EXERCISE 5

Tests for Cerebellar Functions in the Human Being

The cerebellum functions to maintain coordination, posture, and gait. After asking the subject to perform each of the following, check off whether he was able to perform the task (+) or was not able to perform it (−).

Test	+	−
a. With eyes closed, touches index finger of each hand to nose		
b. With eyes open, touches examiner's fingers		
c. Moves hands and fingers fast		
d. Looking straight ahead, moves the heel of one foot down the shin of the other leg		
e. Looking straight ahead, touches his outstretched hand with corresponding toe		
f. Stands with feet together and eyes closed without losing balance		
g. While walking, arms swing slightly		
h. While looking straight ahead, walks in tandem (heel to toe) without losing balance		

UNIT X

Nervous System

DISCUSSION

Multiple Choice

_____ 1. The structural and functional unit of the nervous system is the:
    a. Schwann cell
    b. neuroglia
    c. neuron
    d. telodendria

_____ 2. The ventral root of a spinal nerve:
    a. conducts motor impulses to effectors
    b. is synonymous with the anterior root
    c. conducts impulses that are voluntarily controlled
    d. all of these

_____ 3. The trochlear nerve (IV) functions in:
    a. facial movements
    b. eye movements
    c. tongue movements
    d. hearing

Photo Identification

Use the photograph of the human brain (superior aspect) to identify the structures numbered 4 and 5.

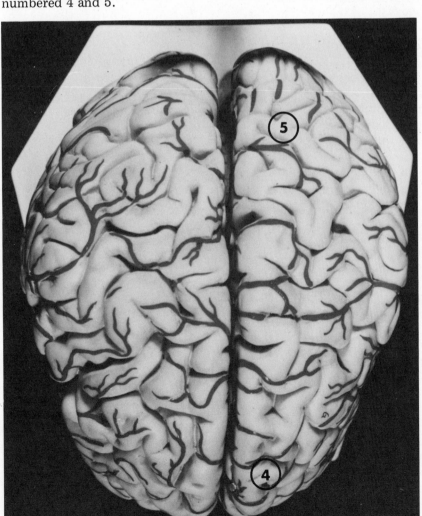

4. The lobe numbered 5 serves as the "motor area" for the body:
   a. occipital
   b. parietal
   c. frontal
   d. temporal

5. This posterior lobe (number 4) is encased and protected by a bone bearing the same name:
   a. occipital
   b. parietal
   c. frontal
   d. temporal

6. During the generation of the nerve impulse, ... and potassium move across the membrane during the action potential.
   a. calcium
   b. magnesium
   c. sodium
   d. chloride

7. The cell body of a neuron is known as:
   a. the axon
   b. the perikaryon
   c. neurofibril
   d. dendrite

8. Identify the correct superior to inferior sequence of spinal nerves:
   a. cervical, thoracic, lumbar, sacral, coccygeal
   b. thoracic, lumbar, cervical, sacral, coccygeal
   c. coccygeal, sacral, lumbar, thoracic, cervical
   d. lumbar, sacral, coccygeal, cervical, thoracic

9. In the spinal cord, the funiculi are:
   a. divisions of the gray matter
   b. divisions of the white matter
   c. same as ventral horns
   d. same as dorsal horns

10. The trochlear nerve functions in:
    a. hearing
    b. tongue movements
    c. facial movements
    d. eye movements

11. Which of these is a mixed cranial nerve (i.e., both sensory and motor)?
    a. trigeminal
    b. acoustic
    c. olfactory
    d. hypoglossal

12. Which meninge is the most external?
    a. pia mater
    b. dura mater
    c. arachnoid

13. A term that is synonymous with pituitary gland is:
    a. infundibulum
    b. optic chiasma
    c. hypophysis
    d. pons

14. The arbor vitae are located in the:
    a. cerebellum
    b. cerebrum
    c. diencephalon
    d. medulla oblongata

15. The brachial plexus consists of which nerves?
    a. cervical only
    b. cervical and thoracic
    c. cervical and lumbar
    d. thoracic

16. Which of the following cranial nerves influences maintenance of equilibrium?
    a. glossopharyngeal
    b. acoustic
    c. olfactory
    d. vagus

_____ 17. Which spinal reflex protects the delicate eye tissues by causing blinking?
     a. photo-pupil reflex
     b. Babinski reflex
     c. accommodation reflex
     d. corneal reflex

_____ 18. Which of these divides the cerebrum into left and right hemispheres?
     a. longitudinal fissure
     b. corpus callosum
     c. central sulcus
     d. transverse fissure

_____ 19. If the anterior root of a spinal nerve were cut, how would the regions served by the nerve be affected?
     a. they would be painful
     b. there would be loss of sensation
     c. there would be loss of movement
     d. there would be loss of sensation and movement

_____ 20. Which of these nerve fibers would be capable of regenerating?
     a. facial nerve
     b. cerebral tracts
     c. spinal tracts
     d. none could regenerate

_____ 21. Which cranial nerves innervate the extrinsic eye muscles?
     a. I, II, and III
     b. III, IV, and V
     c. III, IV, and VI
     d. II, III, and VI

_____ 22. Which of these is not necessary for a reflex to occur?
     a. afferent spinal nerves
     b. efferent spinal nerves
     c. cerebral cortex
     d. effectors

_____ 23. Inability to touch the index finger to the nose with the eyes closed probably indicates a disorder of the:
     a. cerebrum
     b. basal ganglia
     c. eighth cranial nerve
     d. cerebellum

_____ 24. Where are cell bodies of afferent neurons located?
     a. dorsal root ganglia
     b. pyramidal cells of the cerebrum
     c. ventral horns of the spinal cord
     d. at the effector

_____ 25. Injury to which part of the brain would most likely cause death?
     a. left cerebral hemisphere
     b. medulla
     c. corpus callosum
     d. cerebellum

DISCUSSION QUESTIONS (Optional)

1. Give reasons why mature neurons are not replaced after destruction.

2. Give the function of the following:

    a. olfactory bulbs

b. pyramidal tracts

c. optic chiasma

d. thalamus

e. hypothalamus

f. cerebellum

g. arbor vitae

4. What is a spinal reflex? Give an example.

5. Which ions are involved in impulse transmission?

6. Attach and interpret your recorded results from Exercise 2 of the physiology portion of this unit (Selected Characteristics of Nerve Impulse Transmission) here.

7. Complete the following chart for cranial nerves:

Number	Name and Branches	Sensory or Motor	Function

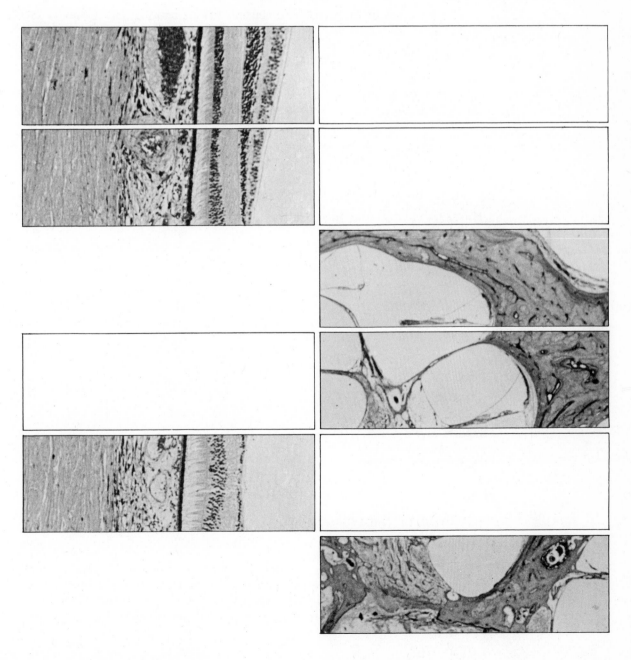

Special Senses

■ *Purpose*

The purpose of Unit XI is to familiarize you with the basic anatomy of the major sense organs and with the physiology of the eye, ear, nose, taste buds, and cutaneous receptors.

■ *Objectives*

In order to complete Unit XI, you must be able to do the following:

1. Identify the structural components of the eye.
2. Perform physiological exercises related to vision.
3. Identify anatomical features of the ear.
4. Perform physiological exercises related to hearing.
5. Perform physiological exercises related to olfactory discrimination.
6. Perform physiological tests to indicate various taste discriminations.
7. Identify histological characteristics of cutaneous receptors.
8. Perform exercises illustrating physiological characteristics of cutaneous receptors.

One commonly thinks of five senses—vision, hearing, taste, smell, and touch. The senses also include other sensations—those of warmth, cold, and pain.

The sense organs, which include all the sensory neurons, are involved in experiencing or reacting to any of these senses or sensations. These organs serve two functions—that of sensation and that of reflex—both of which result from the stimulation of sense **receptors**. Sensory receptors are classified into three main groups:

1. *Exteroceptors:* Surface receptors (dendrites) in the skin, eye, and ear.
2. *Visceroceptors:* Receptors in blood vessels, stomach, and intestines.
3. *Proprioceptors:* Receptors in muscles, tendons, joints, and the internal ear (to detect body positioning and movement).

The nerve endings primarily contained in the skin are referred to as **cutaneous receptors**.

EXERCISE 1

The Eye

Materials

Dissecting kits
Preserved sheep or cow eyes
Snellen charts
Tape measures
Color-blindness charts

Procedure

DISSECTION OF THE EYEBALL

(*see* **Color Plate 32**)
Make a coronal section through the preserved specimen of the eyeball and gently place it on your dissecting tray. It is important that you keep the contents relatively intact to facilitate identification of structures. Locate the three layers, or coats, covering the eyeball: the **sclera, choroid,** and **retina.**

These three layers are beneath the **extrinsic rectus** muscles. Identify the extrinsic muscles: the *superior rectus, inferior rectus, medial rectus, lateral rectus, superior oblique,* and *inferior oblique.* The sclera is often referred to as the white of the eye. Note that the anterior portion, or **cornea,** is transparent and covers the **iris,** or pigmented portion, of the eyeball. The choroid layer contains the *ciliary body,* which is located between the anterior margin of the retina and posterior margin of the iris. Attached to the ciliary body is the *suspensory ligament,* which appears as transparent threads and holds the **lens** in place. The lens is a hard (but flexible in the living state) kernellike structure that refracts light. The *iris* is part of the choroid layer. The inner layer, the retina, is located only in the posterior portion of the eyeball. It appears to be a thin, brownish layer of tissue containing photoreceptor neurons—the **rods** and **cones** (*see* Figures 171–174).

The eyeball also contains two cavities: the **anterior cavity** that is filled with clear and watery *aqueous humor;* and the *posterior cavity* that is larger and contains **vitreous humor,** a gelatinous viscid substance that functions to prevent the eyeball from collapsing (*see* Figures 171 and 172).

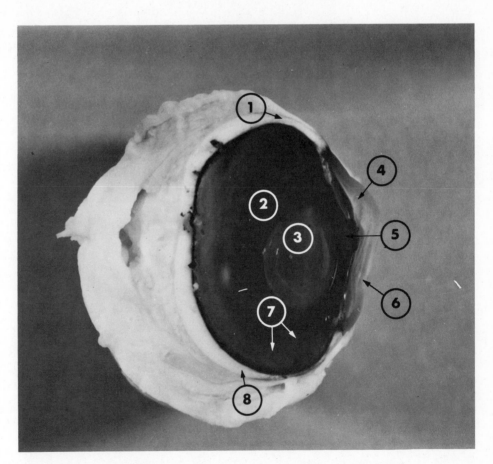

FIGURE 171 *Sheep eye, midsagittal section, showing gross internal detail*

1. Sclera
2. Retina
3. _____
4. _____

5. _____
6. Conjunctiva
7. Vitreous Humor
8. _____

FIGURE 172 *Sheep eye, coronal section, showing internal detail posterior to lens*

1. _____
2. Fovea (Macula Lutea)
3. Retina
4. _____

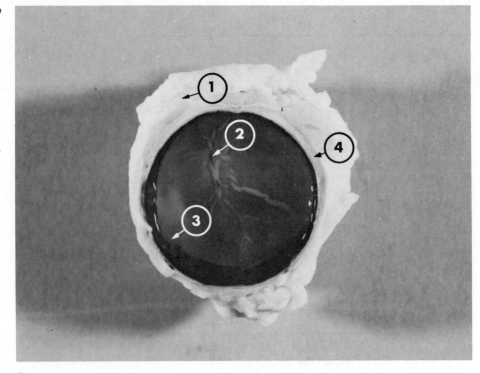

FIGURE 173 *Fovea centralis area of retina*

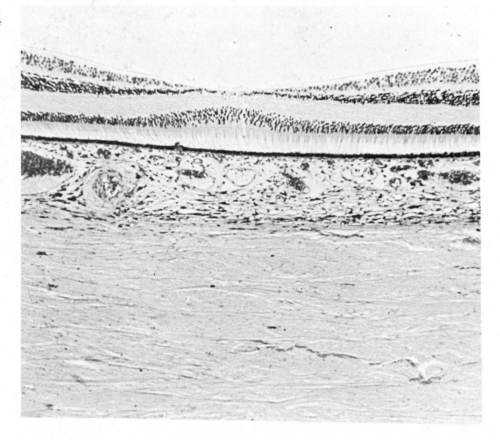

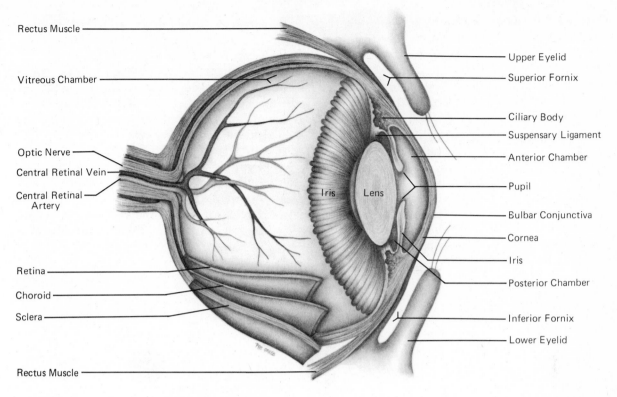

Rectus Muscle

Vitreous Chamber

Optic Nerve

Central Retinal Vein

Central Retinal Artery

Retina

Choroid

Sclera

Rectus Muscle

Iris   Lens

Upper Eyelid

Superior Fornix

Ciliary Body

Suspensary Ligament

Anterior Chamber

Pupil

Bulbar Conjunctiva

Cornea

Iris

Posterior Chamber

Inferior Fornix

Lower Eyelid

FIGURE 174   *Human eye, midsagittal section*

VISION

The eyes function in the image-formation aspect of vision and in the stimulation of nerve impulses that are conducted to visual areas of the cerebral cortex. The image upon the retina is formed by refraction of light rays as they move through the eyeball from the cornea to the retina. Rods and cones are dense toward the center of the retina where vision is most acute. This region is called the **fovea**, as seen in Figure 173. Toward the periphery, the photoreceptors become less dense and visual acuity decreases. At the point where the *optic nerve* enters the retina, there is a complete absence of rods and cones. This region is referred to as the **blind spot**. Examine prepared microscope slides of the eye (*see* Figure 173).

The following exercises exemplify some characteristics of the eye and vision:

1. Following your instructor's directions, read the letters on the Snellen chart. By covering each eye in turn, you can determine the approximate strength of vision in your right and left eye. *RECORD.* Normal vision is 20/20, which means that an individual being tested can read letters at a distance of 20 feet that a person with normal vision can see at 20 feet. If vision were 20/50, the person being tested would see at 20 feet what an individual with normal vision could see at 50 feet. The measurements 20/50, 20/20, and so on, are specific to either the right or the left eye.

2. You can determine the approximate distance from your eye that an image falls onto the blind spot of each eye by doing the following exercises.

a. To determine the blind spot in your left eye, hold this page 18 inches away from you and, closing your right eye, focus on the circle. Slowly bring the page closer, still focusing on the circle. The point at which the cross disappears is the blind spot where the optic nerve enters your right eye. *MEASURE* this distance and *RECORD* it in the following table.

b. To determine the blind spot in your right eye, hold this page 18 inches away from you and, closing your left eye, focus on the cross. Slowly bring the page closer, still focusing on the cross. The point at which the circle disappears is the blind spot where the optic nerve enters your right eye. *MEASURE* this distance and *RECORD* it in the table opposite.

3. Examine the various color-blindness charts available. Be sure to follow the instructor's directions in reference to each chart.

TABULATION OF CLASS NORMS FOR COMPARISON

Person Tested	Distance from Eye at Which Blind Spot Becomes Evident:	
	Right Eye	Left Eye
Class Average:		

EXERCISE 2
The Ear

Materials

Stopwatch   Tuning forks: (1) 512 hertz (2) 2000 hertz
Rubber reflex hammer   (cycles per second)

Procedure

The auditory apparatus is comprised of the ears, auditory nerves, and auditory areas of the temporal lobes of the cerebrum. Each ear consists of (1) an external ear; (2) a middle ear; and (3) an internal ear, as seen in Figure 175. The external ear has two divisions: (1) the **auricle**, or *pinna*—the flap of tissue on the outside of the ear; and (2) the **external**, or **auditory**, **acoustic meatus**—a canal terminating at the **tympanic membrane**, or *eardrum*. The tympanic membrane is a tense membrane of circular and radial fibers that separates the external ear from the middle ear. This membrane transforms sound waves

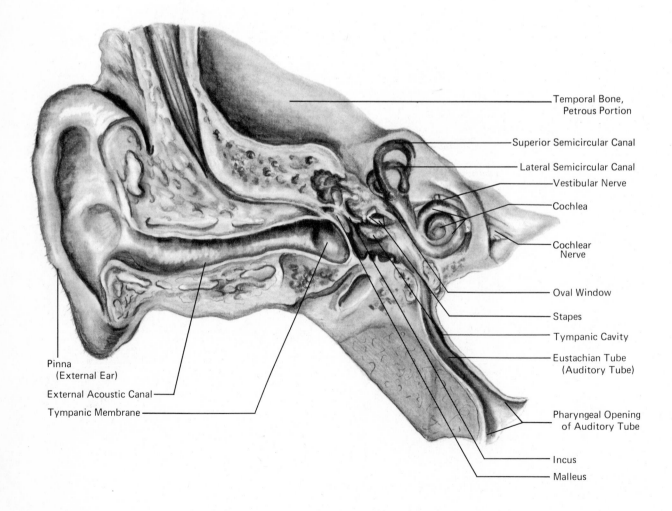

Temporal Bone, Petrous Portion

Superior Semicircular Canal

Lateral Semicircular Canal

Vestibular Nerve

Cochlea

Cochlear Nerve

Oval Window

Stapes

Tympanic Cavity

Eustachian Tube (Auditory Tube)

Pharyngeal Opening of Auditory Tube

Incus

Malleus

Pinna (External Ear)

External Acoustic Canal

Tympanic Membrane

FIGURE 175 *Frontal section of right ear showing external, middle and internal structures*

into mechanical vibrations that are transmitted into the middle and inner ear. The tympanic membrane also serves as a barrier that protects the delicate auditory ossicles of the middle ear.

The *middle ear*, or **tympanic cavity**, contains three tiny bones called *auditory ossicles*. They are the **malleus,** or hammer; the **incus,** or anvil; and the **stapes,** or stirrup. These bones transfer and intensify sound waves to the fluid within the cochlea of the inner ear.

The **inner ear** is concerned with both hearing and the sense of equilibrium. The hearing organ itself is the **organ of Corti** and is contained within the **cochlea** portion of the inner ear (*see* Figures 176 and 177). Hearing results from stimulation of the auditory area of the cerebral cortex, located in the temporal lobe of the brain. Equilibrium, or balance, is sensed by the flow of fluid within the three **semicircular canals** in the inner ear.

The ear transmits patterns of sound vibrations, their intensities, and directions of origin to the temporal lobe. Sound waves have two major characteristics—*frequency*, or wavelength, which determines pitch; and *intensity*, or *amplitude*, which determines loudness. The following exercises are designed to illustrate characteristics of hearing.

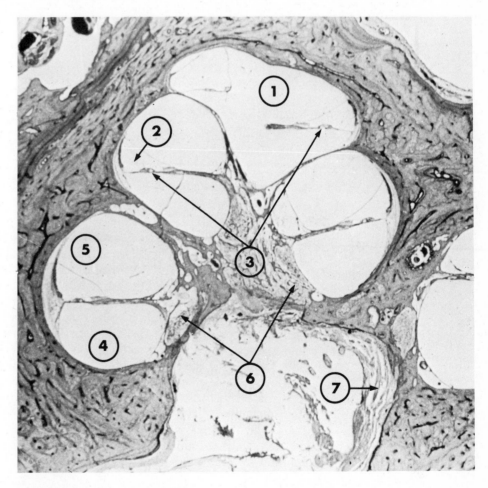

FIGURE 176 *Section of cochlea*

1. Helicotrema
2. Scala Media
3. Organ of Corti
4. Scala Tympani
5. Scala Vestibuli
6. Spiral Ganglion
7. Cochlear Nerve

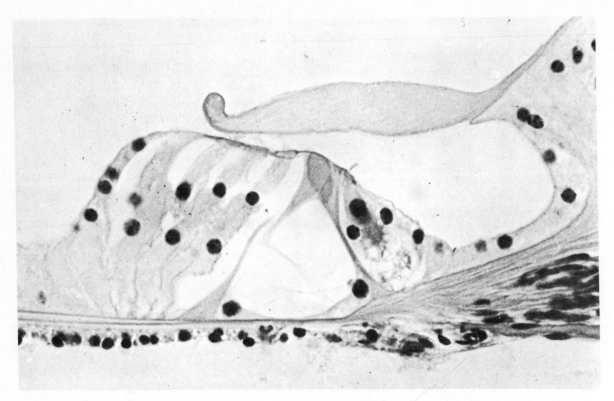

FIGURE 177   *Organ of Corti*

a. *Watch-tick Test*

Work with a partner. Hold a stopwatch very close to your partner's ear. Slowly move it away from his or her ear until the tick is no longer audible. *REPEAT* with the other ear. *CHART* your results in inches after measuring when the sound becomes inaudible. *TABULATE* class data for comparison.

b. *Weber and Rinne Tests*

A tuning fork can be stimulated to vibrate with a rubber reflex hammer and be used to compare hearing in both ears. It is also possible to differentiate between conductive and perceptive (or sensineural) deafness using a tuning fork. To do the *Weber* test, strike the tuning fork with the hammer and place the handle on your forehead in a medial position. If the tone is heard in the middle of the head, you have equal hearing or loss of hearing in both ears. If nerve deafness is present in one ear, the tone is heard in the other ear. In conduction deafness, the sould is heard in the ear in which there is hearing loss. In the *Rinne* test, strike the tuning fork and place the handle on the mastoid process. Then place the tines parallel to the external auditory meatus. Alternate the fork in the two positions a number of times. If the tone is heard equally well in both places, hearing loss is probably mixed. If hearing is normal, the tone is louder and heard longer at the side of the ear (air conduction). If it is louder through the mastoid process (bone conduction), a conductive hearing loss may be indicated.

c. *Examination of Prepared Slides*

Examine prepared microscope slides that show the middle and inner ear structures.

EXERCISE 3
Olfactory Discrimination

Materials

Samples of 10 spices, such as nutmeg, thyme, cinnamon, sage, vanilla extract, allspice, mustard powder, ginger, rosemary, garlic powder, pepper, oregano, clove, paprika, and slices of apple and potato.

Procedure

Receptors for the sense of smell are located high in the interior of the nose between the median **septum** and in the region of the superior turbinate. This area is referred to as the **olfactory cleft.** Olfactory nerves pass superiorly through the cribriform plate of the ethmoid, which is lateral and inferior to the crista galli, and from there pass to the nasal interior (*see* Figure 144).

The nasal passages are lined with mucous membranes and cilia, which condition the air in the breathing process. When one has a cold, the membranes tend to swell and secrete excessive mucus that covers the lining of the cavity and, therefore, impairs the sense of smell.

The following exercises are designed to test your ability to discriminate different smells. The sense of smell is complex and abilities to determine various smells are individual. Test your ability to discriminate and compare it with your classmates'.

1. Class groups of 10 students are suggested for this exercise. Your instructor has selected several different spices. You are to smell each with your eyes closed and identify the spice. *Keep a list* of the numbered spice samples and your responses.
2. Pinch your nostrils together, close your eyes, and have your partner place a piece of either potato or apple in your mouth. Is it apple or potato? *Explain.*

EXERCISE 4
Taste

Materials

Prepared slides of tongue cross sections	Sugar, lemon, burnt almonds, salt, pepper, mustard
Menthol eucalyptus	Ice cubes
Cotton-tipped applicators	Hot water
PTC paper	

Procedure

Taste, like smell, is a result of sensory neuron stimulation by chemicals. The tongue, which is covered with papillae (*see* Figures 104 and 105), is the principal organ of taste. Taste buds (*see* Figure 103) are spread over the surface of the tongue and are sensitive to substances dissolved in water. The four tastes are *salty*, stimulated by metallic cations; *sour*, stimulated by hydrogen ions; *sweet*, stimulated by a hydroxyl group in sugar or alcohol; and *bitter*, triggered by alkaloids.

1. Your instructor will have various solutions in jars that you can apply to your tongue systematically. Dip the applicator in one solution and gently

touch the applicator to the tongue in the following positions: (1) the apex, or tip; (2) posterior medial region; (3) anterior lateral sections; and (4) posterior lateral sections. *MAP on the drawing* provided where you can most readily identify the solutions. *REPEAT* the applications with other solutions, mapping your results. Compare your mapped responses with other class members'.

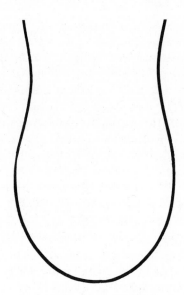

2. Hold an ice cube on your tongue for 1 minute. Apply a sugar solution to your tongue. What is the result?

3. Hold hot water in your mouth so the tongue is exposed to the hot water. Remove the water and apply your sugar solution to the tongue. What is the result?

4. Place a piece of PTC paper in your mouth. The ability to taste the chemical in the paper is inherited. Compare your ability to taste the chemical with your classmates'.

   What percentage of the class are "tasters"? . . . .

   What percentage of the class are "nontasters"? . . . .

5. Examine prepared microscope slides of cross sections of the tongue showing taste buds.

EXERCISE 5

Cutaneous Receptors

Materials

Prepared slides of skin and sense
  receptors
Powdered charcoal
Hot water
Brushes with firm bristles

Pins and needles
Ice
Centimeter rulers
Microscopes

Procedure

The following exercises will familiarize you with a variation of sensations
related to the skin.

a. *Touch*

Review the cross section of the skin, noting the layers of the epidermis and
their histological characteristics. Note the depth and inclusions of the dermis
as well. Now observe prepared slides of sensory neurons found in the skin,
such as *Pacinian* and *Meissner's corpuscles*. What are the various sensations
stimulated by these neurons?

1. With a pen, mark off a square area 2 cm on a side on the palm of your
    hand and one on the inner surface of your wrist. Use 0.5-cm markings to
    make grids in these square areas on your palm and wrist.
2. With your eyes closed, have your partner use light touch to explore the
    marked-off areas with a bristle or needle, using the same intensity of
    stimuli over 16 different spots within the square areas on your palm and
    wrist. If a sensation is felt, mark a *T* for touch in the appropriate grids.
3. Have your partner place a small piece of cork on the skin of your forearm
    of the same arm that was used in the above steps, and determine how long
    it takes for the initial sensation of pressure to give way to an indifferent
    sensation. *REPEAT* and *RECORD* the average time required to reach this
    accommodation to sensory stimuli.

b. *Temperature Receptors*

With your eyes closed, have your partner map the cold and hot receptors in
the grid areas of your palm and inner wrist. Use a dissecting needle or finishing nail that has been cooled in ice water, wipe the needle dry, and again explore in the same manner 16 different spots within the grids. Mark *C* for
every cold receptor located within the grid. After mapping the wrist and
palm areas for cold receptors, wait 5 minutes before exploring for hot receptors. *REPEAT* the previous procedure, this time using a dry, hot finishing

nail or dissecting needle (be careful not to burn yourself!). Again, record the hot receptors with *H* in the grid areas of the palm and wrist areas.

Repeat procedures *a* and *b*, using the calf of the leg or sole of the foot. Compare the differences in receptors of the palm, wrist, calf, and/or foot according to function.

c. *Discrimination of Stimuli*

Close your eyes and have your partner apply two pins, needles, or bristle points to the palm of your hand and determine how far apart the two points must be before you can discriminate two distinct points of stimulation. Your partner should validate your response by occasionally touching only a single point to the skin. *REPEAT* the above procedure in the following areas: index finger (palmar surface), forearm, back of the neck, and calf or sole of the foot. *RECORD* your results for the above areas of stimulation in terms of millimeters of separation. In which area could you discriminate between the points most easily?

d. *Distribution of Sweat Glands*

In order to demonstrate the distribution of these glands, the palm and wrist of your hand must be towel dried. Your partner will now lightly dust your forearm and the palm of your hand with powdered charcoal. *Use the hand that has the grid marked on the palmar surface.* After dusting with the charcoal powder, wait 15 seconds and, with a sharp puff, blow the excess powder from the hand and wrist. Where sweat glands are present, the powder will adhere to the moisture produced by the glands. Count the dots in the grid areas of the palm and wrist. *RECORD.* Are there any differences in the numbers of glands present? If time permits, you may wish to do other areas of the body in this same manner, such as the sole of the foot.

Special Senses

DISCUSSION

Matching

Column A

_____ 1. cornea

_____ 2. rhodopsin

_____ 3. iris

_____ 4. rectus muscles

Column B

a. controls the amount of light entering the eye

b. eyeball movement

c. refracts light

d. visual pigment

Multiple Choice

_____ 5. A person with 30/20 vision would:
    a. see exceptionally well
    b. have slightly blurred vision
    c. have average vision
    d. see very poorly

_____ 6. Which of these structures contains the largest number of rods and cones?
    a. lens
    b. fovea
    c. iris
    d. aqueous humor

_____ 7. Ringing in the ears (tinnitus) may indicate injury to which cranial nerve?
    a. III
    b. V
    c. VIII
    d. X

_____ 8. A *cataract* refers to opacity of the:
    a. retina
    d. aqueous humor
    c. cornea
    d. lens

_____ 9. Which of these is *not* a refractive medium of the eye?
    a. cornea
    b. vitreous humor
    c. lens
    d. retina

_____ 10. In the normal eye, images are focused:
    a. on the retina
    b. in front of the retina
    c. behind the retina

_____ 11. Why do middle ear infections commonly follow a sore throat?
    a. both areas are contained within the oropharynx
    b. the mucosa of the throat is continuous with the mucosa of the eustachian tube and middle ear
    c. sneezing spreads bacteria from the throat into the cochlea of the ear
    d. rupture of the tympanic membrane allows fluid to drain from the middle ear into the throat

_____ 12. Fusion of the ossicles within the middle ear results in which type of deafness?
    a. conductive
    b. sensineural

13. Hearing is localized in the .... lobe of the brain.
    a. frontal
    b. parietal
    c. temporal
    d. occipital

14. Which of these substances would be tasted at the tip of the tongue?
    a. sour and sweet
    b. salty and bitter
    c. bitter and sour
    d. salty and sweet

15. The .... transmit the impulses for the sense of taste and smell.
    a. baroreceptors
    b. proprioceptors
    c. chemoreceptors
    d. exteroceptors

16. The middle vascular layer of the eye is the:
    a. cornea
    b. choroid
    c. sclera
    d. retina

17. The structural portion of the inner ear that is responsible for hearing is (are) the:
    a. utricle
    b. organ of Corti
    c. auditory ossicles
    d. tympanic membrane

18. The ear ossicles are located in the:
    a. eardrum
    b. inner ear
    c. middle ear
    d. external ear

19. The sensory receptor for cold is:
    a. Krause's corpuscle
    b. Meissner's corpuscle
    c. Pacinian corpuscle
    d. Ruffini's corpuscle

20. The sensory receptor(s) for pain is(are):
    a. Meissner's corpuscle
    b. naked nerve endings
    c. Ruffini's corpuscle
    d. Meissner's corpuscle

21. The inner ear is important because it aids in:
    a. hearing
    b. maintenance of equilibrium
    c. both a and b
    d. neither a nor b

22. The lay term, *white of the eye*, refers to the:
    a. choroid
    b. leukoma
    c. eyelid
    d. sclera

23. Conjunctiva is a membrane lining the:
    a. inner surfaces of the eyelids
    b. naso-lacrimal duct
    c. cornea
    d. pupil

24. The protective coating of the eye is called the:
    a. sclera
    b. retina
    c. ciliary body
    d. choroid

25. The olfactory cleft is located:
    a. in the cerebral cortex of the frontal lobe
    b. around the anterior nares
    c. in the infundibulum
    d. in the region of the superior turbinate and median septum

DISCUSSION QUESTIONS (Optional)

1. Which structures of the eyeball contribute to refraction?

2. How are the Weber and Rinne tests diagnostic?

3. Is the ability to taste approximately the same for all individuals? Explain.

4. What conditions can influence variations in the ability to sense touch?

5. What types of color blindness are the most common?

Urinary System

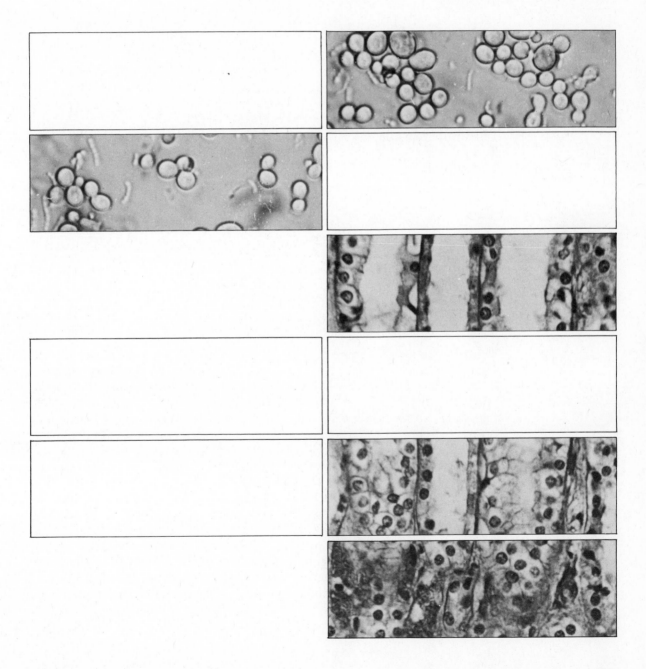

- *Purpose*

The purpose of Unit XII is to familiarize you with the general anatomy and physiology of the urinary system and to enable you to understand the basic tests included in a routine urinalysis.

- *Objectives*

In order to complete Unit XII, you must be able to do the following:

1. Identify the anatomical structures of the urinary system in a male and female cat.
2. Identify gross anatomical features of the kidney.
3. Using a microscope, identify the parts of a nephron.
4. Demonstrate accuracy in performance of basic tests included in a routine urinalysis.

- *Materials*

Sheep or beef kidneys	Clinistix	no. 2 filter paper
Preserved male and female cats	Clinitest	Petri plates
Microscope slides of renal tissue	Albustix	Nitric acid (conc.)
Urinometers	Ictotest	Microscope
Blank microscope slides	Ketostix	
Cover slips	Hemastix	
Nitrazine paper	Combistix	
Unknown urine samples	hot plate	
Test tubes	10% acetic acid	
400-ml beakers	Benedict's reagent	

- *Procedure*

EXERCISE 1

Examination of Urinary Organs of the Male and Female Cat

This exercise may be completed on your own specimen, but be sure to observe a specimen of the opposite sex as well. The urinary organs are located in the pelvic region. The kidneys are covered and attached to the dorsal body wall by a tough serous membrane termed the *peritoneum*. Remove or pull aside the small intestine, colon, and pancreas in order to provide an unobstructed view of the urinary apparatus. Carefully remove the fat, connective tissue, and peritoneum that cover the ventral aspect of the kidneys and blood vessels (renal arteries and veins, descending aorta, and postcava) that supply this organ (*see* Figure 178).

The field should be sufficiently exposed to make a detailed examination of the urinary apparatus. The urinary system is comprised of a pair of **kidneys**, a pair of **ureters**, a **urinary bladder**, and a **urethra** (*see* Figures 178 and 179). Examine and identify the following surfaces of each kidney: superior, inferior, medial, lateral, dorsal and ventral. Identify the creamy orange **adrenal glands** that lie on the superior and medial borders of each kidney. These are endocrine glands that secrete various hormones that influence metabolism, cardiovascular output, and electrolyte and water balance. The hormones secreted by these glands have physiological influences on the kidneys but are not a functional component of the urinary system. Identify the medial border of each kidney and note the depressed, or indented, portion known as the **renal hilum**. This is the point where the renal blood vessels, nerves, and ureter enter and exit from the kidney. Identify the **ureters**, which exit from the hilum and lie against the dorsal body wall. Trace the ureters inferiorly from the kidneys to the point of attachment to an elongated, collapsed sac that lies between the umbilical arteries. This sac is the **urinary bladder**. Dorsally, the bladder narrows to form a short canal, the **urethra**, that will carry urine from the urinary bladder to the outside environment.

Carefully lift up the right kidney and, with your scalpel, slit the kidney with a longitudinal cut from the lateral border to the medial border. Remove the cut half of the kidney and observe the position with the vessels still attached. What are the vessels? Identify the hilar region where the ureter exits from the kidney. Trace the **renal artery** (it may have several branches) from the abdominal aorta to the renal hilum. Follow the **renal vein** from the postcaval vein to the renal hilum.

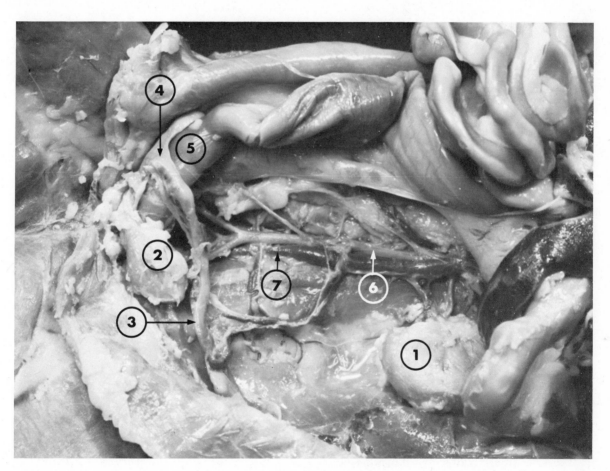

FIGURE 178   *Intestine and colon transected to show urogenital organs*

1. Kidney, Left
2. Urinary Bladder
3. Horn of Uterus
4. Body of Uterus
5. Descending Colon
6. Descending Abdominal Aorta
7. Inferior Vena Cava

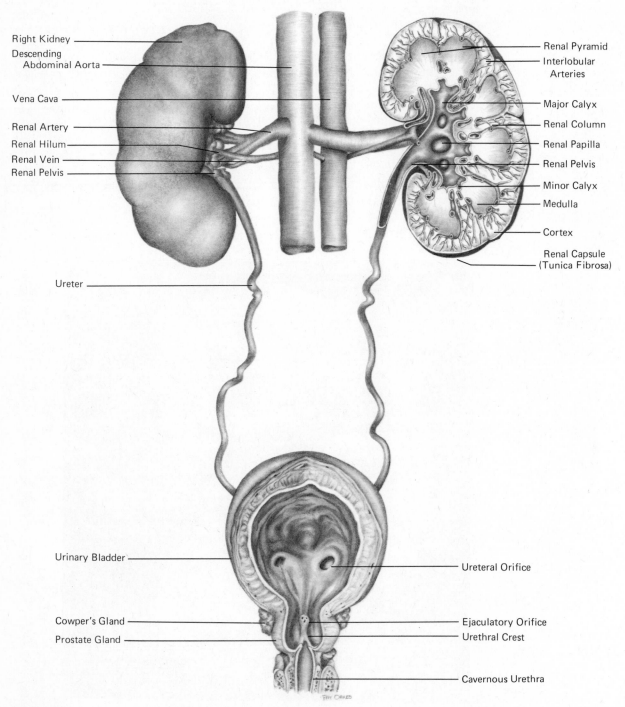

Right Kidney

Descending
 Abdominal Aorta

Vena Cava

Renal Artery

Renal Hilum

Renal Vein

Renal Pelvis

Ureter

Renal Pyramid

Interlobular
 Arteries

Major Calyx

Renal Column

Renal Papilla

Renal Pelvis

Minor Calyx

Medulla

Cortex

Renal Capsule
 (Tunica Fibrosa)

Urinary Bladder

Cowper's Gland

Prostate Gland

Ureteral Orifice

Ejaculatory Orifice

Urethral Crest

Cavernous Urethra

FIGURE 179 *Frontal view of human male urinary system*

EXERCISE 2

Observation of the Gross Anatomy of a Kidney

Use a preserved beef or sheep kidney for this exercise. Carefully remove the excessive adipose tissue. Be careful not to cut the ureter or any attached renal arteries or veins. The adipose tissue holds the kidney in position in a living animal and is very important. Observe the fibrous connective tissue covering that surrounds and protects the kidney. It is called the *renal capsule*. Also, note the **hilum**, which is the indentation where the ureter and blood vessels enter. Make a midsagittal section through the kidney starting at the outer convex region. Open the kidney halves and observe the following structures: **renal cortex, renal medulla** including the **pyramids, medullary rays, papillae, calyces,** and **columns;** and the **renal pelvis** (*see* Figure 180 for reference). Note that the renal pelvis is formed by a wall of thick, white fibrous tissue and forms the orifice of the proximal end of the ureter.

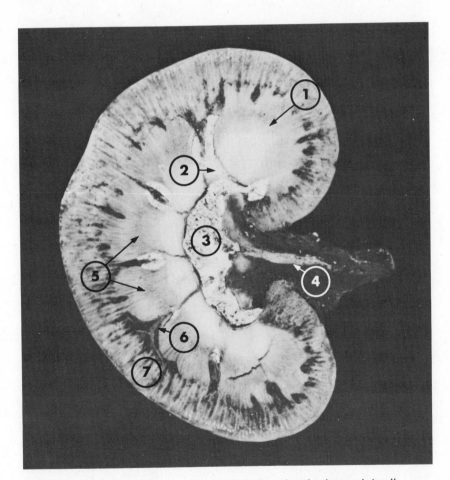

FIGURE 180 *Sheep kidney, midsagittal section, showing internal detail*

1. Medullary Region
2. Calyx
3. Renal Pelvis
4. _____

5. Renal Pyramids
6. Renal Column
7. Cortex

EXERCISE 3

Microscopic Examination of Renal Tissue

(*see* **Color Plates 4 and 30**)
Focus the slide of renal tissue under high, dry power. Observe a renal **cor-puscle**, which includes a tight network of capillaries—a **glomerulus**—and a **Bowman's capsule**, which is a single epithelial cell layer surrounding the glomerulus. Follow a Bowman's capsule as it narrows and extends from the renal corpuscle. This twisted tubule is the **proximal convoluted tubule.**

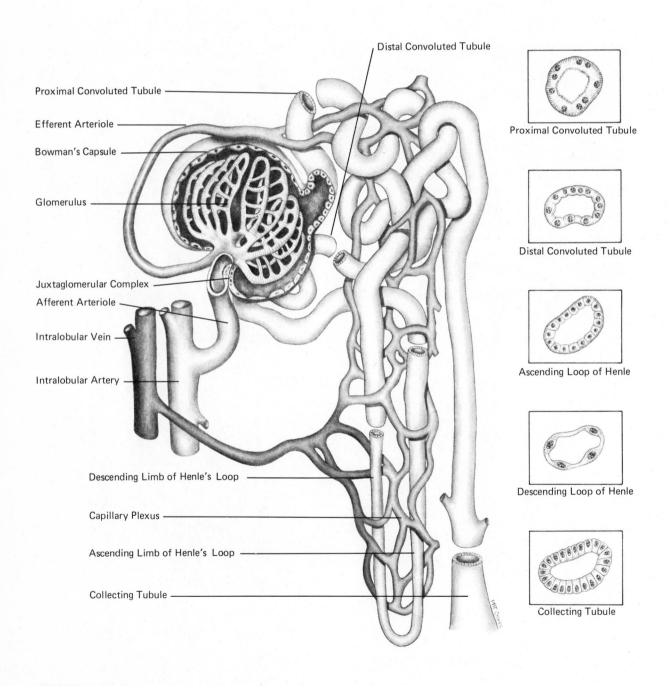

FIGURE 181 *Nephron unit showing histological detail of tubules*

Note that the tubule continues in an inferior direction and then turns superiorly, forming a loop. This tubule section is referred to as the **loop of Henle.** As the loop of Henle continues and extends from the renal corpuscle, it is referred to as the **distal convoluted tubule.** As the distal convoluted tubule continues, note that the walls become slightly thicker. This thicker-walled, straight tubular extension is called a **collecting tubule.** A **nephron**—the structural and functional unit of the kidney—is comprised of a glomer-ulus, a Bowman's capsule, a proximal convoluted tubule, a loop of Henle, and a distal convoluted tubule. Two or more distal convoluted tubules merge into a straight collecting duct. (*See* Figures 181–184.)

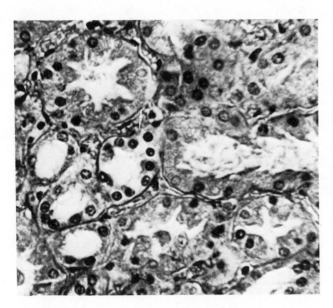

FIGURE 182 *Kidney tubules, cortical region*

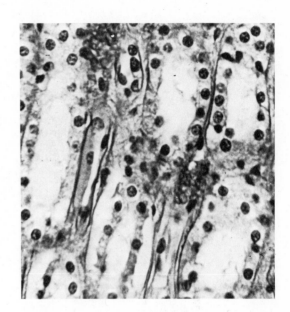

FIGURE 183 *Kidney tubules, medullary region*

FIGURE 184 *Glomerulus of kidney*

1. Glomerulus
2. Bowman's Capsule
3. Bowman's Space

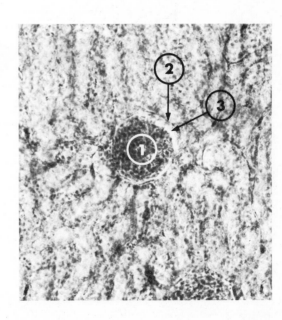

EXERCISE 4

Routine Urinalysis

Urine is approximately 95% water plus the following dissolved substances: pigments, electrolytes (for example, sodium and potassium), hormones such as estrogens, and the nitrogenous wastes—urea, uric acid, creatinine, and ammonia. During illness or disease, other substances such as bacterial toxins can be found, as well as blood, sugar, and protein (*see* Figures 185 and 186). Urine is, therefore, a good general indicator of an individual's state of health. More complete urinalyses are done with a 24-hour urine collection, which is normally about 1200–1500 ml in volume, but routine examinations are done on samples of lesser amounts.

After cleansing the area surrounding the urinary meatus, collect a small sample of urine—approximately 30–50 ml—from yourself. The first urination of the day is best, but it is often difficult to collect in a school situation. Observe the physical characteristics of your urine and of an "unknown" urine sample provided by your instructor and record in the table provided.

In order to determine specific gravity, use the urinometer or hydrometer and follow the directions specific to it. In order to observe the urine microscopically, make a smear on a clean slide and observe. You may see some epithelial cells sloughed off from the lining of the urinary tract, some white

PHYSICAL CHARACTERISTICS OF URINE

Characteristic	Normal	Your Urine	Unknown
Color	Amber to straw colored; can vary with diet, medication, or state of hydration		
Odor	Fresh—no odor; ammonia odor develops after standing due to bacterial breakdown and formation of ammonia substances		
Turbidity	Clear; becomes cloudy upon standing; heavier particles settle to bottom		
Specific gravity	1.003–1.030 in normal adult; can be higher if any early morning specimen; specific gravity is temperature dependent, so add 0.001 for each 3°C above 20°C; subtract 0.001 for each 3°C below 20°C		
Microscopic examination	Can include epithelial cells, leukocytes, pigment, or casts		

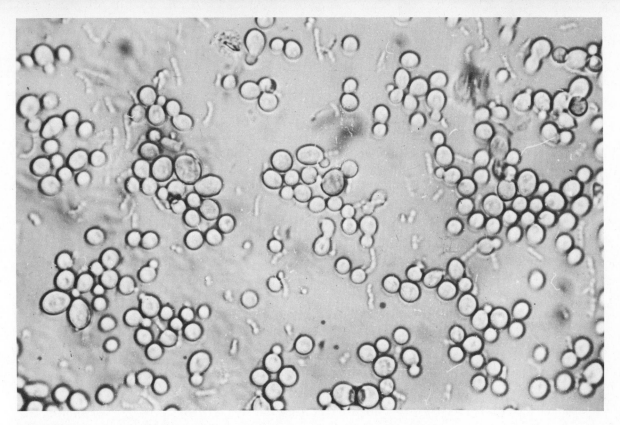

FIGURE 185 *Yeast cells and bacteria as seen in sediment of urine from urinary tract infection (100X).* (Courtesy Ames Company)

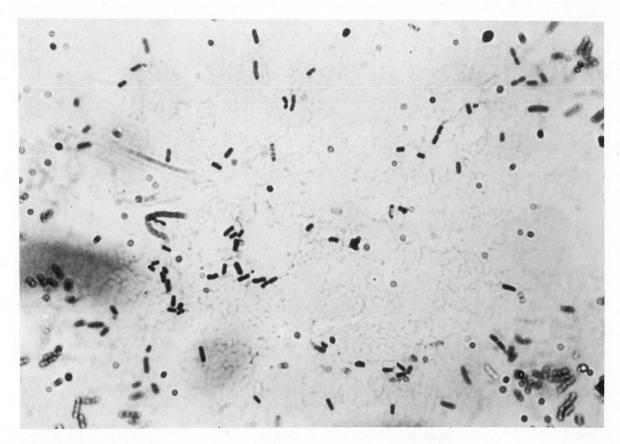

FIGURE 186 *Bacteria, high-power view, as seen in urinary tract infection. The bacteria are shown as small rod-shaped structures. The background consists of cellular and non-cellular debris. The dark circular areas represent out-of-focus material.* (Courtesy Ames Company)

blood cells, some parasites, and/or casts. *Casts* are collections of substances such as inorganic materials or mucus that have hardened and have assumed the shape of the tubule in which they were formed.

Also included in a routine urinalysis are tests to determine the absence or presence of substances that are abnormal constituents of urine. These tests can be completed accurately and effectively by use of various dipsticks that are on the market. Directions must be followed for each of the specific tests and the results must be compared with the available charts on either the bottle or box. In order to complete the following common tests, dip each treated strip of paper into your specimen, wait the required time, and record the results in the table provided. For the Clinitest and Ictotest, the procedure is different. Follow the directions on the label for these tests. Your instructor will again provide you with an unknown urine specimen. Analyze both your own and the unknown at the same time. Abnormal components may have been added to the unknown.

TESTS FOR ABNORMAL CONSTITUENTS OF URINE

| | | Results | |
| | | Your Urine (+: presence of substance; –: absence) | Unknown |
Indicator	*Substance Indicated*		
Clinistix	Glucose	———	———
Albustix	Albumins, (protein)	———	———
Ketostix	Ketones	———	———
Hemastix	Blood	———	———
Combistix	More than one substance	———	———
Ictotest	Bilirubin	———	———
Clinitest	Glucose	———	———
Nitrazine paper	pH (normal range pH is 4.8–8.0; average is pH 6; may vary with diet and medication	(give pH) ———	———

EXERCISE 5

Urine Screening Tests

a. *Test for Urine Protein*

Normal adult urine may contain a trace of plasma proteins (2–8 mg/100 ml). Plasma proteins are normally prevented from entering the glomerular filtrate by the glomerulus itself and by their large molecular weights. An increase in protein of 0.5 g to 4 g per day may be evidence of glomerular disease.

Pour 10 ml of urine into a test tube. Put the test tube into a beaker of tap water and onto a hot plate. Bring the contents of the tube to a boil. As soon as the urine boils, add three drops of 10% acetic acid to the test tube. If a precipitate results, then albumin and globulins are present: this indicates a positive test for proteins.
RESULTS?

List and describe three diseases that may be indicated by *proteinuria*.

1.

2.

3.

b. *Benedict's Qualitative Method for Detecting Urinary Glucose*

Benedict's is a copper reduction test that is used for the qualitative screening of glucose in urine. If glucose is present, it reduces Benedict's reagent from blue alkaline copper sulfate to a precipitate of red cuprous oxide. If a yellow, green, orange, or brick red precipitate forms, a reducing sugar (e.g., glucose) is present in the specimen. The particular color produced depends upon the concentration of reducing sugar present. The brick red color indicates a maximum concentration. A disadvantage of this test is that any reducing sugar will give a positive test, not only glucose.

Place 0.5 ml, or eight drops, of urine into a clean, dry test tube. Add 5 ml of Benedict's reagent. Mix by agitating the tube. Place the test tube in a beaker of tap water and onto a hot plate. *Let boil for 5 minutes.* Remove from the boiling water and observe immediately for any color change.

Color Interpretation	*Concentration*	*Results*
a. Clear blue or green with no precipitate	0–100 mg/100 ml	0
b. Green with a yellow precipitate	100–500 mg/100 ml	1+
c. Yellowish green with a yellow precipitate	500–1400 mg/100 ml	2+
d. Brownish orange with a yellow precipitate	1400–2000 mg/100 ml	3+
e. Orange to brick red precipitate	2000 or more mg/100 ml	4+

RESULTS:

List and describe three diseases that may be indicated by *glycosuria.*

1.

2.

3.

c. *Test for Urinary Bilirubin*

Bilirubin results from the breakdown of hemoglobin within the liver. Free bilirubin within the plasma normally does not pass through the glomerulus and is not found in the glomerular filtrate. A positive test for urinary bilirubin may indicate hepatic damage.

Place a small strip of no. 2 filter paper in a Petri dish or watch glass. Add two drops of urine and one drop of concentrated nitric acid (HNO_3). *PLEASE USE CAUTION.* A green color is positive for biliverdin. Bilirubin is not stable in urine and is oxidized to biliverdin. Other colors are negative. *RESULTS?*

List and describe three disorders that may be indicated by high levels of urinary bilirubin and/or biliverdin.

1.

2.

3.

d. *Test for Urea*

Normal adult urine contains large amounts of urea. The major portion of dietary nitrogen is excreted in the form of urea in the urine.

Place two drops of urine on a clean, dry microscope slide. *Carefully,* add two drops of concentrated nitric acid (HNO_3). Slowly warm the slide on a slide warming tray or hot plate. *Do not allow it to boil.* Cool. When crystals begin to form, examine under the low power of your microscope. The crystals that form are urea nitrate. *DRAW* your results.

List, describe, and microscopically identify three kinds of crystals that may be found in normal acidic or alkaline urine.

1.

2.

3.

Urinary System

DISCUSSION

Multiple Choice

_____ 1. Which of these is the correct sequence of vessels through which blood normally passes going to and from a nephron?
   a. intralobular vein, capillary plexus, efferent arteriole, afferent arteriole, glomerulus
   b. glomerulus, afferent arteriole, capillary plexus, efferent arteriole, intralobular vein
   c. afferent arteriole, glomerulus, efferent arteriole, capillary plexus, intralobular vein
   d. glomerulus, Bowman's capsule, proximal convoluted tubule, loop of Henle, distal convoluted tubule, collecting tubule

_____ 2. Which of these structures is located in the cortex of the kidney?
   a. glomerulus
   b. loop of Henle
   c. calyx
   d. collecting tubule

_____ 3. Who would be more prone to urinary tract infections?
   a. men
   b. women

_____ 4. Why is it advantageous that kidney tubules be only one cell layer thick?
   a. they occupy less space
   b. molecules and ions may be transported more readily
   c. diffusion of gases takes place more quickly
   d. to provide more surface area for water absorption

_____ 5. If a person has neither eaten nor drunk anything for eight hours, what would you expect the specific gravity of his or her urine to be?
   a. high
   b. low
   c. state of hydration has no effect on specific gravity

_____ 6. Which of these pituitary hormones influences the volume of urine excreted?
   a. oxytocin
   b. parathyroid hormone
   c. antidiuretic hormone
   d. aldosterone

_____ 7. A diagnostic test for diabetes mellitus is the presence of . . . . in the urine.
   a. protein
   b. sodium
   c. bilirubin
   d. glucose

_____ 8. Which of these may be found in the urine if fats are not completely metabolized?
   a. ketones
   b. glucose
   c. bilirubin
   d. urea

_____ 9. In most people, urine pH:
   a. is acidic
   b. is basic
   c. varies
   d. is neutral

10. The volume of glomerular filtrate produced per day is normally:
   a. about the same as urine
   b. about 100 times greater than urine
   c. about 100 times less than urine
   d. there is no relationship between the two volumes

11. Which of the following represents the correct sequence of structures through which glomerular filtrate and urine must pass?
   a. Bowman's capsule, proximal tubule, glomerulus, distal tubule, Henle's loop, ureter
   b. glomerulus, Bowman's capsule, proximal tubule, Henle's loop, distal tubule, ureter
   c. Bowman's capsule, glomerulus, proximal tubule, Henle's loop, distal tubule, ureter
   d. ureter, Henle's loop, proximal tubule, Bowman's capsule, distal tubule, glomerulus

12. Which of the following would be considered abnormal constituents of urine?
   a. protein
   b. glucose
   c. hemoglobin
   d. all of the above

13. The structural and functional unit of the kidney is the:
   a. ureter
   b. glomerulus
   c. nephron
   d. calyx

14. Which of the following is *not* a function of the kidney?
   a. maintains acid-base balance
   b. maintains H_2O balance
   c. elimination of by-products of catabolism
   d. all of the above are functions

Photo Identification

Refer to the photograph of the kidney for Questions 15–18.

15. If high concentrations of uric acid are allowed to build up in regions 2 and 3 of the kidney, ... may result.
   a. anuria
   b. kidney stones
   c. polyglomerular nephritis
   d. nephritis

16. The nonvascular tubular structure that exits from the hilum as indicated by 4 on the photograph is the:
   a. renal artery
   b. renal vein
   c. ureter
   d. urethra

17. The region identified by 7 contains *mostly*:
   a. collecting tubules
   b. glomeruli
   c. neither of the above
   d. both of the above

18. The region identified by 1 contains mostly:
   a. collecting tubules
   b. loops of Henle
   c. glomeruli
   d. both a and b

19. If blood is present in the urine as a result of an acute inflammation of the urinary tract, which of the following tests would be used to determine its presence?
   a. Ketostix
   b. Hemastix
   c. Clinistix
   d. nitrazine paper

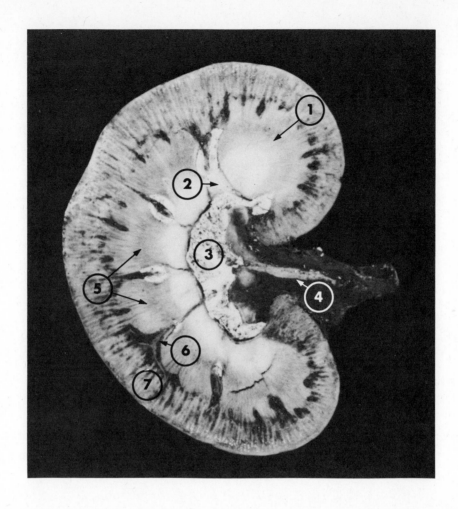

20. Water reabsorption in the kidney depends upon the system's need for it and upon which of the following hormones secreted by the posterior lobe of the pituitary gland?
    a. oxytoxin
    b. antidiuretic
    c. insulin
    d. none of the above

21. The average daily amount of urine excreted is:
    a. 700 ml
    b. 100 ml
    c. 1000 ml
    d. 1500 ml

22. The muscle of the bladder is the:
    a. Myometrium
    b. Rectus abdominis
    c. Detrusor
    d. Levator ani

23. From the renal pelvis, urine flows to the:
    a. glomerulus
    b. calyces
    c. ureter
    d. bladder

24. In which of the following parts of the nephron are water, glucose, and salt primarily reabsorbed into the blood?
    a. proximal convoluted tubule
    b. Henle's loop
    c. distal convoluted tubule
    d. collecting tubule

25. The body's blood supply initially comes into contact with the nephron at the:
    a. glomerulus
    b. pyramids
    c. papillae
    d. calyces

DISCUSSION QUESTIONS (Optional)

1. Why is it advantageous that Bowman's capsule be one cell layer in thickness?

2. How does the urinary system of the male differ from that of the female?

3. Turbidity of urine and specific gravity are related. Why?

4. Why is urinalysis considered to be a valuable diagnostic test?

5. Which nerves and hormones influence urine formation?

Acid-Base Balance

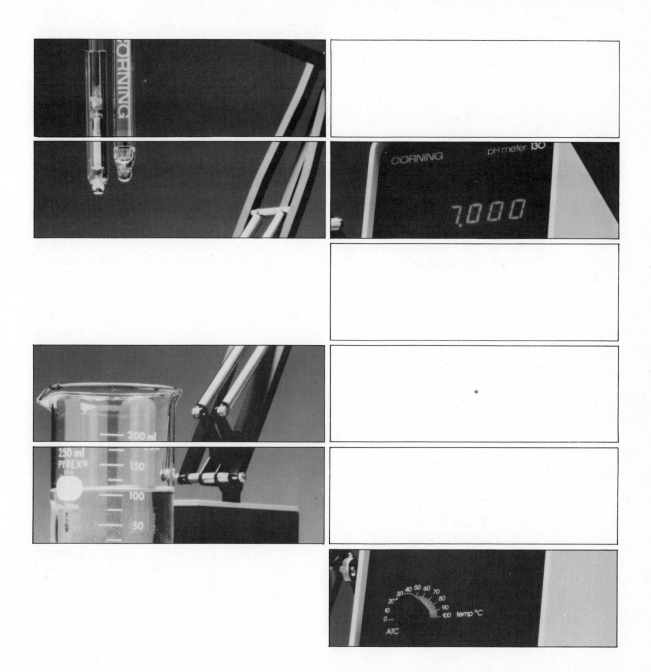

■ *Purpose*

The purpose of Unit XIII is to enable you to understand the concepts of acidity, alkalinity, and buffer systems in common solutions and biological fluids, and to understand their importance in maintaining homeostasis within the body.

■ *Objectives*

In order to complete Unit XIII, you must be able to do the following:

1. Define *acid*, *base*, *pH*, and *buffer system*.
2. Use litmus paper, pH paper, and a pH meter to determine the acidity or alkalinity of several common solutions and biological fluids.
3. Demonstrate the action of buffer systems on pH change.
4. Explain how the respiratory and renal systems function to maintain acid-base balance in the body.
5. Define and identify the types of acidosis and alkalosis occurring in certain disease states.

■ *Procedure*

In order for the body to maintain a state of *homeostasis*, or a stable internal environment, mechanisms exist for controlling the various amounts of acids and bases within body fluids.

Before beginning the exercises, it will be necessary to define the following terms:

1. **Acid.**   A hydrogen ion (H^+) donor.

2. **Base.**   A hydroxyl ion (OH^-) donor. In biological systems, bases are often called *alkalies* and are defined as negatively charged ions that combine with hydrogen ions in solution.

3. **pH.**   A measure of the acidity (amount of acid, or H^+) or alkalinity (amount of base, or OH^-) of a solution. Mathematically, pH is equal to the negative logarithm of the hydrogen ion concentration, or $-\log[H^+]$. It can also be represented as $\log 1/[H^+]$.

4. **Buffer system.**   A combination of two or more chemicals (usually a weak acid and its salt) that minimize the pH change of a solution when a strong acid or base is added to it.

EXERCISE 1

Measuring the pH of Common Solutions

The pH scale ranges from 0 through 14. The lower the pH, the greater the hydrogen ion concentration, and, therefore, the greater the acidity of a solution. The higher the pH, the lower the hydrogen ion concentration, which indicates an alkaline, or basic, solution. A pH of 7 is midway between 0 and 14 and represents neutrality (where the number of hydrogen ions equals the number of hydroxyl ions) and, therefore, the solution is neither acidic nor basic, but neutral. A convenient notation for indicating the pH of a solution is the **pH scale** (*see* Figure 187).

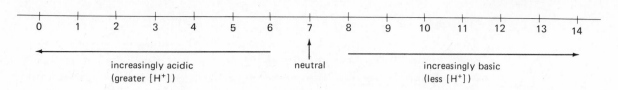

FIGURE 187 *pH scale*

It is important to understand that a shift of one pH unit represents a tenfold increase or decrease in hydrogen ion concentration. For example, a solution of pH 4 has 10 times the concentration of H^+ as a solution of pH 5. A solution of pH 4 has 100 times the concentration of H^+ as a solution of pH 6. Conversely, a solution of pH 10 has 1/10 the hydrogen ion concentration of a solution of pH 9 and 1/100 the hydrogen ion concentration of a solution of pH 8.

Materials

Red litmus paper	Milk
Blue litmus paper	Coffee
pH meter	Cola
Wash bottles with distilled water	Lemon juice
pH 7 buffer solution	Baking soda solution
pH paper (or nitrazine paper)	Distilled water
	Beakers

a. *Determination of pH Using Litmus Paper*

Dip a 2-inch strip of red, then blue, litmus paper into each of the following solutions: milk, coffee, cola, lemon juice, baking soda solution, and distilled water. Observe for a color change, using fresh strips for each solution. Acidic solutions will turn blue litmus red. Basic, or alkaline, solutions will turn red litmus blue. *RECORD* your results in Table 1.

TABLE 1   Determination of pH of Common Solutions Using Litmus Paper

Solution	Red Litmus	Blue Litmus	Conclusion
Milk			
Coffee			
Cola			
Lemon juice			
Baking soda			
Distilled water			

b. *Determination of pH Using pH Paper*

Dip a 2-inch strip of pH paper into each of the following solutions: milk, coffee, cola, lemon juice, baking soda solution, and distilled water. Compare the colors that appear on the chart that accompanies the pH paper container. Record the pH of each solution and *RECORD* in Table 2.

TABLE 2 Determination of pH of Common Solutions Using pH Paper

Solution	pH	Conclusion
Milk		
Coffee		
Cola		
Lemon juice		
Baking soda		
Distilled water		

c. *Determination of pH Using a pH Meter*

Plug in and turn on the pH meter at this time, because some models may take up to 30 minutes to warm up. (Your instructor will give you any additional information about the particular instrument you will be using.)

There are several types of pH meters, however, they all operate by a similar principle. Observe the pH meter you will be using. Your instructor will demonstrate its use, but, in general, pH meters have the following components that you should be able to identify (*see* Figure 188):

FIGURE 188 *pH meter with electrodes* (Courtesy Corning Glass Works, Corning, New York)

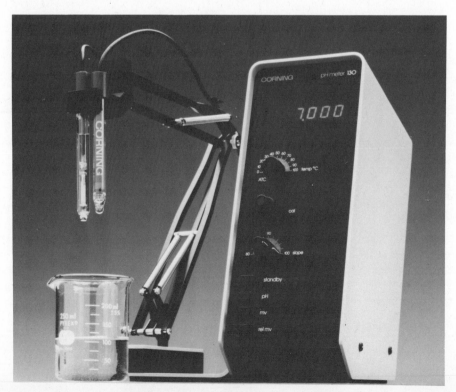

1. *Electrodes.* The two electrodes project in a downward direction and are attached to the body of the pH meter by cables. The **glass electrode** consists of an internal sealed tube with a metallic tip. Around the internal tube is an external tube containing a standard solution. A pH-sensitive glass bulb protrudes from the external tube and will be immersed in the solution being tested. The **reference electrode** appears similar to the glass electrode but may be slightly longer. It is comprised of a metallic internal element that is housed within an outer tube containing an electrolyte solution such as KCl. Minute amounts of the electrolyte solution flow into the sample solution when the pH meter is in operation, thus completing an electrical circuit. (*Note: In some pH meters, the glass and reference electrodes are combined into a single electrode. Also, when storing a pH meter, always have the electrodes immersed in a buffer solution. Do not allow them to become dry.*)

2. *Dial or Digital Readout.* The pH value of the solution to be tested is read from a dial imprinted with the pH scale or from a digital display readout.

3. *Temperature.* You must adjust for the temperature of the experimental solution with the temperature control knob.

4. *Calibrate, pH, and Standby Buttons or Knobs.* These are usually located on the pH meter and will be explained during its operation.

You will now determine the pH of each of the solutions that you previously worked with by using the pH meter. Mix each solution well before use and do the following:

1. Calibrate the instrument by placing the electrode(s) in the beaker containing a prepared pH 7 buffer. The electrodes should be immersed at least 2 cm. Adjust the pH meter until the pH value of the buffer (7) appears on the dial or readout.

2. Depress the standby button and carefully remove the electrodes from the buffer solution by raising the bar that holds the electrodes. Avoid scratching the electrodes or touching them with your hands. Rinse the electrodes with distilled water from a wash bottle, allowing the rinse water to drain into an empty beaker.

3. Immerse the electrode tips in each solution to be tested. (Be sure to rinse the electrodes with distilled water draining into a separate beaker after each solution is tested.) Release the standby button, press the pH button, and wait for the dial or readout to stabilize before taking a final reading. *RECORD* your results for each solution tested in Table 3.

TABLE 3 Determination of pH of Common Solutions Using a pH Meter

Solution	pH	Conclusion
Milk		
Coffee		
Cola		
Lemon juice		
Baking soda		
Distilled water		

EXERCISE 2

Determining the pH of Biological Solutions Using pH Paper

Materials

Saliva
Urine
Blood plasma (obtained by centrifuging fresh, whole blood for 5 minutes at
   1800 rpm)
pH paper (or nitrazine paper)

Procedure

The pH range in arterial blood that is compatible with life is relatively narrow, ranging from about 6.8 to 7.8. In the healthy adult, the pH is between
7.35 and 7.45. The pH range for aterial blood may be represented as follows:

| Death | Acidosis | Normal | Alkalosis | Death |

6.80   7.35   7.45   7.80

Using specimens of saliva, freshly voided urine, stale urine (urine that has
been standing for several hours or overnight), and blood plasma, determine
the pH of each solution using pH or nitrazine paper. *RECORD* the pH of
each of these in Table 4.

How do you account for the fact that saliva and urine may have a pH
below 6.8 or above 7.8 in a healthy individual?

Was there a difference in pH of the freshly voided urine and the urine
that was left to stand? *Explain.*

TABLE 4   Determination of pH of Biological Solutions Using pH Paper

Solution	*pH*	*Conclusion*
Saliva		
Freshly voided urine		
Stale urine		
Plasma		

EXERCISE 3
Buffer Systems

Materials

pH meter	Dropper bottles of 0.01M NaOH
Two beakers with 100 ml each of:	Saliva
distilled water	Urine
pH 7 buffer solution	Plasma
Dropper bottles of 0.05M HCl	pH paper
Dropper bottles of 0.05M NaOH	Beakers
Dropper bottles of 0.01M HCl	Distilled water

a. *Effect of Buffers on Acidic and Alkaline Solutions*

In this exercise, you will determine the effect of buffers on the extent of pH change in acidic and alkaline solutions.

Procedure

1. Immerse the electrodes in 100 ml of distilled water and determine the pH. *RECORD.*
2. Slowly add 0.05M HCl dropwise to the distilled water, gently swirling the beaker after each drop. *RECORD* the number of drops of HCl added to change the pH one unit. In which direction did the pH change? . . . .
3. Remove the electrodes from the solution and rinse with distilled water. Immerse in the beaker containing 100 ml of pH 7 buffer solution. Add 0.05M HCl dropwise to the buffer, gently swirling after each drop. How many drops of HCl were added to change the pH of the buffer one unit?

    . . . .
4. After rinsing the electrodes with distilled water, immerse them in the second beaker of distilled water. Add 0.05M NaOH dropwise to the water, gently swirling the beaker after each drop. How many drops of NaOH were added to change the pH one unit? . . . . In which direction did the pH change? . . . .
5. Again, remove the electrodes from the solution and rinse with distilled water. Immerse the electrodes in the second beaker of pH 7 buffer. Add 0.05M NaOH dropwise to the buffer, again, gently swirling after each drop. How many drops of NaOH were added to change the pH one unit?

    . . . .

What conclusion can you draw about the effect of buffers on pH change?

b. *Effect of Buffers in Biological Fluids on pH Change*

The body utilizes circulating chemical buffers, the lungs, and the kidneys to keep the pH of the blood within narrow limits. Three chemical acid-base buffer systems are found in the blood:

1. *Bicarbonate Buffer System.* This consists of carbonic acid (H_2CO_3) and sodium bicarbonate ($NaHCO_3$), which are present in extracellular and intracellular fluids. When minute amounts of a strong acid such as HCl are added to blood in the body, the following reaction takes place:

$$HCl \ + \ NaHCO_3 \longrightarrow H_2CO_3 \ + \ NaCl$$

strong acid   buffer in   weak acid;   salt
           plasma   formed in
                          plasma

If a strong base such as NaOH is added to the bicarbonate buffer system, it will react with the carbonic acid component of this system:

$$NaOH \ + \ H_2CO_3 \longrightarrow NaHCO_3 \ + \ H_2O$$

strong base   carbonic acid;   weak base;   water
          buffer in plasma   formed in plasma

2. *Phosphate Buffer System.* This system is also present in intraceullar and extracellular body fluids. It primarily functions to regulate the H^+ concentration in the nephrons. The following reactions occur if a strong acid or strong base, respectively, is added to the system:

$$HCl \ + Na_2HPO_4 \longrightarrow NaH_2PO_4 + NaCl$$

strong acid   buffer   weak acid   salt

$$NaOH \ + NaH_2PO_4 \longrightarrow Na_2HPO_4 + H_2O$$

strong base   buffer   weak base   water

3. *Protein Buffer System.* This buffer system consists of plasma proteins and hemoglobin. Recall that amino acids contain carboxyl groups ($-COOH$) and amino groups ($-NH_2$).

Free carboxyl groups may function as acids by releasing a hydrogen ion:

$$-COOH \longrightarrow -COO^- + H^+$$

If there is excess acid in the blood, ($-COO^-$) may accept a hydrogen ion and become a free carboxyl group ($-COOH$).

Likewise, free amino groups may accept hydrogen ions in the presence of excess acid:

$$-NH_2 + H^+ \longrightarrow NH_3^+$$

Hydrogen ions may also be released in the presence of excess base, thus making the blood more acidic by the following reaction:

$$-NH_3^+ \longrightarrow NH_2 + H^+$$

Procedure

1. Obtain two beakers of distilled water and two 5-cc specimens each of saliva, freshly voided urine, and blood plasma.

2. Using pH paper, determine the pH of one of the beakers of distilled water and one specimen of each of the above.

3. Add 0.01*M* HCl dropwise to each of the above until the pH changes one unit. *RECORD* the number of drops required. . . . .

4. Add 0.01*M* NaOH dropwise to another beaker of distilled water and the second specimen of each of the above until the pH changes one unit. *RECORD* the number of drops required. . . . . Did

you notice a buffering effect in both Exercises 3a and 3b? . . . . Explain.

EXERCISE 4

Maintaining Acid-Base Balance in the Body

Materials

Small paper bags	250 ml beakers
pH meter	Straws
Distilled water	Clock or watch with second hand

Two body systems assist in the regulation of blood pH—the urinary (renal) system and the respiratory system. The urinary system functions to retain or excrete bicarbonate ions in the kidney tubules, which results in an increase or decrease in blood pH, respectively. The kidney tubules also retain or secrete hydrogen ions, which result in a decrease or increase in blood pH, respectively. The respiratory system functions to retain or release (through exhalation) carbon dioxide (CO_2), which results in a decrease or increase in blood pH, respectively. The term "pCO_2" or "$paCO_2$" refers to the partial pressure of free carbon dioxide dissolved in arterial blood. An increased pCO_2 results in a low (acidic) blood pH, a decreased pCO_2 results in a high (basic) blood pH.

Procedure

a. *Effect of pCO_2 Levels on the Respiratory Rate*

1. Observe your lab partner breathing normally. *RECORD* his respiratory rate (number of breaths per minute). . . . .

2. Instruct your lab partner to take rapid, deep breaths, until he tires. After he stops hyperventilating, *RECORD* his respiratory rate. . . . .

3. Instruct your lab partner to breathe in and out of a paper bag through his nose and mouth until breathing becomes difficult. Remove the bag and *RECORD* his respiratory rate. . . . .

Explain the differences in rates in Steps 2 and 3.

b. *Effect of pCO_2 Changes on pH*

1. Using the pH meter, determine the pH of a beaker of distilled water. Pinching your nostrils shut, exhale into the bottom of the beaker through a straw. Have your lab partner determine and *RECORD* the pH of the water. *REPEAT*, blowing into the water five additional times, taking a deep breath each time. *RECORD* the pH after each exhalation in Table 5.

TABLE 5 Effects of pCO$_2$ Changes on pH

Solution	pH After Breathing Room Air	pH After Breathing Into Paper Bag
Original distilled H$_2$O		
H$_2$O after exhalation 1		
H$_2$O after exhalation 2		
H$_2$O after exhalation 3		
H$_2$O after exhalation 4		
H$_2$O after exhalation 5		
H$_2$O after exhalation 6		

2. Rinse the electrodes and determine the pH of the water in the other beaker. Breathe into a paper bag until breathing becomes difficult. Remove the bag and, pinching your nostrils shut, exhale through the straw into the water. Have your lab partner *RECORD* the pH after the first exhalation and after each of five additional normal breaths of room air in Table 5.

EXERCISE 5

Disorders of Acid-Base Balance in the Body: Acidosis and Alkalosis

In a healthy person, the pH of the blood is maintained within narrow limits, as described in Exercise 2, by chemical and physiological buffers. In certain pathological conditions, however, the pH may increase, resulting in alkalosis, or decrease, resulting in acidosis. Fortunately, the respiratory and renal systems are able to adjust for these changes to some extent by a process known as **compensation**, thus returning the pH to within normal or nearly normal limits.

a. *Acidosis*

Acidosis occurs when the pH of arterial blood drops below 7.35. It may result from an accumulation of acids or a loss of bases. There are two types of acidosis:

1. **Respiratory acidosis.** In this condition, there is too much dissolved carbon dioxide in the plasma. Thus, the pCO$_2$ will be *elevated*. Disease states that may cause this condition include injury to the respiratory center in the brainstem, which results in decreased rate and depth of breathing; emphysema, in which movement of air within the alveoli is impeded; and pneumonia. Respiratory acidosis is often compensated for by metabolic alkalosis, in which the kidneys conserve bicarbonate ions, thus raising the pH of the blood.
2. **Metabolic acidosis.** In this condition, the pH of arterial blood drops as a result of the accumulation of nonrespiratory acids or the loss of bases.

Disease states that may lead to this condition include *uremia*, in which there is a failure of the kidneys to excrete acids formed through metabolic processes; prolonged diarrhea, in which there is an excessive loss of alkaline intestinal secretions; and diabetes mellitus, in which keto acids accumulate in the blood and bicarbonate ions are excessively excreted in the urine. Metabolic acidosis is often compensated for by respiratory alkalosis, in which the lungs eliminate excess CO_2 through rapid, deep breathing, thus raising the pH of the blood.

b. *Alkalosis*

Alkalosis occurs when the pH of arterial blood rises above 7.45. It may result from a loss of acids or an accumulation of bases. There are two types of alkalosis:

1. **Respiratory alkalosis.** In this condition, there is too little dissolved carbon dioxide in the plasma. Thus, the pCO_2 of arterial blood will be *decreased* due to hyperventilation. Conditions that may cause hyperventilation and, thus, respiratory alkalosis include anxiety, early stages of salicylate poisoning, or high altitudes. Light-headedness, agitation, tingling sensations of the extremities, and **vertigo** may result. Respiratory alkalosis is often compensated for by metabolic acidosis, in which the kidneys excrete more bicarbonate ions, thus lowering the pH of the blood.
2. **Metabolic alkalosis.** In this condition, the pH of arterial blood rises as a result of an excessive loss of hydrogen ions or a gain in bases. Disease states that may lead to this condition include vomiting or gastric suctioning, and excessive intake of alkaline drugs such as sodium bicarbonate (baking soda) to relieve heartburn or other gastric distress. Metabolic alkalosis is often compensated for by respiratory acidosis, in which there is decreased rate and depth of breathing, which will raise the arterial pCO_2 and, thus, lower the pH of the blood.

By analyzing arterial blood gases, it is possible to determine whether a person is in a state of acidosis or alkalosis and to determine the type (metabolic or respiratory) of acidosis or alkalosis that results in a pH shift. Normal values and the result of increased or decreased values are as follows:

Increased	Alkalosis		Acidosis		Alkalosis	
Normal	pH	7.35–7.45	pCO_2	40 mm Hg	HCO_3^-	24 mEq/liter
Decreased	Acidosis		Alkalosis		Acidosis	

When the pH shift is primarily reflected by pCO_2 values, it is respiratory in nature. When the pH shift is primarily reflected by HCO_3^- values, it is metabolic in nature. This pH shift is usually resisted or compensated for by an increase or decrease in respiratory center activity or by chemical buffer activity. If compensation does not occur, death may result.

Acid-Base Problems

Solve the following acid-base balance problems with respect to:

a. type of acidosis or alkalosis
b. presence or absence of compensation

1. pH   7.48
   pCO_2 48
   HCO_3^- 32

2. pH   7.31
   pCO_2 50
   HCO_3^- 32

3. pH   7.30
   pCO_2 41
   HCO_3^- 18

4. pH   7.50
   pCO_2 30
   HCO_3^- 24

UNIT XIII
Acid-Base Balance

DISCUSSION

Matching

Select the letter that best describes the acid-base imbalance for each condition listed.

_____ 1. Diarrhea

_____ 2. Oral intake of excessive NaHCO$_3$

_____ 3. Emphysema

_____ 4. Early stage of overdose of aspirin

_____ 5. Vomiting

_____ 6. Diabetes mellitus

_____ 7. Pneumonia

_____ 8. Anxiety

a. respiratory acidosis

b. metabolic acidosis

c. respiratory alkalosis

d. metabolic alkalosis

Multiple Choice

_____ 9. The most accurate method of determining pH is to:
a. use litmus paper
b. use pH or nitrazine paper
c. use a pH meter
d. taste with your tongue

_____ 10. A solution that has the same number of hydrogen ions and hydroxyl ions is:
a. neutral
b. acidic
c. basic
d. salty

_____ 11. If one solution is 1000 times as acidic as another, how many pH units would there be between the two solutions?
a. one
b. two
c. three
d. four

_____ 12. A person with an arterial blood pH of 7.62 would be in a state of:
a. acidosis
b. alkalosis
c. neutrality
d. death

_____ 13. The two organs primarily responsible for maintaining acid-base balance in the body are:
a. liver and spleen
b. liver and lung
c. kidney and heart
d. lung and kidney

_____ 14. Which of these would be formed if a strong acid were added to a weak base in a buffered solution?
a. weak acid
b. water
c. strong base
d. weak base and strong acid

_____ 15. Which of these acids is the strongest?
   a. 0.01M HCl   c. 1M NaOH
   b. 1M HCl   d. 0.05M HCl

_____ 16. What effect does breathing in and out of a paper bag have on a person's arterial pCO$_2$ level?
   a. the level decreases   c. it has no effect
   b. the level increases

_____ 17. The strength of an acid refers to:
   a. the number of hydrogen ions in   c. the degree to which its molecules
      each molecule   ionize in water
   b. the type of inorganic salt it forms   d. the concentration of acid molecules

_____ 18. The most common buffer in plasma and intercellular fluid is:
   a. bicarbonate   c. hemoglobin
   b. phosphate   d. sodium and potassium

True or False

_____ 19. The pH of blood may vary considerably, depending on diet.

_____ 20. The pH of venous blood is usually lower than that of arterial blood.

_____ 21. Generally speaking, buffer solutions neutralize acidic or alkaline solutions.

_____ 22. Phosphate buffers may act as either acids or bases.

_____ 23. More drops of acid are required to change the pH of a buffered solution than a nonbuffered solution.

_____ 24. The respiratory system is the major compensatory mechanism for respiratory acidosis.

_____ 25. A base is a hydroxyl ion acceptor.

DISCUSSION QUESTIONS (Optional)

1. Define acids, bases, and buffers by writing chemical equations to show the type of ions each produces in water.

2. List and describe three general properties of acids and bases.

3. Explain how the kidneys and respiratory center regulate acid-base balance in the body.

Reproductive System

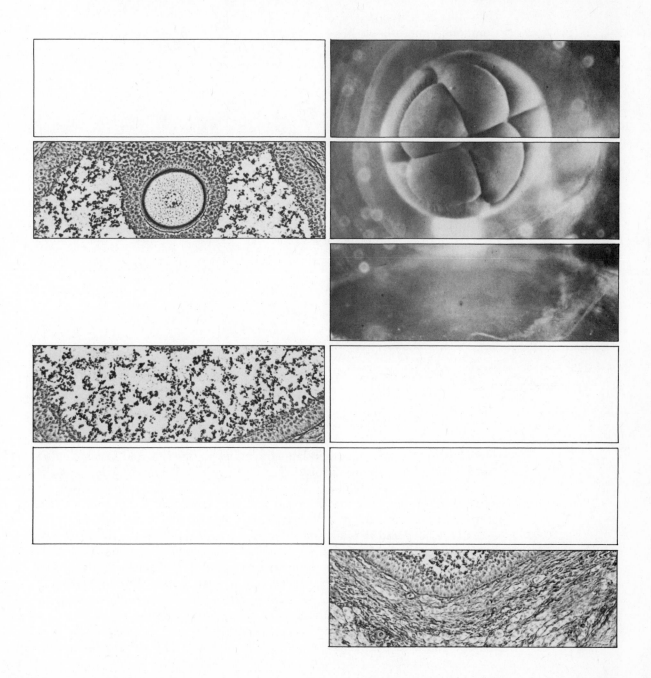

■ *Purpose*

The purpose of Unit XIV is to enable you to understand the anatomy and physiology of the male and female reproductive systems.

■ *Objectives*

In order to complete Unit XIV, you must be able to do the following:

1. Identify the structures of the male and female cat reproductive systems.
2. Observe specimens of human or bovine **gonads** (the ovary and testis), penis, placenta, and uterus with fetus in situ.
3. Identify the stages in oogenesis and spermatogenesis.
4. Using a microscope, recognize sections of penis, uterus, ovary, and testis.
5. Understand the basic principles of reproductive physiology.

■ *Materials*

Dissecting microscope
Compound microscope
Pituitary extract (optional)
Dissecting instruments
Mature male and female live
    frogs (optional)
Holtfreter's solution
    (optional)
Gross specimens of prepared ovary, testis,
    penis, uterus with fetus in situ, and
    placenta
Microscope slides of ovary, testis, penis,
    uterus, and sperm smears
Petri dishes (optional)
Preserved male and female cats
Fertilized frog eggs (optional)

■ *Procedure*

EXERCISE 1

Cat Reproductive Organs

a. *Female Reproductive System*

In order to observe the reproductive organs of a female cat (*see* Figures 189 and 190), reflect the small and large intestines to one side and locate a small cream-colored oval-shaped ovary lying just below each kidney. Each **ovary** is connected to a short coiled **oviduct (fallopian tube)** that conveys **ova**, or eggs, to the **uterus**. The uterus in the cat is bipartite, that is, made up of two **uterine horns** that are continuous with the oviducts. The uterine horns extend medially and caudally and then merge to form the **body of the uterus**, dorsal to the base of the urinary bladder. The human uterus has no horns. Cut completely through the pubic symphysis and follow the body of the uterus caudally. The **vagina** is dorsal to the urethra that extends from the bladder to the **vestibule**, into which the urethra and vagina merge. Caudal to the vestibule is the **urogenital sinus** that opens to the outside.

To observe the vagina and uterus internally, make a ventral incision into the opening of the urogenital sinus and extend it to the horn of the uterus.

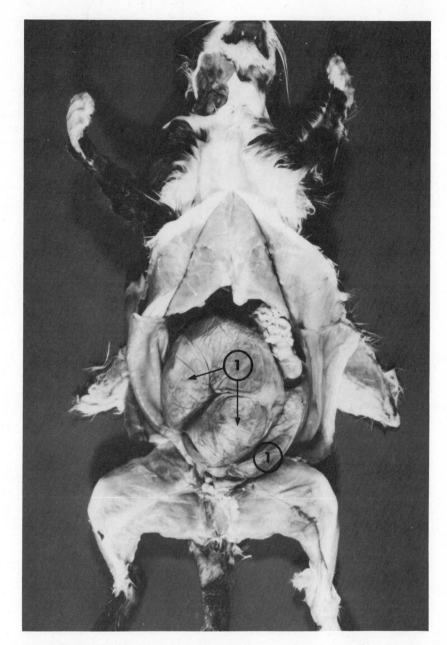

FIGURE 189 *Ventral aspect of abdominal cavity of a full-term cat exposed to show fetuses within uterine horn* (From M. J. Timmons, *A Visual Guide to Dissection: Cat Anatomy Slides,* Courtesy J. B. Lippincott Company)

1. Uterine Horns Containing Fetuses

Locate the **cervix** of the uterus, which is a hard knot of tissue separating the body of the uterus from the vagina. *IDENTIFY: ovaries, vagina, urogenital sinus, cervix, oviducts, uterine horns,* and the *body of the uterus.*

b. *Male Reproductive System*

Observe the reproductive organs of a male cat. Immediately caudal to the pelvic region, locate the **scrotum,** which is a sac covered with skin. Carefully cutting through the skin and connective tissue at the caudal ventral portion

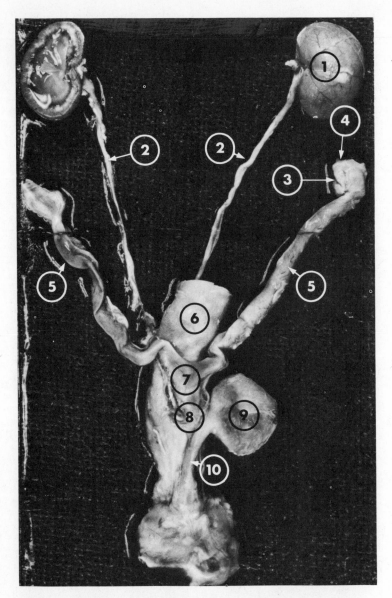

FIGURE 190 *Female cat urogenital organs* (From M. J. Timmons, *A Visual Guide to Dissection: Cat Anatomy Slides,* courtesy J. B. Lippincott Company)

1. Kidney, Left
2. Ureter
3. Ovary, Left
4. Fallopian Tube
5. Uterine Horn
6. Rectum
7. Body of Uterus
8. Vagina
9. Urinary Bladder
10. Urethra

of the scrotum, expose the paired **testes**. Remove the peritoneum covering the testes and observe the **epididymis**—a large coiled duct on the anterior lateral surface of each testis. Extending from the epididymis is an ascending tube, the **spermatic** *cord*, that consists of the **vas deferens (ductus deferens),** blood vessels, and a nerve. Trace the spermatic cord and notice that it enters the body cavity through an opening—the **inguinal canal**. When the spermatic cord reaches the level of the urinary bladder, the vas deferens leaves the spermatic cord and makes a sharp bend medially and caudally over the **ureter** and continues caudally until it penetrates the urethra at the level of the **prostate gland**. Cut through the pubic symphysis and trace the urethra caudally until it enters the **penis**, an external structure. The opening at the anterior end of the penis where urine and semen are released is the **urogenital opening**. *IDENTIFY: scrotum, testes, epididymis, spermatic cord, vas deferens, inguinal canal, urethra, prostate gland,* and *penis. (See* Figures 191 and 192).

FIGURE 191 *Male cat urogenital organs* (From M. J. Timmons, *A Visual Guide to Dissection: Cat Anatomy Slides,* courtesy J. B. Lippincott Company)

1. Kidney, Left
2. Ureter, Left
3. Rectum
4. Urinary Bladder
5. Urethra

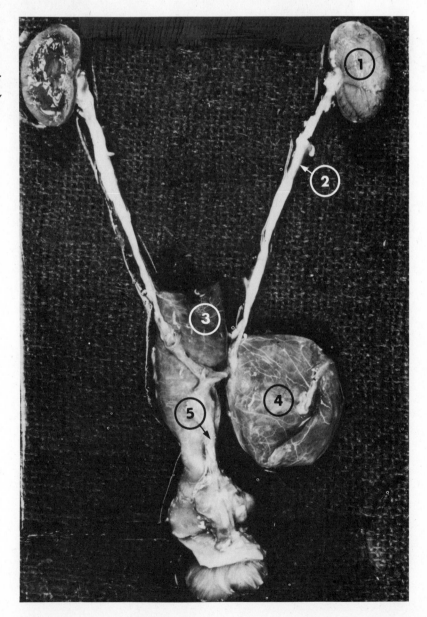

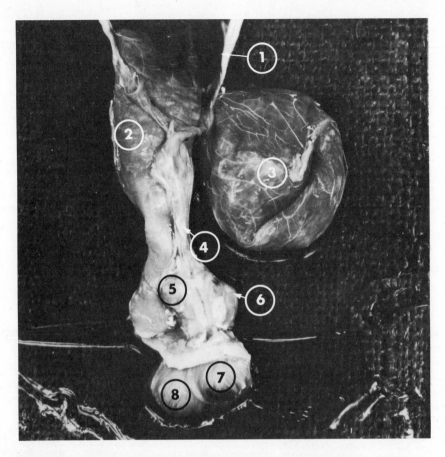

FIGURE 192 *Male cat urogenital system showing both external and internal detail* (From M. J. Timmons, *A Visual Guide to Dissection: Cat Anatomy Slides,* courtesy J. B. Lippincott Company)

1. Ureter, Left
2. Rectum
3. Urinary Bladder, Reflected
4. Urethra
5. Prostate Gland
6. Bulbourethral Gland
7. Penis
8. Testis

EXERCISE 2

Observation of Gross Specimens

Your instructor will provide you with specimens of **ovary**, **testis**, **penis**, uterus, and **placenta** for observation. (*See* Figures 193 and 194 for views of the internal viscera and external genitalia of the human male and female.)

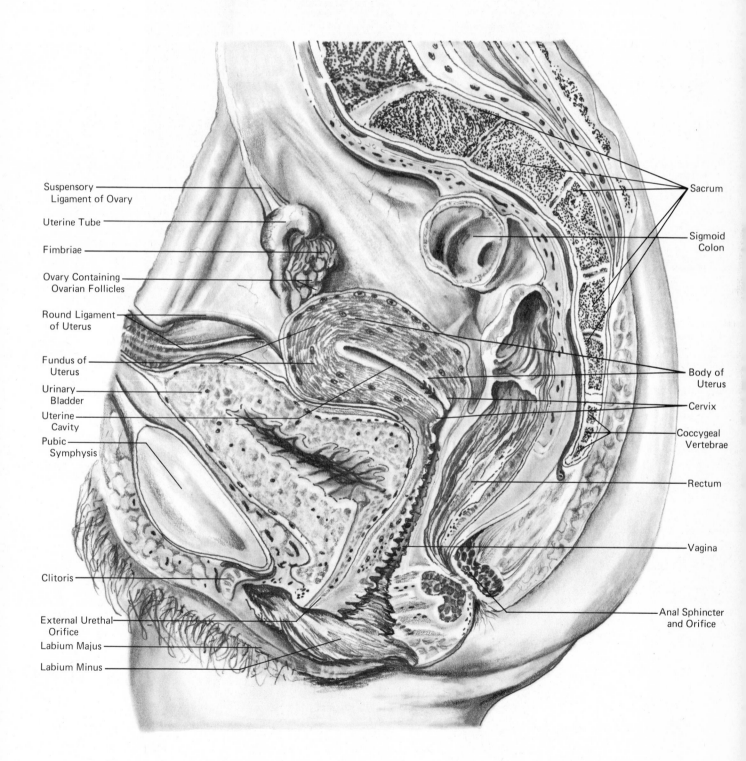

FIGURE 193 *Midsagittal section of female pelvis showing internal viscera and external genitalia*

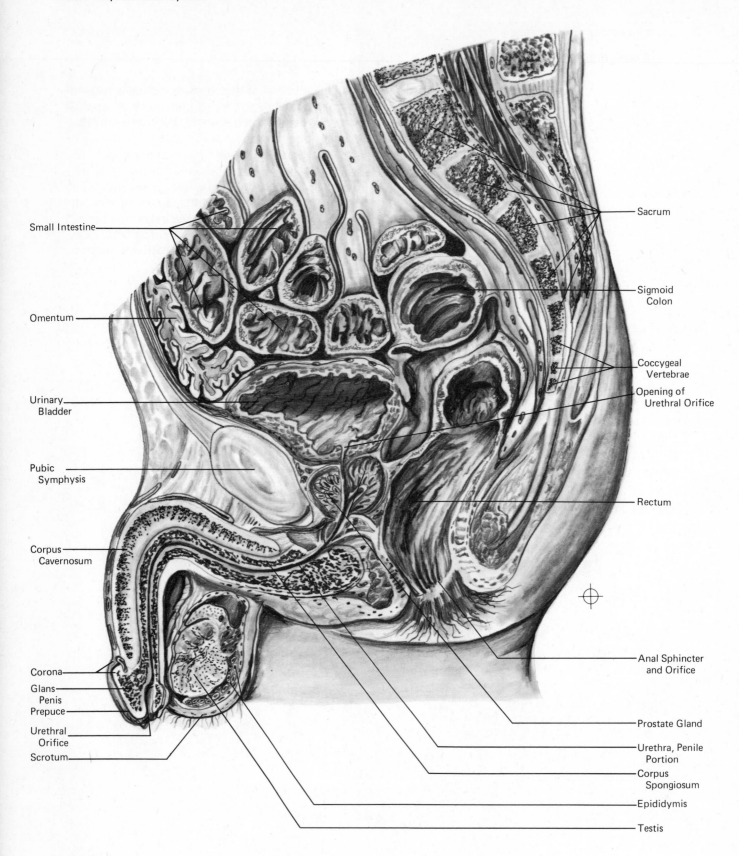

Small Intestine

Omentum

Urinary
Bladder

Pubic
Symphysis

Corpus
Cavernosum

Corona

Glans
Penis

Prepuce

Urethral
Orifice

Scrotum

Sacrum

Sigmoid
Colon

Coccygeal
Vertebrae

Opening of
Urethral Orifice

Rectum

Anal Sphincter
and Orifice

Prostate Gland

Urethra, Penile
Portion

Corpus
Spongiosum

Epididymis

Testis

FIGURE 194 *Midsagittal section of male pelvis showing internal viscera and external genitalia*

EXERCISE 3

Microscopic Study of an Ovary

The sexually mature ovary contains millions of undifferentiated cells termed **oogonia** that, like the spermatogonia, undergo many mitotic divisions in order to increase their number. As the oogonia mature, they develop into larger **primary oocytes**. The primary oocytes are large cells that contain yolk. These cells divide by **meiosis** into two cell types—the **secondary oocyte**, which contains all yolk, and the **first polar body**. Meiosis continues as the secondary oocyte divides to become a large **ootid** and a smaller **second polar body**. The ootid undergoes further minor changes until it is a mature **ovum** (egg). The polar bodies are formed in order to eliminate the extra number of chromosomes. Thus, the ovum contains half the number of chromosomes as the diploid oogonium (*see* Figure 195). The process by which the mature ovum is formed is called **oogenesis**.

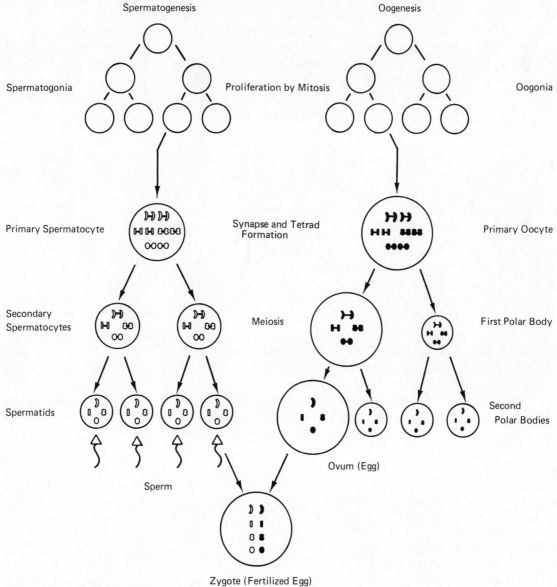

FIGURE 195 *Gametogenesis*

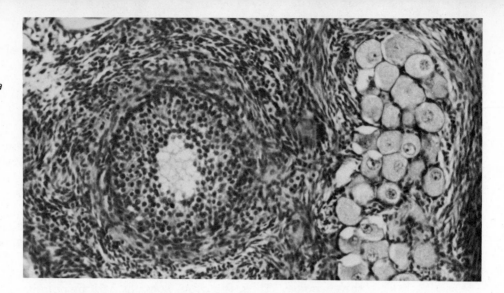

Observe a slide of a sectioned ovary showing **Graafian follicles** (*see* Figure 196). Notice the *cuboidal epithelium* around the periphery of the section. Medial to the cuboidal epithelium you may see **primary follicles**. After puberty, under hormonal influence, several follicles begin to grow prior to ovulation and become primary oocytes, secondary oocytes, and, later, Graafian follicles. Graafian follicles occur only in mammals and are hollow sacs containing an ovum surrounded by **follicular fluid** (or liquor folliculi). The cavity containing the follicular fluid is the *antrum*. Immediately underlying the ovum within the Graafian follicle is a mound of cells known as the **cumulus oophorus** (*see* Figure 197). *SKETCH* what you see and *LABEL* the various structures.

Observe a slide of a sectioned ovary showing corpora lutea. A **corpus luteum** is a collapsed follicle that has expelled its ovum. Corpora lutea first appear yellow and then regress into scar tissue known as the **corpus albicans**. The active corpus luteum produces *progesterone*—a hormone essential to pregnancy (*see* Figure 198). *SKETCH* in the corpus luteum and corpus albicans in the above drawing and *LABEL* these structures.

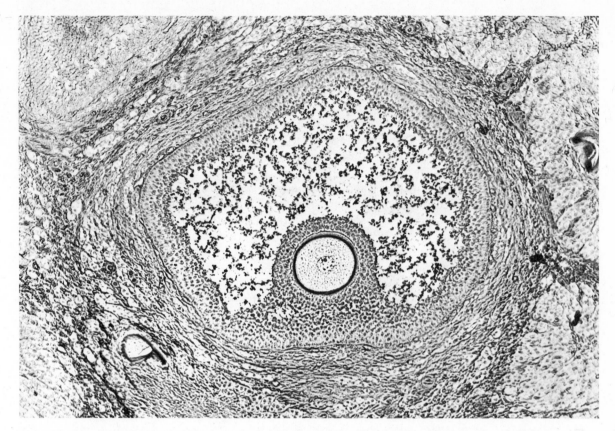

FIGURE 197 *Graafian follicle showing cumulus oophorus* (Courtesy Carolina Biological Supply Company)

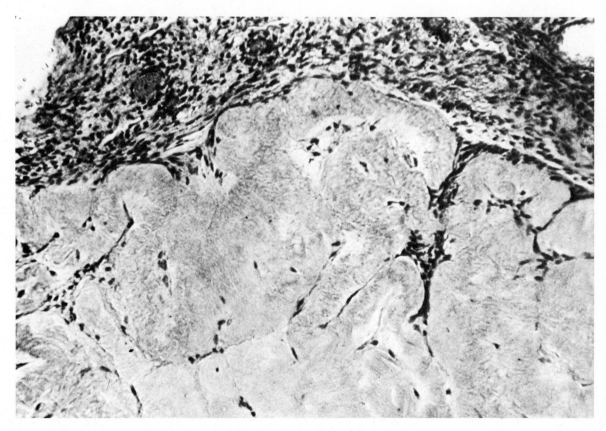

FIGURE 198 *Corpus albicans*

EXERCISE 4

Microscopic Study of a Testis

The testis consists of thousands of coiled tubules that develop millions of sperm. These tubes are lined with undifferentiated germ cells termed **spermatogonia**. These spermatogonia divide by the process of mitosis to increase their number. At puberty, the spermatogonia cells begin to undergo **spermatogenesis**. The first stage of this process is the growth of these cells into much larger cells termed **primary spermatocytes**. These cells in turn divide by meiosis into two cells of equal size termed **secondary spermatocytes**. Meiosis continues as the secondary spermatocytes immediately divide into four equal-sized cell types termed **spermatids**. Spermatids contain half the number of chromosomes as diploid spermatogonia and require further minor changes before becoming functional **spermatozoa** (*see* Figure 200).

Observe a slide containing a section of a testis. Within the testis, you will find numerous **seminiferous tubules** (*see* Figure 199). Under high power, focus on a cross section near the periphery of a seminiferous tubule. Starting at the periphery and moving toward the center, you should be able to see the maturation of sperm (spermatogenesis) from spermatogonia to primary spermatocytes to secondary spermatocytes to spermatids to mature **sperm** in the center of the tubules (*see* Figure 200). Mature sperm can be recognized by the presence of tails. Between the seminiferous tubules are the **interstitial cells of Leydig**, which produce and secrete the male hormone **testosterone**. *DRAW and LABEL* a cross section of a testis.

Observe a slide of sperm smear. *IDENTIFY:* the head and tail of a spermatozoan.

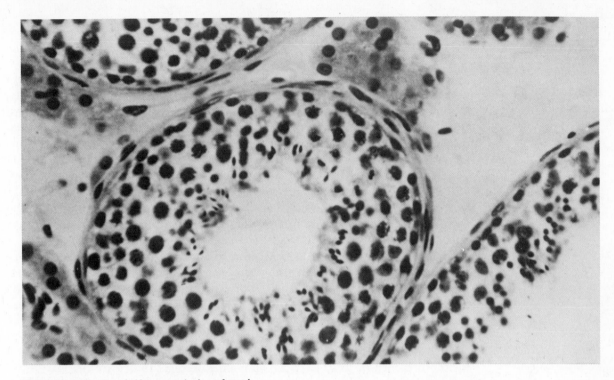

FIGURE 199   *Seminiferous tubules of testis*

FIGURE 200   *Detail of seminiferous tubule of rat testis showing spermatogenesis*

EXERCISE 5

Microscopic Study of a Penis

Obtain a slide of a cross section of a penis. This slide will be best observed under a dissecting microscope. *IDENTIFY:* two **corpora cavernosa**—corpus cavernosum urethrae (also called **corpus spongiosum**) and the *urethra*, which is found in the corpus spongiosum. Also, observe the connective tissue, muscle, sinuses, and blood vessels in this section. *SKETCH and LABEL.*

EXERCISE 6

Microscopic Study of a Uterus

Obtain a section of uterus (any phase) and identify the smooth muscle area and endometrial lining. In the **endometrium**, which is sloughed off during menstruation, look for arteries and uterine glands. *SKETCH and LABEL* this slide.

EXERCISE 7

Induction of Ovulation in the Frog (Optional)

In this exercise, you will inject frogs with pituitary extract containing **gonadotropic hormones** in order to induce ovulation. Your instructor will provide you with instructions and materials for doing this.

EXERCISE 8

Determination of Human Pregnancy Using Human Chorionic Gonadotropin as an Indicator

The first laboratory diagnosis of pregnancy originated with the Ascheim and Zondek test in 1928. This test was the first to be based on the presence of **Human Chorionic Gonadotropin (HCG)**, a hormone that is produced by the placenta and excreted in the urine of pregnant women. When this urine was injected into mice, Ascheim and Zondek discovered that the HCG caused premature maturation of ovarian follicles in the mice. In 1929, Friedman developed a similar test using a rabbit, which took less time. Several years later, an even more rapid test was developed, using a frog. More recently, tests have been developed that eliminate the need for living animals, and use only the urine of the female to be tested.

The hormonal changes of pregnancy chiefly involve the placenta. The developing placenta secretes human chorionic gonadatropin (HCG) that stimulates the corpus luteum during pregnancy to continue its large output of estrogen and progesterone, which are essential for the maintenance of early pregnancy. A significant level of the hormone HCG is found only in urine of pregnant women. It is the presence of this urinary HCG that constitutes the basis of current pregnancy test methods and the basis for this exercise.

The Early Pregnancy Test (E.P.T.®)* is an early indicator of pregnancy that is considered to be 97% accurate. This test can detect HCG in a fresh, first-voided morning urine sample as early as nine days following the expected date of the first missed menstrual period. At this time, there is sufficient concentration of HCG in a woman's first-voided morning urine to provide an accurate indication of pregnancy. The E.P.T. test simply verifies whether or not HCG hormone is present in urine.

The principle of this pregnancy test is based on the ability of HCG to cause antibody formation. The HCG is the *antigen* and the *antibodies* are proteins, known as *anti-HCG*. The antibody has the property of being able to combine with the HCG antigen to form an *antigen-antibody reaction*, or simply to neutralize the HCG antigen.

The neutralization of HCG by anti-HCG serus is not visible to the naked eye. Therefore, to make the reaction visible, HCG must be absorbed, or coated, onto the surface of sheep red blood cells. When anti-HCG is added to a suspension of HCG-coated sheep red blood cells, the cells are agglutinated (clumped) and become visible upon standing. They are *seen as a yellow-red deposit* in the bottom of the test tube in which the reaction takes place.

This test is based on the ability to accurately measure the amount of HCG and anti-HCG serum so there is an optimum amount of each to react with the other with no excess of either being present. In other words, one can standardize the amount of antigen and antibody. If HCG is added from another source, as in urine from a pregnant woman, the HCG in the urine will react with the available anti-HCG serum. There will be no anti-HCG serum left to react with the HCG absorbed on the sheep red blood cells, and no agglutination can occur. Therefore, the unagglutinated cells will settle to the bottom of the test tube in a *dark brownish ring pattern* that indicates a *positive test* for pregnancy. If there is no HCG or an insufficient amount of HCG to react with the anti-HCG serum, the interaction between the HCG-coated red blood cells and the anti-HCG serum will proceed as follows. Agglutination will occur and the agglutinated cells will settle to the base of the test tube as a *yellow-red deposit* rather than in a *ring pattern*. This is interpreted as a *negative test*.

* Registered trademark of the Warner/Chilcott Company, Morris Plains, New Jersey.

Materials

Early Pregnancy Test (E.P.T.) Kit (*see* Figure 201) that consists of:

  a. A test tube containing a lyophilized (freeze-dried) mixture of antigen (from sheep red blood cells coated with HCG)

  b. An antibody (from blood serum of HCG-sensitized rabbits)

  c. A vial containing distilled water

  d. A medicine dropper

  e. A clear plastic support with an angled mirror at the bottom

First-voided morning urine sample (collected in a freshly washed and rinsed receptacle)

Experimental urine samples (provided by your instructor)

Procedure

1. Place three drops of your urine (or sample provided by your instructor) into the test tube that already contains the test reagents.
2. Completely empty the plastic vial of distilled water into the test tube.
3. Press the rubber stopper of the dropper into the test tube and *shake vigorously for 10 seconds*.
4. Place the test tube back into its original position in the holder and **LEAVE IT COMPLETELY UNDISTRUBED FOR EXACTLY 2 HOURS.** *Note:* The test must be completely undisturbed and free from jarring or vibrations for 2 hours because the basis of the test is the settling of the unagglutinated red cells into a ring at the bottom of the tube. Even the slightest jarring can interfere with this process. The test must be read in *exactly* 2 hours.)

FIGURE 201 *Early Pregnancy Test (E.P.T.) performed by student* (Courtesy Warner-Lambert Company)

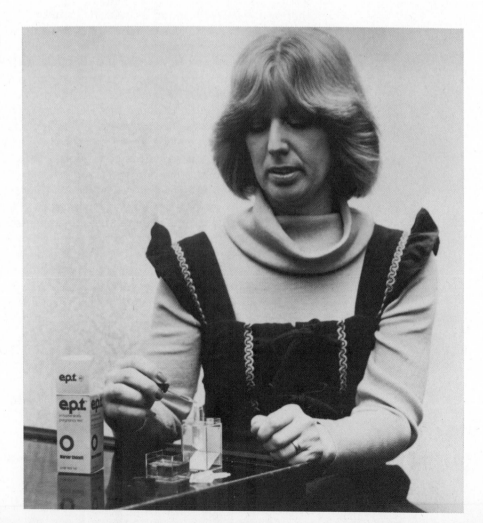

FIGURE 202 *Positive results for HCG pregnancy hormone as indicated by the dark doughnut-shaped ring that has formed in the bottom of the test tube* (Courtesy Warner-Lambert Company)

5. Interpretation of the test (*see* Figure 202):
   a. *Positive test results.* After 2 hours, if a dark brown doughnut-shaped ring is visible in the mirror under the test tube, it indicates that the urine sample contains the hormone HCG. (*Note:* False positive tests may be caused by residues of detergent used to clean the urine sample container, blood in the urine, or high concentrations of protein in the urine.)
   b. *Negative test results.* When there is no ring, or doughnut pattern, but only a yellow-red deposit, the reaction is negative. This means no pregnancy hormone has been detected in the urine sample. (*Note:* False negative tests may be caused by a low level of HCG in the urine, the test being performed too soon after a missed period, or the test tube being jarred during the 2-hour waiting period. In an **ectopic pregnancy** —a pregnancy that takes place outside the uterus, generally in the fallopian tube—inadequate placental tissue formation results in low levels of HCG. In the case of this unnatural development, the test will be negative.)

Questions

1. Why is a first-voided morning urine specimen advised for this test?

2. To what errors of observed interpretation is this test subjected?

3. Briefly explain the principle of the HCG agglutination test.

4. Why does a brown ring formed in the bottom of the test tube indicate a positive test?

EXERCISE 9

Embryology of the Frog (Optional)

1. First, transfer several unfertilized eggs (from Exercise 7) to a Petri dish containing Holtfreter's solution (*see* Solution Appendix), which should completely cover the eggs. Place a few eggs on a slide, cover with water, and observe the **fertilization membrane** under low power. Examine the remaining eggs under a dissecting microscope and *DRAW* a few eggs, paying particular attention to their pigmentation. The animal pole of the egg is highly pigmented, whereas the vegetable pole is not.

2. Observe fertilized frog eggs in various stages of embryological development. Identify the following developmental stages of the **embryo**: cleavage (2-, 4-, 8-, and 16-cell stages), blastula, gastrula, and **neurula** (*see* Figures 203–216). *DRAW* and label the stages of development.

FIGURE 203 *Fertilized frog egg*

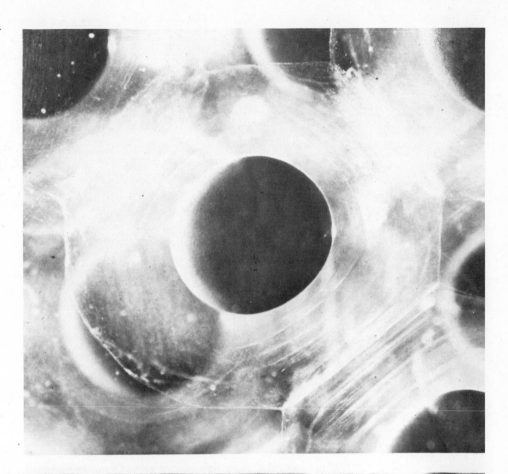

FIGURE 204 *Frog egg, 2-cell stage*

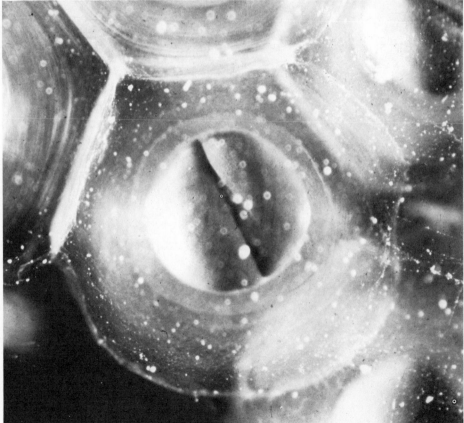

Figures 203–216 courtesy Carolina Biological Supply Company

FIGURE 205 *Frog egg, 4-cell stage*

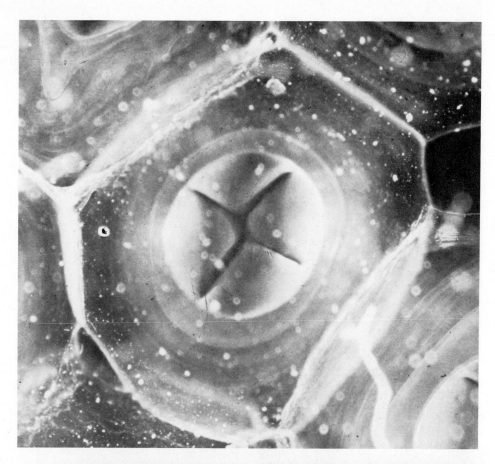

FIGURE 206 *Frog egg, 8-cell stage*

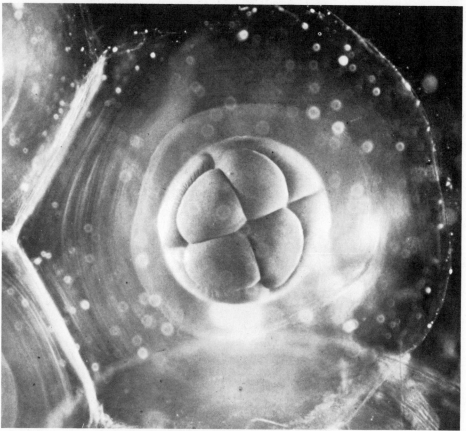

FIGURE 207 *Frog egg, 16-cell stage*

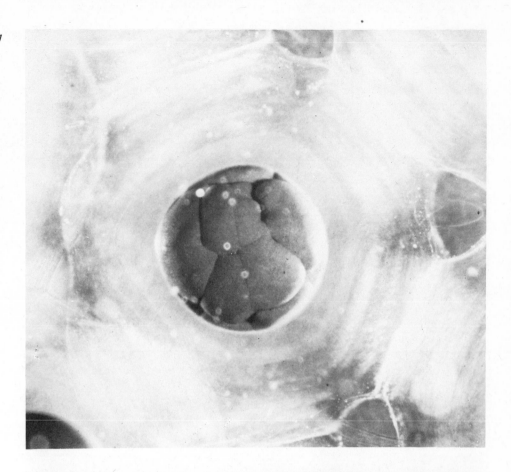

FIGURE 208 *Frog egg, 32-cell stage*

FIGURE 209 *Frog egg, early blastula stage*

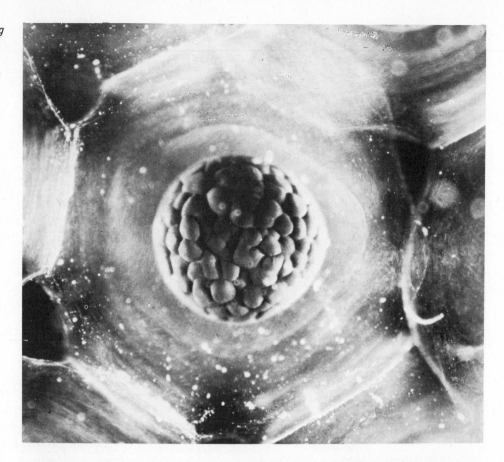

FIGURE 210 *Frog egg, late blastula stage*

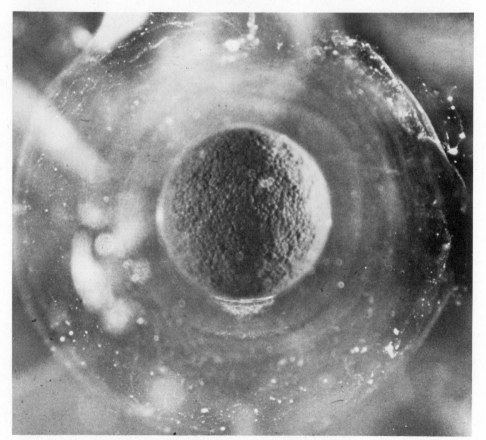

FIGURE 211 *Frog egg, early yolk plug stage*

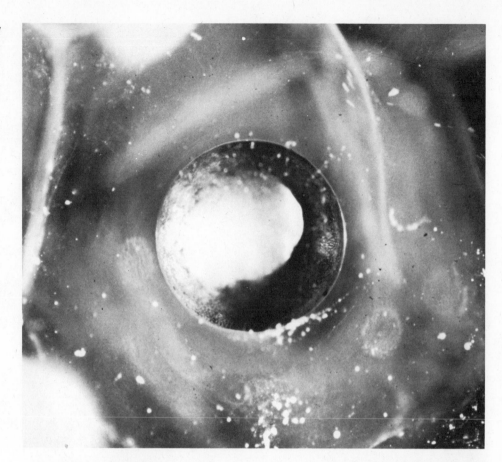

FIGURE 212 *Frog egg, late yolk plug stage*

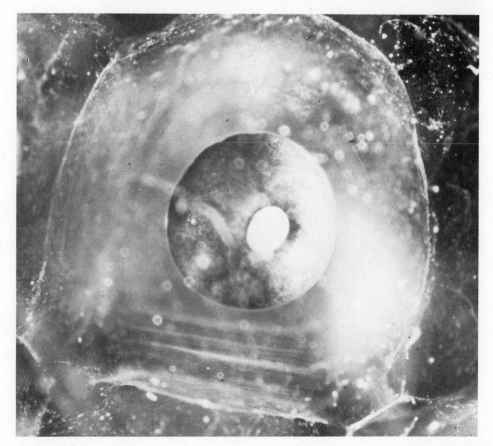

FIGURE 213 *Frog egg, neural plate stage*

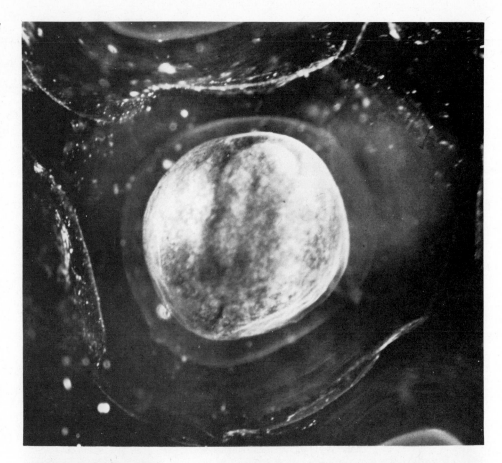

FIGURE 214 *Frog egg, early neural groove stage*

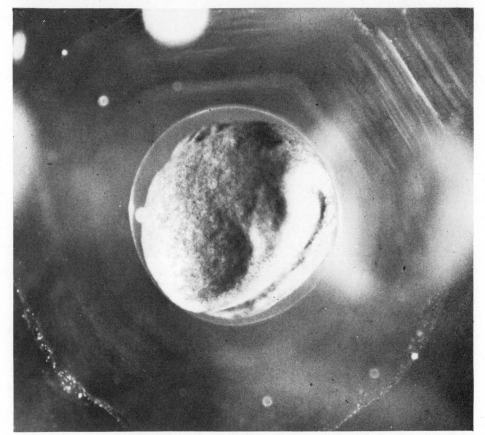

FIGURE 215 *Frog egg, late neural groove stage*

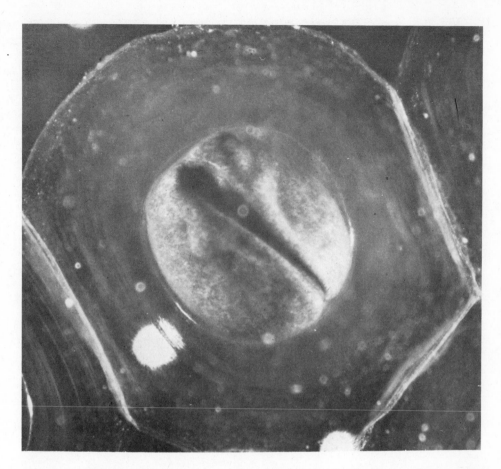

FIGURE 216 *Frog egg, neural tube stage*

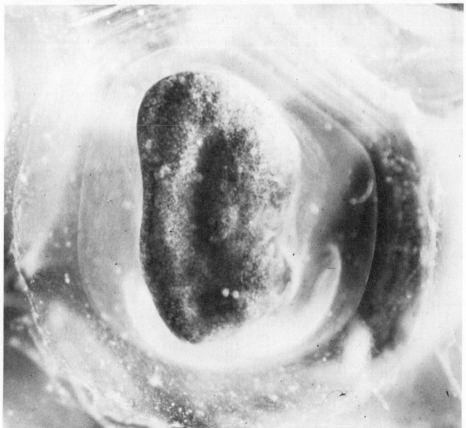

UNIT XIV

Reproductive System

DISCUSSION

Multiple Choice

_____ 1. Which of these is the correct sequence for oogenesis?
a. oogonium, primary oocyte, ovum, secondary oocyte, ootid
b. primary oocyte, secondary oocyte, oogonium, ootid, ovum
c. oogonium, primary oocyte, secondary oocyte, ootid, ovum
d. ovum, primary oocyte, oogonium, secondary oocyte, ootid

_____ 2. How many spermatozoa are normally produced from one primary spermatocyte?
a. one
b. one plus three polar bodies
c. two
d. four

_____ 3. How many chromosomes does a human ovum normally contain?
a. 23
b. 23 pairs
c. 46
d. 46 plus 2 sex chromosomes

_____ 4. Hormones that promote ovulation are secreted by the:
a. pituitary gland
b. ovary
c. adrenal glands
d. uterus

_____ 5. Which of these correctly describes development of a human fertilized ovum?
a. zygote, blastula, gastrula, morula
b. gastrula, zygote, blastula, morula
c. zygote, morula, blastula, gastrula
d. blastula, gastrula, morula, zygote

_____ 6. Spermatogenesis begins:
a. immediately prior to ejaculation
b. at puberty
c. within the first year of life
d. during embryonic development

_____ 7. What do human ova and spermatozoa have in common?
a. they contain the same number of chromosomes
b. they secrete the same hormones in equal quantities
c. they are approximately the same size
d. they are produced in equal numbers

_____ 8. Which of these could not occur within the body if the fallopian tubes were cut and ligated?
a. fertilization
b. ovulation
c. sexual intercourse
d. menstruation

_____ 9. What is the purpose of having the testes suspended within the scrotum?
a. to protect the sperm from becoming contaminated with urine
b. to provide additional surface area for sperm maturation
c. to facilitate the passage of sperm through the vas deferens
d. to protect the sperm from higher temperatures within the abdominal cavity

10. Which of these hormones maintains male secondary sex characteristics?
   a. corticosterone
   b. FSH
   c. testosterone
   d. progesterone

11. Which of the following structures is not contained in the testes?
   a. seminiferous tubules
   b. rete testes
   c. vas deferens
   d. interstitial cells of Leydig

12. Which of the following glands manufacture(s) semen?
   a. seminal vesicles
   b. prostate
   c. Cowper's
   d. all of the above

13. The major anatomical feature contributing to penile erection is:
   a. bulbourethral contraction
   b. urethral distention
   c. retention of blood in corporal cavities
   d. scrotum distention

14. In the female, the structure immediately anterior to the urethral opening is the:
   a. hymen
   b. clitoris
   c. mons pubis
   d. vaginal orifice

15. In a vasectomy, the duct that is cut and ligated is the:
   a. epididymis
   b. vas deferens
   c. ejaculatory duct
   d. seminiferous tubule

16. Human fertilization normally takes place in the:
   a. ovary
   b. fallopian tubes
   c. uterus
   d. vagina

17. The uterine cycle, or menstruation, is triggered by:
   a. lack of progesterone
   b. FSH production
   c. corpus luteum formation
   d. ovulation

Photo Identification

Refer to the photograph of a seminiferous tubule for Questions 18-20.

18. The cell type identified by M is a:
   a. spermatogonium
   b. sperm cell
   c. primary spermatocyte
   d. secondary spermatocyte

19. The first meiotic division produces the cell type identified by O, which is:
   a. a spermatid
   b. a sperm cell
   c. a secondary spermatocyte
   d. none of the above

20. Of the following cell types, which is not found in the seminiferous tubule photograph?
   a. secondary spermatocyte
   b. primary spermatocyte
   c. interstitial cell of Leydig
   d. spermatid

21. Which of the following does *not* secrete some substance that is ejaculated by the male?
   a. testis
   b. vas deferens
   c. seminal vesicle
   d. prostate gland

22. The spermatic cords enclose:
   a. the seminal ducts
   b. blood vessels
   c. nerves
   d. all of the above

_____ ° 23. The hymen is structurally part of the:
    a. cervix   c. vagina
    b. clitoris   d. labia minora

_____ 24. To which of the following structures does *fimbriated* refer?
    a. ovary   c. uterus
    b. Fallopian tube   d. vagina

_____ 25. Which of the following is exclusively found in the female?
    a. parotid gland   c. prostate gland
    b. Bartholin's gland   d. Cowper's gland

DISCUSSION QUESTIONS (Optional)

1. Trace the passage of sperm from the seminiferous tubules to the point of ejaculation.

2. a. Trace the passage of an egg from its release from a Graafian follicle through implantation.

   b. What is the fate of a Graafian follicle after ovulation?

3. Indicate whether each of the following contains the diploid or haploid number of chromosomes:

   a. sperm . . . .

   b. primary oocyte . . . .

   c. secondary spermatocyte . . . .

   d. Leydig cell . . . .

   e. ovum . . . .

4. In the human, which hormone(s) is(are) responsible for inducing ovulation?

5. In terms of the number of germ layers, how does a blastula differ from a gastrula?

Endocrine System

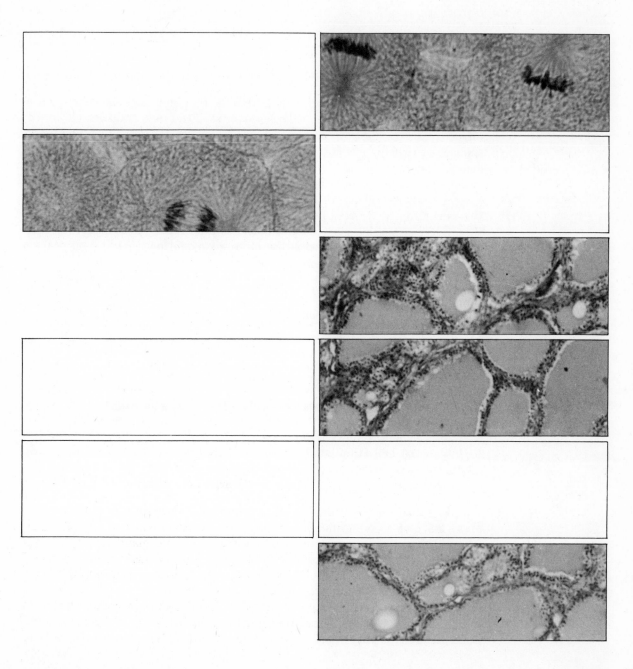

■ *Purpose*

The purpose of Unit XV is to enable you to understand the physiology of and to microscopically recognize the major endocrine glands and to identify their locations.

■ *Objectives*

In order to complete Unit XV, you must be able to do the following:

1. Understand the basic physiology of selected endocrine glands and their hormones.
2. State the location of each endocrine gland in the body.
3. Using a microscope, recognize sections of pituitary, thyroid, parathyroid, pancreas, and adrenal glands.
4. List various pathological conditions associated with **hypersecretion** and **hyposecretion** of hormones by endocrine glands.

■ *Materials*

microscope slides of the following endocrine glands: pituitary, thyroid, parathyroid, pancreas, and adrenal

■ *Procedure*

Endocrine glands are ductless glands that secrete chemical agents known as **hormones.** Hormones from each endocrine gland are transported in the bloodstream to another part of the body where they evoke systemic responses or adjustments by acting on target tissues or organs. The endocrine glands together with the nervous system integrate functions of organs and systems in the body (*see* Figure 217 in Exercise 1).

The **pituitary gland,** or **hypophysis,** (*see* **Color Plate 31**) is located within the sella turcica of the sphenoid bone. It was at one time considered to be the master endocrine gland in the body. It does regulate the secretory functions of other endocrine glands, but it is now believed that the pituitary is regulated by the **hypothalamus** of the brain.

The anterior lobe of the pituitary is known as the **adenohypophysis (or pars distalis).** The adenohypophysis secretes several hormones, some of which are known as **tropic hormones** because they exert their effects indirectly by stimulating the functional activities of other endocrine glands or body cells. The major tropic hormones and the endocrine glands they affect (target glands) include the following:

1. **Thyroid stimulating hormone (TSH)** ——→ thyroid gland
2. **Adrenocorticotropic hormone (ACTH)** ——→ adrenal cortex
3. **Follicle stimulating hormone (FSH)** ——→ ovaries and testes
4. **Luteinizing hormone (LH)** ——→ ovaries
5. **Interstitial cell stimulating hormone (ICSH)** ——→ testes (LH and ICSH are the same hormone)
6. **Luteotropic hormone (LTH)** ——→ ovaries and mammary glands

After a tropic hormone has been released by the pituitary gland, it stimulates its target endocrine gland to secrete hormone(s). For example, TSH (thyroid stimulating hormone) stimulates the thyroid gland to produce the hormone thyroxine that regulates basal metabolism in the human. The hypothalamus regulates the amount of TSH secreted by the adenohypophysis in response to the level of thyroxine in the bloodstream.

Other hormones secreted by the adenohypophysis are **growth hormone (GH)**, and **melanocyte stimulating hormone (MSH)**. GH is a general metabolic hormone having a variety of actions. It enhances growth of the cartilaginous portion of bone, causes an increase in muscle mass, decreases glomerular filtration rate, and liberates free fatty acids from adipose tissue. Growth hormone is now classified as a tropic hormone. Its secretion seems to be regulated by blood protein and glucose levels. MSH exerts its action on melanocytes, the pigment-producing cells in the skin. In the infant, MSH is found in the intermediate lobe of the pituitary gland. Later in life, the intermediate lobe becomes part of the adenohypophysis and continues MSH production.

The posterior lobe of the pituitary is known as the **neurohypophysis (or pars nervosa)**. The neurohypophysis secretes two hormones. The hormone **oxytocin** functions in stimulating uterine contractions and milk ejection. **Antidiuretic hormone** (ADH) promotes reabsorption of water from the distal convoluted tubules and collecting tubules in the kidney. ADH, when administered by injection or nasal spray, is known as **vasopressin**. These two hormones are normally secreted by neurosecretory cells in the hypothalamus and transported by means of nerve fibers to the neurohypophysis where they are stored.

The **thyroid gland,** which secretes two hormones, is located anterior to the trachea and inferior to the larynx. Thyroid hormone, or **thyroxine** regulates metabolic rate and is a derivative of **thyroglobulin,** a glycoprotein. The biosynthesis of thyroxin from thyroglobulin is relatively complex and involves several steps. **Thyrocalcitonin (or calcitonin)** is a hormone that lowers serum calcium levels and functions as an antagonist to parathyroid hormone.

The **parathyroid glands,** of which there are usually four, are located on the posterior surface of the lateral lobes of the thyroid. The parathyroids secrete **parathyroid hormone** (PTH) that increases serum calcium levels by mobilization of calcium from bone in response to lowered calcium levels in blood. If the serum calcium level falls too low, a sometimes fatal condition known as **tetany** develops.

The **pancreas** has an endocrine function in addition to its **exocrine** function of secreting digestive enzymes, which you studied in Unit VII. Interspersed among the acinar regions of the pancreas are areas known as **islets of Langerhans** that contain two cell types—**alpha,** which produce the hormone **glucagon,** and **beta,** which produce the hormone **insulin.** Insulin secretion regulates cellular uptake of glucose and blood glucose level. A lack of insulin results in **hyperglycemia** (elevated blood glucose level), mobilization of depot fat in adipose tissue into the blood, a decrease in protein synthesis, and dehydration. These symptoms characterize the disease **diabetes mellitus.** Glucagon raises the blood glucose level and is thought to promote glycogen breakdown to glucose in the liver. The major stimulant for the release of either insulin or glucagon is a change in the circulating blood glucose level. An increase in circulating blood glucose levels results in insulin release. Correspondingly, a decrease in blood glucose results in glucagon release.

There are two **adrenal glands (or suprarenal glands),** one positioned on the superior border of each kidney. Each gland consists of a **cortex,** which secretes steroid hormones, that surrounds the **medulla,** which secretes the **catecholamines epinephrine** and **norepinephrine.**

Steroids of the adrenal cortex include (1) **mineralocorticoids,** which increase sodium reabsorption by the distal convoluted tubules in the kidney, (2) **glucocorticoids,** which influence the metabolism of glucose, protein, and fat, and (3) **androgens,** which produce masculinization and are present in both sexes. The most abundant mineralocorticoid is **aldosterone.** Glucocorticoids include **cortisol** and **corticosterone.** The androgens, or male sex hormones, include the *17-keto steroids.* There is evidence that **estrogens,** which are female hormones, and **androgens** are present in the adrenal cortex of both sexes.

EXERCISE 1
Location of Endocrine Glands

LABEL the endocrine glands in Figure 217.

EXERCISE 2
Pituitary Gland (Hypophysis)

(*see* **Color Plate 31**)
Examine a section of pituitary gland. *DRAW* the section and *LABEL* the following: **infundibular stalk, pars distalis, chromophil cells, pars nervosa,** and **pars intermedia.** First observe this section under a dissecting microscope to see its entirety.

EXERCISE 3
Thyroid Gland

In this section, you will see cuboidal epithelium surrounding **colloid,** which stains pink and contains **thyroglobulin,** the precursor and storage form of thyroxine (*see* Figure 218). *DRAW and LABEL:* colloid, cuboidal epithelium.

FIGURE 217 *Endocrine system*

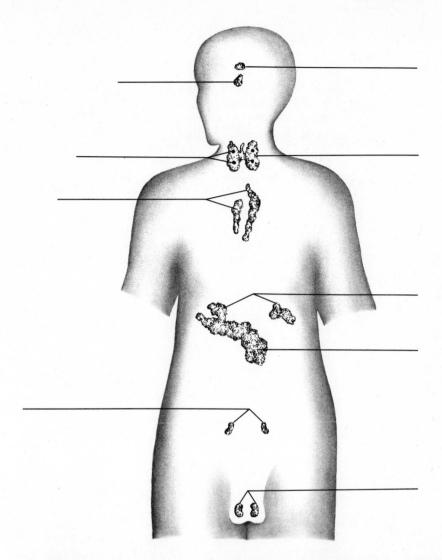

FIGURE 218 *Thyroid gland*

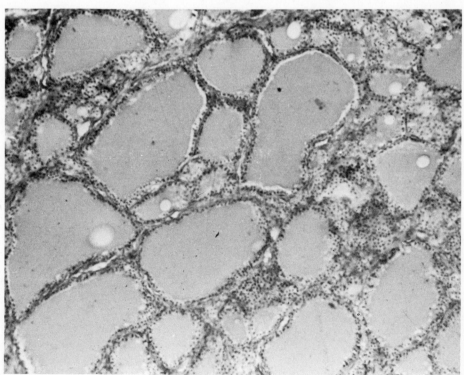

EXERCISE 4
Parathyroid Glands

Parathyroid tissue in the human adult is comprised of two principal types of cells—**chief** (or **principal**) cells and **oxyphilic** cells. Chief cells occur in masses, whereas oxyphilic cells occur singly or in small groups. Oxyphils are not present in young children. *DRAW* a section of parathyroid tissue *and LABEL:* oxyphilic cells, chief cells, and connective tissue.

EXERCISE 5
The Pancreas

(*see* **Color Plate 16**)
Observe the scattered **islets of Langerhans** among the acinar tissue (Figures 219 and 220). Recall that the islets produce two hormones—glucagon from alpha cells and insulin from beta cells. *DRAW* a section of pancreas *and LABEL:* the acinar and islet portions.

FIGURE 219 *General view of pancreas*

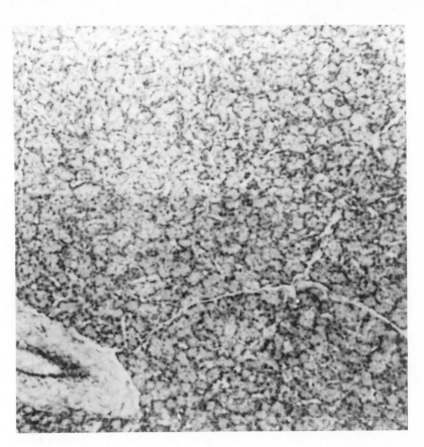

FIGURE 220 *Pancreas, islet of Langerhans*

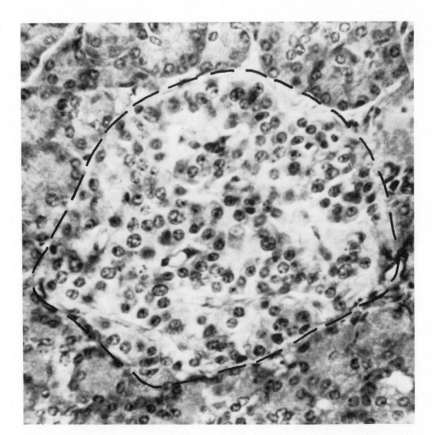

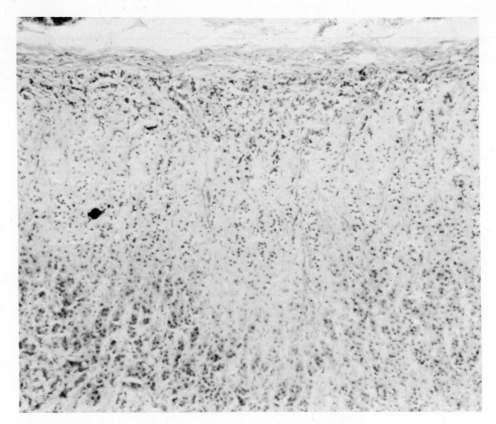

FIGURE 221 *View of the cortex and medulla of suprarenal gland*

EXERCISE 6
Adrenal Glands

Each adrenal gland has an outer **cortex** surrounding the **medulla.** The cortex consists of three zones— the outer **zona glomerulosa,** which secretes aldosterone; and the middle **zona fasciculata** and the inner **zona reticularis,** which both secrete glucocorticoids and sex hormones (*see* Figure 221).

Endocrine System

DISCUSSION

Matching

    a. pituitary
    b. parathyroid
    c. thyroid
    d. pancreas
    e. adrenal

_____ 1. This gland is both an endocrine and an exocrine gland.

_____ 2. This gland contains the following structures: pars distalis, pars nervosa, and pars intermedia.

_____ 3. Chief and oxyphilic cell types comprise this gland.

_____ 4. There are usually four of these glands.

_____ 5. This gland produces numerous tropic hormones.

_____ 6. This gland produces steroid hormones in its cortex and epinephrine and norepinephrine in its medulla.

_____ 7. There is evidence that this gland produces androgens and estrogens in both sexes.

_____ 8. This gland produces TSH (thyroid stimulating hormone).

Multiple Choice

_____ 9. Which of these hormones can be used to induce labor in a pregnant woman?
    a. luteinizing hormone   c. sodium chloride solution
    b. oxytocin   d. progesterone

_____ 10. Which of these hormones is secreted in response to an elevated blood glucose level?
    a. insulin   c. epinephrine
    b. glucagon   d. thyroxine

_____ 11. The chemical element necessary for normal functioning of the thyroid gland is:
    a. cobalt   c. zinc
    b. iodine   d. calcium

_____ 12. When there is an insufficient amount of thyroxine in the circulating blood, the anterior pituitary normally increases its output of thyroid stimulating hormone. This is an example of:
    a. reverse homeostasis   c. decreased metabolic rate
    b. the all-or-none law   d. negative feedback

_____ 13. Which of these is *not* true of epinephrine?
     a. it increases liver glycogenolysis   c. it is necessary for life
     b. it elevates blood pressure   d. it increases the heart rate

_____ 14. Calcitonin is to the thyroid gland as cortisol is to the:
     a. neurohypophysis   c. adrenal cortex
     b. pancreas   d. parathyroid gland

_____ 15. What is the mechanism of action of insulin?
     a. it increases the rate of glucose   c. it increases urinary excretion of exces-
         oxidation within cells   sive blood glucose
     b. it facilitates transport of glucose   d. it increases the basal metabolic rate
         across the cell membrane in skeletal
         muscle and adipose tissue

_____ 16. Which of the following would *not* develop as a result of a hypophysectomy (removal of the pituitary gland)?
     a. sterility   c. hypothyroidism
     b. diabetes mellitus   d. diabetes insipidus

_____ 17. Females may develop masculinization of features due to a tumor of the:
     a. adrenal cortex   c. pancreas
     b. thyroid   d. thymus

_____ 18. Glucagon is secreted by which of these glands?
     a. adrenal   c. ovary
     b. pancreas   d. pituitary

_____ 19. The adrenal glands are located:
     a. in the neck   c. above each kidney
     b. in the pelvis   d. in the chest cavity

_____ 20. . . . . stimulates follicle maturation in the ovary.
     a. thyroid stimulating hormone   c. luteinizing hormone
     b. insulin   d. progesterone

_____ 21. . . . . is a hormone that lowers serum calcium levels and functions as an antagonist to parathyroid hormone.
     a. glucagon   c. insulin
     b. calcitonin   d. growth hormone

22. The zona glomerulosa of the adrenal gland secretes:
     a. sex hormones   c. glucocorticoids
     b. aldosterone   d. glucagon

_____ 23. A lack of insulin results in a condition of:
     a. hypothyroidism   c. hypoglycemia
     b. hyperthyroidism   d. hyperglycemia

True or False

_____ 24. Hormones influence only the process of metabolism and have no functional control over other bodily processes such as growth and development.

_____ 25. Hormones are secreted directly into the circulatory system.

DISCUSSION QUESTIONS (Optional)

1. Using your text or reference books, name the disease that may result from:

   a. removal of the parathyroid glands ....

   b. hypersecretion of GH in an adult ....

   c. hyposecretion of insulin ....

   d. hypersecretion of corticosteroids ....

   e. hyposecretion of corticosteroids ....

   f. deficiency of thyroxine in an adult ....

   g. insufficient iodine in the diet ....

   h. hypothyroidism present from birth ....

2. Does the human pituitary have a distinct pars intermedia?

3. a. How does thyroglobulin differ from thyroxine functionally?

   b. If a thyroid gland were overactive, would you expect the follicular epithelium to retain its cuboidal shape? .... Explain.

Glossary

A band *p. 142* Anisotropic band; myosin and actin filaments in a myofibril.

Abduct Movement away from body midline or from body part.

Acapnia Condition represented by a reduction of carbon dioxide in the blood.

Achilles tendon *p. 134* The tendon posterior to the heel that connects the Gastrocnemius and Soleus muscles with the tuberosity of the calcaneus.

Acid *p. 334* A hydrogen ion (H^+) donor.

Acidophil (Eosinophil) *p. 201* A cell somewhat larger than a neutrophil with an irregularly shaped and partially constricted bilobed nucleus. Acidophils constitute 2%–4% of the normal leukocyte population.

Acidosis *p. 342* A condition that occurs when pH of arterial blood drops below 7.35.

Adduct Movement toward body midline or toward another body part.

Adenohypophysis *p. 382* *See* Pars distalis.

Adhesion Process by which parts come together on close approximation.

Adipose *p. 44* Fat tissue.

Adrenal glands *p. 314* Suprarenal glands; one located on superior border of each kidney.

Adrenalin *p. 236* Epinephrine.

Adrenocorticotrophic hormone (ACTH) *p. 382* A hormone produced by the adenohypophysis, having its effect on the adrenal cortex.

Adventitia *p. 165* The outermost fibrous coat of the digestive tract.

Afferent *p. 265* Movement or conveyance toward a central point. A component leading toward a central point (e.g., afferent arteriole).

Agglutination (coagulation) *p. 222* The clumping together of erythrocytes.

Albustix *p. 322* A urine test for the presence of albumin.

Aldosterone *p. 383* A mineralocorticoid produced by the adrenal cortex.

Alkalosis *p. 342* A condition that occurs when pH of arterial blood rises above 7.45.

Alpha cells *p. 383* Cells in the islets of Langerhans that produce the hormone glucagon.

Alveolus *p. 254* Saclike structure found in lung and bone.

Amenorrhea Condition of absence of or cessation of menses.

Amnion Thin extraembryonic membrane surrounding each fetus in utero.

Anabolism Process by which simple substances are converted to complex compounds by living systems.

Analgesia Absence of sensitivity to pain.

Anaphase *p. 28* Mitotic phase in which homologous chromosomes move apart to opposite poles.

Anaphylaxis Hypersusceptibility to foreign protein.

Anastomosis A joining, or union, between separate structures (e.g., the joining of blood vessels).

Androgen *p. 383* A class of steroid hormones that produces masculinization.

Anesthesia Condition represented by loss of sensation.

Anion *p. 36* A negative ion.

Anoxia Condition of tissues caused by oxygen reduction below normal physiological levels.

Antibody (agglutinin) *p. 222* A protein produced in response to an invasion by a foreign substance.

Antidiuretic hormone *p. 383* *See* Vasopressin.

Antigen (agglutinogens) *p. 222* Any foreign substance that, when introduced into the body, will cause the production of an antibody that reacts specifically with the foreign substance.

Aperture A body opening, or entrance.

Aphasia Condition represented by loss of and/or comprehension of speech or symbols.

Aponeurosis A sheath of connective tissue.

Apoplexy Condition represented by acute hemorrhage of the brain; this condition may also result from thrombosis and embolism.

Arbor vitae *p. 268* White matter of the cerebellum assuming a treelike shape.

Articulation *p. 101* Union of bones in the body; a joint.

Ascending colon *p. 158* Portion of large intestine between the cecum and transverse colon.

Ataxia Condition represented by defective muscular coordination.

ATP *p. 144* Adenosine triphosphate—a nucleotide contained in all cells; serves as energy reserve.

Atria *p. 215* The upper chambers of the heart.

Atrophy A condition represented by a decrease in size and function of cells, tissues, organs, and organ parts.

Barfoed's test *p. 181* Test for presence of monosaccharides.

Base *p. 334* A hydroxyl ion (OH^-) donor.

Basement membrane A supporting membrane underlying mucosal epithelium.

Basophil *p. 201* A granular leukocyte.

Benedict's test *p. 181* A test for presence of reducing sugars.

Beta cells *p. 383* Cells in the islets of Langerhans that secrete the hormone insulin.

Bicuspid *p. 155* A tooth posterior to the cuspid that has two points, or cusps; premolar.

Bicuspid valve *p. 218* Valve consisting of two flaps, or cusps, of tissue between the left atrium and left ventricle; *mitral* valve.

Bile *p. 160* An alkaline fluid secreted by the liver and concentrated and stored in the gallbladder that facilitates the emulsification of fats.

Biuret reaction *p. 179* A reaction specific for compounds containing two peptide bonds united by a nitrogen or carbon atom.

Blastula *p. 367* A hollow ball of cells formed from a cleaving egg; a stage in embryological development.

Body *p. 157* The central portion of an organ.

Bolus A rounded mass; food that is swallowed or is in the superior part of the digestive tract.

Bowman's capsule *p. 318* First collection unit of a nephron, shaped as a hemisphere with single-layered epithelial walls.

Brainstem *p. 269* Inferior section of the brain consisting of the midbrain, pons, and medulla oblongata.

Broad ligament One of the uterine ligaments.

Brunner's glands *p. 169* Mucous glands in the submucosa of the duodenum.

Buffer *p. 334* A fluid substance that tends to reduce changes in hydrogen ion concentration.

Buffer system *p. 334* A combination of two or more chemicals that minimizes the pH change of a solution when an acid or base is added to the solution.

Bursa Saclike pouch that reduces friction between bones and tendons.

Calcitonin *p. 383* Thyrocalcitonin; hormone of the thyroid gland.

Calorie Unit of heat; amount of heat required to elevate 1 g of water 1°C.

Calyx *p. 317* Cup-shaped structure for urine collection in the renal medulla.

Canaliculus A small passageway; found in Haversian system of bone.

Cancellous Spongy; "soft" bone tissue.

Canine *p. 155* A conical pointed tooth located between the lateral incisor and first premolar; cuspid.

Carbohydrate *p. 180* An aldehyde or ketone derivative or a polyhydric alcohol; contains carbon, hydrogen, and oxygen.

Carboxyl group *p. 340* COOH—a radical group.

Carcinoma A malignant growth of epithelial cells.

Cardiac *p. 44* Pertaining to the heart; a muscle type occurring in the heart; portion of the stomach nearest the esophagus.

Cartilage *p. 44* A firm, flexible supporting connective tissue.

Catecholamine *p. 383* A derivative of catechol (e.g., epinephrine and norepinephrine).

Cation *p. 36* A positive ion.

Caudal *p. 2* Inferior in human anatomy; tending toward tail.

Cecum *p. 155* A blind pouch; first section of the large intestine.

Cell *p. 26* The structural and functional unit of the body.

Cementum *p. 161* A bony layer surrounding the dentin of the root of a tooth.

Centriole *p. 26* Structures within a centrosome that divide and function in spindle formation during cell division.

Centrosome A cellular organelle located outside the nucleus that contains centrioles.

Cerebellum *p. 269* Second largest brain division; concerned mainly with coordination of movements.

Cerebrum *p. 269* Largest brain division of man; consists of two hemispheres connected by the corpus callosum.

Cervix *p. 351* Neck, or necklike structure; the inferior end of the uterus.

Chiasma X-shaped intersection, as of the fibers of the optic nerve on the ventral brain surface.

Chief cells *p. 166* Cells in the cardiac and fundic region of stomach mucosa that secrete pepsinogen; cells in parathyroid tissue.

Chorion An extraembryonic membrane surrounding the fetus in utero.

Chordae tendineae *p. 215* Slender fibers attached to papillary muscles and cusps of the bicuspid and tricuspid valves. The papillary muscles and chordae tendineae serve to prevent the cusps of the valve from being forced open during systole.

Choroid *p. 296* Middle layer or tunic of the eye.

Chromatin *p. 26* Strands of DNA and protein in the nucleus of the cell that condense into chromosomes during cell division.

Chromophil cells *p. 384* Readily staining cells with granular cytoplasm in the adenohypophysis.

Chyme Semifluid material resulting from gastric digestion.

Cleavage *p. 367* Initial series of mitotic divisions of a fertilized egg.

Clinistix *p. 322* Dipstick indicator for glucose used in urinalysis.

Clinitest *p. 322* A urine test for the presence of glucose.

Clitoris A small erectile body, the size of a pea, located at anterior union of the vulva in the female; analogous to the penis in the male.

Cochlea *p. 303* A spiral-shaped portion of the inner ear.

Cohnheim's areas *p. 142* Areas within a muscle fiber that contain myofibrils, seen in cross section; Cohnheim's fields.

Collecting tubules *p. 319* Thick-walled straight tubules extending from nephron into the renal medulla.

Colloid A state of matter usually consisting of a liquid medium and a solute. The solute particles in contrast to crystalloids will not permeate a dialyzing membrane.

Colon *p. 158* The large intestine; bowel.

Colostrum Substance secreted by mammary glands as a function of pregnancy and parturition.

Columnar *p. 44* Tall, having greater length than width; epithelial type.

Columns *p. 317* Region between pyramids in the renal medulla.

Coma An abnormal state of unconsciousness and reduced sensitivity.

Combistix *p. 322* Simple clinical method for indicating the presence of a number of substances not found in normal urine.

Common bile duct *p. 160* A duct formed by the cystic and hepatic ducts, leading to the duodenum.

Compensation *p. 342* The process by which the respiratory and renal systems normally adjust to pH changes.

Condyle *p. 72* Bone marking—a rounded, or knoblike, projection.

Contraction period *p. 145* The time in which a muscle actually contracts or shortens.

Corium *p. 57* Dermis of the skin.

Cornea *p. 297* Anterior transparent area of the sclera of the eye.

Coronal *p. 5* A transverse plane dividing the body or any of its parts into anterior and posterior portions.

Coronal suture *p. 103* An immovable, or synarthrodial, joint that joins the frontal bone with the parietal bones.

Corpus albicans *p. 358* A regressed corpus luteum.

Corpus cavernosum *p. 362* One of two lateral masses of erectile tissue found in the penis.

Corpus luteum *p. 358* A collapsed follicle that has expelled its ovum.

Corpus spongiosum *p. 362* A mass of erectile tissue in the penis through which the urethra passes (corpus cavernosum urethrae).

Cortex *p. 383* The outer portion of a structure.

Corticosterone *p. 383* A glucocorticoid; hormone of adrenal cortex.

Cortisol *p. 383* A glucocorticoid; hormone of adrenal cortex.

Costal Pertaining to the ribs.

Cranial *p. 2* Toward the head end of the body.

Crest *p. 72* Bone marking—an elevation or a ridge that serves as a point for muscle attachment.

Crown *p. 160* The exposed portion of a tooth.

Crystalloid A noncolloidal substance resembling a crystal that dissolves and forms a true solution; will disperse through a dialyzing membrane.

Cuboidal *p. 44* Cube shaped.

Cumulus oophorus *p. 358* A mound of cells immediately surrounding an ovum within a Graafian follicle.

Cuspid *p. 155* A conical pointed tooth situated between the lateral incisor and first premolar; canine.

Cutaneous *p. 296* Having to do with the skin.

Cyanosis Blueness of the skin; indicates hypoxia.

Cystic duct *p. 160* A duct leading to the common bile duct connecting the gallbladder with the hepatic duct.

Cytoplasm *p. 26* The portion of a cell outside the nucleus and within the plasma membrane.

Deamination Removal of an amino group ($-NH_2$) from an amino acid.

Dehydration Removal of water from the body or a tissue.

Dendrite *p. 264* Short process extending from the nerve cell body to a presynaptic neuron.

Dense, fibrous *p. 44* A type of connective tissue consisting mainly of parallel rows of bundles of fibers.

Dentin (Dentine) *p. 160* A collagenous and calcified substance comprising the principal mass of a tooth.

Dermis *p. 57* The layer of dense, vascular connective tissue subjacent to the epidermis.

Descending colon *p. 158* A portion of the large intestine lying in the vertical position on the left side of the abdomen, extending from the stomach to the level of the iliac crest.

Diabetes mellitus *p. 383* A chronic metabolic disease resulting from insufficient insulin production by the pancreas.

Diabetes insipidus A comparatively rare disease of pituitary origin that is characterized by the excretion of large volumes of urine.

Dialysis *p. 32* The separation of crystalloids from colloids utilizing a semipermeable membrane.

Diapedesis The outward movement of formed elements of blood through intact walls of vessels.

Diaphragm *p. 156* A skeletal muscle separating the thoracic cavity from the abdominal cavity; important in breathing.

Diarthroses *p. 104* One of two major groups of joints in the body; movable joints.

Diastole *p. 225* The noncontractile phase of a heart chamber.

Diencephalon Inferior mesial brain section consisting of the thalamus and hypothalamus; primarily of embryological significance.

Diffusion *p. 31* Movement of molecules from a highly concentrated area to a lower concentrated area until equilibrium is achieved.

Digit A finger or a toe.

Dipeptide *p. 179* Two amino acids linked together by a peptide bond.

Disaccharide *p. 180* A sugar formed by the union of two monosaccharides.

Distal *p. 2* Farthest from the trunk or point of origin of a part.

Distal convoluted tubule *p. 319* Last tubular portion of a nephron; continuation of the loop of Henle.

Diurnal Occurring daily.

Diverticulum A pouch, or pocket, off of a tube or passage.

Dorsal *p. 2* Pertaining to the back.

Double pith *p. 145* Destruction of the brain and spinal cord for experimental purposes.

Ductus deferens *p. 353* Vas deferens; sperm duct.

Duodenum *p. 155* The first portion of the small intestine, about 10 inches long.

Ectopic pregnancy A pregnancy that occurs outside the uterus, generally in a Fallopian tube.

Edema Presence of extraordinary amounts of fluid in extracellular tissue spaces.

Effector A nerve that activates muscular contraction or glandular secretion.

Efferent *p. 265* Movement away from a central point. A component leading away from a central structure (e.g., efferent neuron).

Electrocardiogram *p. 238* A tracing of the electric current that initiates cardiac muscle contraction.

Electrolyte A solution that is able to conduct electricity due to the presence of ions.

Embolus *p. 221* A plug of undissolved matter—solid, liquid, or gaseous—that circulates in the bloodstream; frequently it may cause an obstruction within a blood vessel that is termed an *embolism.*

Embryo *p. 367* An animal in early stages of development; a human being from implantation through the eighth week postconception.

Emulsified fat *p. 177* Fat that has been converted into small fat droplets.

Enamel *p. 160* A hard covering of the crown of a tooth composed primarily of calcium salts.

Endocrine *p. 382* Glandular organs that secrete hormones directly into blood.

Endometrium *p. 362* The inner lining of the uterus.

Endoplasmic reticulum *p. 26* A protein and lipid membraneous network in the cytoplasm through which biochemical components of the cell move.

Endothelium *p. 197* Simple squamous epithelium that surrounds the lumen of blood vessels; innermost lining of blood vessels.

Enzyme *p. 184* An organic catalyst.

Eosinophil *p. 201* *See* Acidophil.

Epicondyle Bone marking—a small rounded process above a condyle.

Epidermis *p. 57* The outermost layer of skin; consists of five layers.

Epididymis *p. 353* A long coiled duct on the anterior lateral surface of each testis.

Epigastric *p. 5* An abdominal region lying superior to the stomach.

Epimysium *p. 142* Connective tissue surrounding a muscle.

Epinephrine *p. 383* A catecholamine secreted by the adrenal medulla; commercially known as Adrenalin.

Epiphysis *p. 67* Part of bone connected in embryological stage of life by cartilage to long bone; later becomes end portion of a long bone.

Epithelium *p. 44* A sheet of cells that covers an external surface or lines an internal surface, and functions in protection, absorption, secretion, and filtration.

Erythrocytes *pp. 33, 200* Red blood cells; have the shape of biconcave disks and have no nuclei; normal human males have 5.0 million per cubic millimeter of blood; females have approximately 4.0–5.0 million erythrocytes per cubic millimeter of blood.

Esophagus *p. 155* A muscular tube, about 10 inches long, extending from the pharynx to the stomach.

Estrogen *p. 383* A class of female hormones.

Evagination An out-pocketing of a part.

Exocrine *p. 383* A glandular organ that secretes outwardly through ducts.

External respiration Exchange of gases between the blood and external environment; occurs in the alveoli of the lungs.

Falciform ligament *p. 160* A ligament dividing the liver into right and left lobes, which are then further subdivided.

Fallopian tubes *p. 350* Paired tubes that serve as ducts from the ovaries to the uterus, through which ova pass.

Fascia *p. 114* Dense, fibrous connective tissue covering muscles.

Fascicle, fasciculi (pl.) *p. 142* A subdivision of an entire muscle; a bundle of muscle fibers.

Fat *p. 177* *See* Lipid.

Fatty acid *p. 177* A saturated or unsaturated organic (monocarboxylic) acid.

Fertilization Union of sperm and egg; formation of a zygote.

Fertilization membrane *p. 366* A membrane found elevated from the egg surface at fertilization.

Fetus An unborn vertebrate; an unborn human being from the third month after conception until birth.

Fibrin *p. 221* An insoluble protein forming a mass of tangled strands in which red and white blood cells become enmeshed during coagulation.

Fibrous *p. 44* *See* Dense, fibrous.

First polar body *p. 357* A structure containing chromosomes formed by the meiotic division of a primary oocyte.

Fissure *p. 72* A bone marking—narrow slitlike opening.

Follicle *p. 358* A small excretory sac or gland.

Follicle stimulating hormone (FSH) *p. 382* A gonadotropic hormone whose target organ is the ovary or testis; produced by the anterior pituitary gland.

Follicular fluid *p. 358* Liquid within a Graafian follicle surrounding an ovum.

Fontanels *p. 103* Fibrous areas existing before the four cranial bones completely form and fuse together.

Foramen *p. 72* An opening through a bone for transmission of nerves and/or blood vessels.

Foramen ovale *p. 218* An orifice that develops in the interatrial septum during fetal development. Failure to close after birth is known as *atrial septal defect.*

Fossa *p. 72* A bone marking—a cavity or hollow area in a bone.

Frontal *p. 5* A lengthwise plane dividing the body into anterior and posterior portions.

Fundus *p. 157* Enlarged portion of the stomach to the left of and superior to the cardioesophageal junction.

Gallbladder *p. 155* An accessory organ of the digestive tract that concentrates and stores bile, later ejecting it into the duodenum.

Ganglion A collection of nerve cells.

Gastric pits *p. 166* Narrow depressions containing glands that extend through the full thickness of the gastric (stomach) mucosa.

Gastrula *p. 367* An early developmental stage following the blastula stage, during which the germ layers are formed.

Glass electrode *p. 337* An internal sealed tube with a metallic tip surrounded by an external tube containing a standard solution and a pH-sensitive glass bulb; used to determine pH.

Glomerulus *p. 318* A cluster of blood vessels in close proximity to Bowman's capsule.

Glucagon *p. 383* A hormone produced by alpha cells of the islets of Langerhans; involved in carbohydrate metabolism.

Glucocorticoids *p. 383* A class of adrenal cortical steroid hormones that influence the metabolism of glucose, protein, and fat.

Glyceride An ester (salt) of glycerol with fatty acids.

Glycerol *p. 183* An alcohol, $C_3H_8O_3$, found in neutral fats.

Golgi apparatus An organelle comprised of membranes stacked upon each other, functioning in glycoprotein synthesis and the movement of substances to the external cell environment.

Gonadotropic hormone *p. 362* A type of hormone secreted by the adenohypophysis (pars distalis) that stimulates the gonads to secrete hormones.

Gonads *p. 350* Ovaries and testes; sex organs.

Graafian follicle *p. 358* A hollow fluid-containing sac surrounding a primary or secondary oocyte contained within an ovary.

Greater omentum *p. 156* A large fold of peritoneum covering the small and large intestines.

Growth hormone (GH) *p. 382* Tropic hormone secreted by adenohypophysis, which influences metabolism of proteins, fats, and carbohydrates.

Haustra *p. 158* A sacculation of the wall of the large intestine.

Head *p. 72* Rounded surface of bone connected to the shaft by the neck.

Hemastix *p. 322* A urine test for the presence of blood.

Hemopoietic Pertaining to blood cell formation.

Hemorrhage The escape of a large amount of blood.

Hemostasis *p. 203* An arrest, or stoppage, of escaping blood.

Hepatic duct *p. 160* A duct from the liver that joins the cystic duct to form the common bile duct.

Hepatic flexure *p. 158* The angle formed by the ascending and transverse colons.

Hilum *p. 317* Mesial indentation of kidney and lungs.

Homeostasis A tendency toward maintenance of stability of the internal environment.

Horizontal *p. 5* A plane dividing the body or its parts into superior and inferior portions.

Hormone *p. 382* A chemical messenger secreted by an endocrine gland.

Humerus *p. 90* Longest and largest bone of the upper arm.

Hyaline *p. 44* The most common type of cartilage; glassy in appearance.

Hydrolysis *p. 183* (*see* Formation of triglyceride) The splitting of a compound by the addition of water, in which the hydroxyl group and hydrogen atom are incorporated in the different fragmented compounds.

Hyoid *p. 81* A bone that supports the base of the tongue.

Hyperglycemia *p. 383* Elevated blood glucose level.

Hypersecretion *p. 382* Secretion of greater than normal amounts of a substance.

Hypertonic solution *p. 33* A solution that, in relation to another solution, contains a greater amount of dissolved material (solute).

Hypertrophy An enlargement, or overgrowth, of an organ or organ part.

Hypocalcemia The condition of having a blood calcium level that is too low.

Hypochondrium *p. 5* An abdominal region lying on either side of the epigastric region and above the lumbar regions.

Hypogastric *p. 5* An abdominal region lying inferior to the umbilical region.

Hypophysis *p. 269* Pituitary gland.

Hyposecretion *p. 382* Secretion of less than normal amounts of a substance.

Hypothalamus *p. 269* A portion of the brain lying beneath the thalamus; contains autonomic regulatory centers.

Hypotonic solution *p. 34* A solution that, in relation to another solution, contains less dissolved material (solute).

Hypoxia A condition represented by a reduced oxygen content.

I band *p. 142* Isotropic band; actin filaments in a myofibril.

Ictotest *p. 322* Urine test for presence of bilirubin.

Ileum *p. 158* The third (last) and longest portion of the small intestine.

Iliac *p. 5* Abdominal regions lying on either side of the hypogastric region; relating to the ilium.

Immovable synarthrodial joint *p. 103* A joint type located in the skull between cranial and facial bones.

Incisor *p. 155* One of four front teeth in each jaw adapted for cutting.

Inferior *p. 2* Lower; away from the head.

Inferior nasal turbinates Paired facial bones in the nasal cavity.

Inflammation A physiological response of cells to injury or destruction by physical, chemical, or bacterial agents; symptoms include swelling, redness, heat, and pain.

Infundibular stalk (infundibulum) *p. 269* A portion of the neurohypophysis that extends superiorly from the pars nervosa, connecting it with the hypothalamus.

Inguinal canal *p. 353* Paired canals in the groin through which spermatic cords pass prior to reaching the scrotum.

Insulin *p. 383* A hormone secreted by the beta cells of the islets of Langerhans; regulates cellular uptake of glucose and blood glucose level.

Intercalated disc *p. 143* Prominent markings at which point the ends of cardiac muscle fibers are joined together.

Internal respiration Metabolic reactions that involve oxygen consumption and carbon dioxide release.

Interphase *p. 28* The stage between successive mitotic divisions during which DNA replication takes place.

Interstitial cells of Leydig *p. 360* A group of cells between the seminiferous tubules that secrete testosterone.

Interstitial cell stimulating hormone (ICSH) *p. 382* A gonadotropic hormone whose target organ is the testis; LH in the female.

Intestinal glands *p. 168* Glands in the mucosa of the small and large intestines that secrete digestive enzymes and/or mucus; also called intestinal crypts of Lieberkühn.

Ion *p. 36* An atom that has gained or lost one or more electrons in its outer shell.

Iris *p. 297* Internal pigmented muscle of the eye that regulates the size of the pupil.

Irritability The ability of a tissue or an organism to respond to a stimulus from the environment, such as heat, cold, or pain.

Islets of Langerhans *p. 383* The endocrine portions of the pancreas.

Isotonic *p. 34* A solution that contains an equivalent amount of solute in relation to another solution.

Jejunum *p. 169* The second portion of the small intestine; between the duodenum and ileum.

Joint *p. 103* A union of bones in the body; an articulation.

Keratin *p. 57* A scleroprotein pigment that is a naturally occurring component of the skin, nails, and hair.

Ketone A chemical compound containing a carbonyl group; ketone bodies are compounds synthesized by the liver in the stepwise process of oxidation of fats.

Ketonuria The incomplete oxidation of fatty acids, giving rise to the accumulation of ketone bodies in the urine.

Ketostix *p. 322* A urine test for the presence of ketones, especially acetone.

Kidney *p. 314* A paired organ that functions in the maintenance of normal blood constituents.

Kinesthetic (kinesthesia) The special sense of awareness of one's bodily positions and movements.

Kymograph *p. 280* Physiological recording apparatus consisting of a rotating drum and drive mechanism.

Lactase *p. 176* An enzyme that hydrolyzes lactose to glucose and galactose.

Lactose *p. 176* A disaccharide comprised of glucose and galactose.

Lacuna An anatomical term referring to a small depression or hollow cavity; a depression containing osteocytes.

Lambdoidal suture *p. 103* A joint that unites the parietal bones with the occipital bone.

Lamella *p. 165* A thin bony plate, such as the Haversian lamella.

Lamina propria A connective tissue layer in the mucosa underlying the mucosal epithelium and basement membrane.

Laryngopharynx *p. 154* A portion of the pharynx (throat) posterior to the larynx.

Latent period *p. 145* The shortest period of muscle contraction, occurs between the stimulus and contraction period.

Lateral *p. 2* Toward the side.

Lecithin A chemical compound of natural occurrence found in most animal tissues, especially egg yolk, semen, and nervous tissue; consists of glycerophosphoric acid and esters of stearic, oleic, or other fatty acids, combined with choline.

Lesion An inclusive term describing any damage to a particular group of tissues, such as ulcers, cataracts, eczema, or tuberculosis.

Leukemia *p. 204* A malignant disease of the spleen, bone marrow, and lymphoid tissue; characterized by wild proliferation of immature leukocytes with a subsequent reduction in the number of erythrocytes and blood platelets, which results in anemia and increased susceptibility to hemorrhage and infection.

Leukocytes *p. 200* White blood cells containing various shaped nuclei.

Ligaments Dense fibrous bands of connective tissue that attach bones to bones.

Liminal stimulus *p. 145* Threshold stimulus; the amount of stimulus needed for a muscle to contract.

Line *p. 72* A bone marking—a slight ridge.

Lipase *p. 177* An enzyme that hydrolyzes emulsified fats to fatty acids and glycerol.

Lipid Fat; a class of compounds including phospholipids, steroids, free fatty acids, and triglycerides.

Liver *p. 155* The largest gland in the body, located

immediately inferior to the diaphragm, occupying most of the right hypochondriac region and part of the epigastric region.

Loop of Henle *p. 319* A portion of the nephron between the proximal and distal convoluted tubules that extends from the renal cortex to the medulla and then proceeds superiorly toward the renal cortex.

Loose, areolar *p. 44* Type of connective tissue found in and around internal organs.

Lumbar vertebrae Five vertebrae inferior to the thoracic vertebrae in the spinal column.

Luteinizing hormone (LH) *p. 382* A gonadotropic hormone whose target gland is the ovary in the female; called ICSH in the male.

Luteotropic hormone (LTH) *p. 382* A tropic hormone secreted by the pars distalis that influences ovaries and mammary glands.

Lymphocyte *p. 201* A type of leukocyte that has a large, round dark-staining nucleus surrounded by a thin rim of clear blue cytoplasm. Lymphocytes constitute 20%–25% of the normal leukocyte population.

Lysosome *p. 26* A membrane-bound organelle containing acid hydrolases; found in the cytoplasm.

Maltase *p. 176* An enzyme that hydrolyzes maltose to two glucose molecules.

Maltose *p. 176* A disaccharide comprised of two glucose units.

Matrix The basic background substance in which cells or tissues grow and develop, such as bone matrix.

Meatus *p. 72* A canal running within a bone.

Medial *p. 2* Toward the midline of the body.

Median sagittal *p. 5* A lengthwise plane from front to back dividing the body or any of its parts into equal right and left halves.

Medulla *p. 383* The inner portion of a structure.

Medulla oblongata *p. 269* The most inferior section of the brain containing control centers for vital functions.

Medullary rays *p. 317* Structure within renal pyramids leading toward papillae and calyces; involved in urine formation.

Megakaryocyte An enormous cell in bone marrow, having a lobulated nucleus, that gives rise to blood platelets.

Meiosis *p. 357* A process of nuclear division in which daughter cells are produced that have half the number of chromosomes as the parent cell; this process involves two cell divisions, the second yielding the haploid number of chromosomes; specific to gametes.

Melanin *p. 57* A skin pigment formed in melanocytes in the stratum germinativum of the epidermis.

Melanocyte stimulating hormone (MSH) *p. 382* A hormone secreted by the pituitary gland that stimulates melanin-producing cells in the epidermis.

Meninges *p. 269* Covering of the brain and spinal cord.

Menopause The gradual waning of the menstrual cycle in the human female; during this period, the ovaries gradually cease to function resulting in the cessation of menstruation.

Menstruation *p. 362* The periodic sloughing off of the endometrial lining, accompanied by bleeding.

Mesentery *p. 160* A fold of parietal peritoneum that anchors abdominal organs to the dorsal body wall.

Mesial *p. 2* Toward the midline of the body.

Mesovarium A mesentery connecting the ovaries to the posterior aspect of the broad ligament.

Metabolic acidosis *p. 342* The decrease in pH of arterial blood as a result of the accumulation of nonrespiratory acids or the loss of bases.

Metabolic alkalosis *p. 343* The increase in pH of arterial blood as a result of excessive loss of H^+ ions or gain in bases.

Metabolism The total of all the chemical processes and reactions taking place in a living organism. Metabolism is composed of two phases—*anabolism*, the constructive phase, and *catabolism*, the destructive phase.

Metaphase *p. 28* A stage in mitosis during which the chromosomes are aligned across the equator of the cell.

Midbrain *p. 269* That portion of the brain containing the corpora quadrigemina and cerebral peduncles; also called the mesencephelon.

Millon reaction *p. 179* A reaction specific for the amino acid tyrosine.

Mineralocorticoids *p. 383* A class of steroid hormones secreted by the adrenal cortex that increases sodium reabsorption by the distal convoluted tubules in the kidney.

Mitochondrion *p. 26* A cytoplasmic membrane-bound organelle; contains oxidative enzymes.

Mitosis *p. 26* A process of cell division in which daughter cells are produced that have the same number of chromosomes as the parent cell.

Molar *p. 155* A tooth posterior to the bicuspid that is adapted for grinding; tricuspid.

Molisch reaction *p. 181* A general test for carbohydrates.

Monocyte *p. 201* A granular leukocyte.

Monosaccharide *p. 180* A simple sugar.

Mucosa *p. 164* The innermost lining of the digestive tract.

Mucus The secretion of mucous cells that contains water, mucin, and various inorganic salts.

Muscle fatigue *p. 147* The result of an accumulation of waste products such as lactic acid in a muscle.

Muscle fiber *p. 142* A muscle cell.

Muscle twitch *p. 145* A single muscle contraction.

Muscularis externa *p. 165* A layer of circular and longitudinal muscle beneath the submucosal layer of the digestive tract.

Muscularis mucosae *p. 165* A smooth muscle layer between the mucosa and submucosa.

Myeloid Relating to or derived from bone marrow.

Myofibril *p. 142* A group of myofilaments.

Myofilament *p. 142* A molecular threadlike contractile element within a myofibril; a contractile component of a myofibril.

Nasal Paired facial bones; from the bridge of the nose.

Nasopharynx *p. 154* Portion of the throat located posterior and inferior to the nasal cavity.

Neck *p. 161* The portion of a tooth at the gum line where the crown and root meet; constricted portion below the head of a bone.

Necrosis *p. 57* The death of cells or tissues as a result of disease or injury.

Nephron *p. 319* The structural and functional unit of the kidney.

Neuroglia Supporting cells of the nervous system.

Neurohypophysis *p. 383* The posterior lobe of the pituitary gland.

Neuron *p. 264* Nerve cell responsible for generation, conduction, and transmission of a nerve impulse.

Neurula *p. 367* An early embryonic form marking the development of the neural tube and the appearance of the nervous system; embryological stage following the gastrula.

Neutrophils *p. 201* White blood cells that usually contain three to five irregular oval-shaped lobes connected by thin strands of nuclear material. These cells constitute 65%–70% of the normal leukocyte population.

Ninhydrin reaction *p. 179* A test for alpha amino acids.

Nitrazine paper *p. 322* Chemically treated paper used as an indicator for pH.

Norepinephrine *p. 383* A hormone secreted by the adrenal medulla; also a sympathetic chemotransmitter.

Nuclear membrane A double protein-lipid membrane surrounding the nucleus.

Nucleolus *p. 26* A small dense structure in a nucleus consisting of RNA and protein.

Nucleus *p. 26* A structure containing DNA within a cell; control center of the cell.

Nystagmus *p. 286* Continuous involuntary movement of the eyeball.

Olfactory *p. 269* Relating to the sense of smell. The olfactory nerve (the first cranial nerve) is concerned with the sense of smell.

Oogenesis *p. 357* The process of egg formation by meiosis.

Oogonium *p. 357* An immature diploid ovum.

Ootid *p. 357* A haploid cell resulting from meiotic division.

Ophthalmic Relating to the eye.

Orifice An opening, entrance, or outlet of a body cavity or structure.

Oropharynx *p. 155* The portion of the throat located posterior to the mouth.

Osmosis *p. 32* The diffusion of water through a semipermeable membrane from a region of greater concentration of solvent to one of lesser concentration.

Osseous *p. 44* A supporting connective tissue; bone tissue.

Ossification Pertaining to the process of formation or conversion into bone or osseous tissue.

Ovary *p. 355* A paired primary reproductive organ in the female that produces ova and female sex hormones.

Oviduct *p. 350* Fallopian tube.

Ovum *p. 357* An egg cell; a female gamete.

Oxidation The loss of electrons by atoms; the opposite reaction to reduction.

Oxyphilic cells *p. 386* Cells found in parathyroid tissue that are capable of producing parathyroid hormone.

Oxytocin *p. 383* A hormone stored by the neurohypophysis that stimulates uterine contractions and milk ejection.

Palatine *p. 80* A paired facial bone; forms posterior one-third of roof of mouth.

Pancreas *p. 155* A tubuloacinar gland lying behind and below the stomach in the curve of the duodenum, having the exocrine function of secreting digestive enzymes and the endocrine function of secreting the hormones insulin and glucagon.

Papillae *p. 317* Indentations in the renal medulla at the base of medullary rays.

Papillary muscles *p. 218* Cylindrical muscular bundles attached to the muscular wall of the ventricles.

Paralysis The loss or deterioration of movement of a limb or limbs of the body as a result of any of a wide variety of physical or emotional disorders.

Paranasal sinuses *p. 81* Air-filled or mucus-filled cavities surrounding the nose.

Parathyroid gland *p. 383* Several (usually four) endocrine glands located on the posterior surface of the lateral lobes of the thyroid gland.

Parathyroid hormone (parathormone) *p. 383* A hormone secreted by the parathyroid glands that increases serum calcium levels.

Parietal *p. 74* A paired cranial bone that forms a portion of the sides and roof of the cranium.

Parotid *p. 155* A paired salivary gland located anterior and inferior to the ear.

Pars distalis *p. 382* Anterior lobe of the pituitary gland.

Pars intermedia *p. 384* The intermediate lobe of the pituitary gland; virtually nonexistant in man.

Pars nervosa *p. 383* The neurohypophysis.

Parturition The act of expulsion of a fetus from the maternal uterus.

Peduncle A term designating a large group of nerve fibers traveling between different regions within the central nervous system.

Penis *p. 353* The male copulatory organ and urinary outlet.

Pepsin *p. 176* A protease in the stomach that hydrolyzes proteins to proteoses and peptones.

Pepsinogen *p. 166* The precursor of pepsin.

Peptidase *p. 176* An enzyme that hydrolyzes peptides to amino acids.

Peptone *p. 176* A partially digested protein.

Pericardium *p. 214* A protective sac of fibrous and serous tissue that encloses the heart.

Perimysium *p. 142* A connective tissue surrounding a fascicle.

Peyer's patches *p. 173* An aggregation of lymph nodules in the mucosa of the ileum.

pH *p. 334* A measure of the acidity or alkalinity of a solution.

pH scale *p. 334* A convenient notation for indicating the pH of a solution on a continuum from acidic to basic.

Phagocyte *p. 202* A term applied to a cell that destroys microorganisms or foreign substances.

Pharynx *p. 155* The region posterior to the nose, mouth, and larynx; the throat.

Physiological apparatus A recording apparatus for muscle contraction, electrocardiograms, heart cycles, breathing rhythms, and other physiological functions.

Pithing *p. 145* The destruction of the brain (single pith) or brain and spinal cord (double pith) for experimental purposes.

Pitressin (vasopressin) Antidiuretic hormone; enhances water reabsorption from the distal convoluted and collecting tubules of the kidney.

Pituitary gland *p. 269* An endocrine gland located in the sella turcica of the sphenoid bone.

Placenta *p. 355* A structure attached to the uterine wall that enables the exchange of oxygen and other substances to take place between maternal and fetal blood.

Plasma *p. 200* The straw-colored fluid portion of the blood that constitutes 55% of the blood volume; transports inorganic and organic substances.

Plasma membrane *p. 26* The protein and lipid membrane surrounding a cell.

Platelet (thrombocyte) *p. 200* Small round-to-oval disks averaging 200-400 thousand per cubic millimeter of blood; platelets are associated with the clotting process.

Plexus A term applied to a network or group of veins or nerves.

Plicae circulares *p. 169* Elevations of submucosa in the small intestine; the tallest plicae circulares are in the jejunum.

Polymorphonuclear *p. 202* Refers to a cell that possesses a deeply lobed nucleus that gives the appearance of being multilobed (e.g., a polymorphonuclear leukocyte).

Polypeptide A chain of amino acids connected by peptide bonds.

Pons *p. 269* Inferior brain section used as a bridge from the medulla oblongata and spinal cord to more superior brain divisions.

Premolar *p. 155* A tooth posterior to the canine; bicuspid.

Primary follicle *p. 358* A layer of cells surrounding an oogonium.

Primary oocyte *p. 357* The mature, diploid potential ovum.

Primary sprematocytes *p. 360* The mature, diploid potential sperm.

Principal cells *p. 386* Predominant cells of the parathyroid glands.

Process *p. 71* Bone marking—a general projection from a bone.

Prognosis A term applied to the probable outcome of a disease or disorder.

Prolapse A term applied to the inferior displacement of a part of the viscera.

Prophase *p. 28* A stage in mitosis during which chromatin shortens and thickens, spindle fibers appear, and the nucleolus and nuclear membrane disappear.

Prostate gland *p. 353* A gland inferior to the bladder in the male that secretes an alkaline fluid that forms the greatest portion of seminal fluid.

Protease *p. 176* An enzyme that hydrolyzes proteins to proteoses, peptones, peptides, and amino acids.

Proteose *p. 176* A partially digested protein.

Protuberance *p. 71* A bone marking—a projection or eminence.

Proximal *p. 2* Nearest the trunk or point of origin of a part.

Proximal convoluted tubule *p. 318* The portion of a nephron between the Bowman's capsule and loop of Henle.

Pseudostratified *p. 44* An epithelium with a layer of staggered cells that give the appearance of several layers.

Ptosis *p. 285* Drooping of an organ or part of an organ, especially of a viscus or an upper eyelid.

Pulp *p. 160* A soft vascularized core of tissue medial to the dentin portion of a tooth.

Pulse *p. 225* The alternating expansion and recoil of an artery.

Purkinje cells *p. 268* Large neurons at the border of the cerebellum and medulla.

Pus *p. 203* An opaque and viscid fluid characteristically produced in bacterial infections that contains dead cell debris and tissue breakdown products.

Pyloric spincter *p. 157* An annular muscle that prevents the backflow of food from the small intestine to the stomach.

Pylorus (pyloric) *p. 157* The inferior portion of the stomach adjacent to the duodenum.

Pyramids *p. 317* The collections of medullary rays in the renal medulla of the kidney.

Ramus A bone marking—an armlike branch extending from the body of a bone.

Receptor *p. 296* The distal ends of dendrites of sense (sensory) neurons.

Rectum *p. 158* The last 7 inches of the large intestine.

Reference electrode *p. 337* A metallic internal element that is housed within an outer tube containing an electrolyte solution; completes the electrical circuit while determining pH.

Reflex *p. 278* An involuntary response to a stimulus.

Relaxation period *p. 145* The period of time following the muscle contraction period in which the muscle returns to its former length.

Renal capsule *p. 317* Fibrous connective-tissue covering of the kidney.

Renal corpuscle *p. 318* A glomerulus and Bowman's capsule.

Renal cortex *p. 317* Periphery of the kidney containing nephrons and glomeruli.

Renal medulla *p. 317* The portion of the kidney subjacent to the cortex.

Renal pelvis *p. 317* Upper expanded end of the ureter.

Respiratory acidosis *p. 342* The decrease in pH of arterial blood due to an increase of dissolved CO_2.

Respiratory alkalosis *p. 343* The increase in pH of arterial blood due to a decrease of dissolved CO_2.

Retina *p. 296* Innermost layer of the eye; contains rods and cones.

Rh factor *p. 222* A red-blood-cell antigen.

Ribosome *p. 41* The site of protein synthesis within a cell.

Root *p. 160* The portion of a tooth within the gum.

Rouleaux formation *p. 200* The occurrence of a group of erythrocytes resembling a stack of coins within capillaries.

Rugae *p. 167* Longitudinal folds of the mucosa and submucosa in the stomach.

Sagittal *p. 3* A lengthwise plane from front to back, dividing the body or any of its parts into right and left sides.

Sagittal suture *p. 103* A suture formed by the articulation between the parietal bones.

Sarcomere *p. 142* The unit of histological structure and physiological action of a muscle fiber; extends from Z line to Z line.

Sciatic nerve A spinal nerve from the lumbosacral plexus; found in the posterior region of the leg.

Sclera *p. 296* The outermost layer of the eyeball.

Scrotum *p. 351* A skin-covered pouch in the male containing testes, epididymides, and the lower portion of spermatic cords; located in the perineal region.

Scrotum processus vaginalis Scrotum; protective muscular sac housing and providing thermal regulation for the testes.

Secondary oocyte *p. 357* A cell resulting from the meiotic division of a primary oocyte.

Secondary spermatocytes *p. 360* Cells formed by the meiotic division of primary spermatocytes.

Second polar body *p. 357* A structure containing a haploid number of chromosomes formed by the meiotic division of a secondary oocyte into a large ootid.

Seliwanow's test *p. 182* A test that differentiates between fructose and glucose.

Semilunar valves *p. 197* Valves that occur in the aorta and pulmonary artery; they prevent the backflow of blood from the aorta and pulmonary artery into the left and right ventricles, respectively.

Seminiferous tubules *p. 360* Long coiled tubules within the testes in which sperm production and maturation take place.

Septum *p. 214* The thickened portion of the myocardium that separates the two atria and ventricles; the muscular wall dividing the atria is the interatrial septum, and the wall dividing the ventricles is the ventricular septum; cartilaginous tissue separating the nostrils.

Serosa *p. 165* The outer layer of an organ that is surfaced with a reflection of peritoneum.

Sigmoid colon Portion of the large intestine that extends from the level of the iliac crest to the rectum in the human being.

Simple *p. 44* One cell layer thick.

Single pith *p. 145* Destruction of the brain for experimental purposes.

Sinus *p. 72* A bone marking—an irregularly shaped space often filled with air and lined with mucosal tissue.

Skeletal muscle *p. 44* Pertaining to the skeleton; striated voluntary muscle.

Slightly movable synarthrodial joint (amphiarthrodial joint) *p. 104* A joint located between the ribs and sternum, between the vertebrae, and between certain bones such as the distal end of the tibia, fibula, and talus.

Smooth muscle *p. 44* An involuntary muscle type occurring primarily in the walls of viscera and blood vessels.

Solute A substance dissolved in a liquid.

Solution A liquid consisting of a solute and a solvent.

Solvent A liquid that dissolves another substance.

Spasm An involuntary, abnormal muscular contraction.

Sperm *p. 360* A male gamete.

Spermatids *p. 360* Haploid cells formed by the division of secondary spermatocytes; when mature, they form sperm cells.

Spermatogenesis *p. 360* The process of sperm formation.

Spermatogonia *p. 360* Immature undifferentiated germ cells in the testis.

Spermatozoa *p. 360* Sperm.

Sphincter An annular muscle surrounding and capable of closing a body opening.

Sphygmomanometer *p. 227* An instrument used for the indirect measurement of arterial blood pressure.

Spine A bone marking—a sharp projection that serves as a point for muscle attachment.

Spleen *p. 160* An organ located in the left hypochondrium, the cells of which function in phagocytosis, hemopoiesis, and the storage of blood.

Splenic flexure *p. 158* The angle formed by the transverse and descending colon.

Squamous *p. 44* Flat; scalelike.

Squamous suture *p. 103* An immovable joint formed by the union of the sphenoid, temporal, frontal, and parietal bones.

Staircase phenomenon *p. 146* The appearance of a

recording of muscle contraction in which single stimuli of constant intensity are applied to a muscle, resulting in each twitch being slightly greater than the preceding one; Treppe or summation contraction.

Stimulus *p. 145* An agent that arouses or incites activity.

Stomach *p. 155* An elongated pouchlike structure that, in the human being, communicates superiorly with the esophagus and inferiorly with the duodenum.

Stratified *p. 44* Several cell layers thick.

Stratum corneum *p. 57* The most superficial layer of epidermis; it is keratinized.

Stratum germinativum *p. 57* The deepest layer of epidermis; contains melanocytes; cells undergo mitosis in this layer.

Stratum granulosum *p. 57* A thin epidermal layer beneath the stratum lucidum.

Stratum lucidum *p. 57* A clear, transluenct epidermal layer beneath the stratum corneum.

Stratum spinosum *p. 57* An epidermal layer beneath the stratum granulosum; contains "prickly" cells.

Subcutaneous Under the skin.

Subliminal *p. 145* A stimulus of less than threshold strength; subthreshold.

Sublingual *p. 155* Salivary glands located in the anterior midportion of the mouth under the tongue.

Submandibular *p. 155* Under the mandible; submaxillary gland.

Submaxillary *p. 155* A pair of salivary glands located in the floor of the mouth inferior to the mandible.

Submucosa *p. 165* The layer immediately underlying the mucosa in the digestive tract.

Subthreshold (subliminal) *p. 145* A stimulus of less than threshold strength.

Sucrase *p. 176* An enzyme that hydrolyzes sucrose to glucose and fructose.

Sucrose *p. 182* A disaccharide composed of glucose and fructose.

Superior *p. 2* Higher; toward the head end of the body.

Suprarenal gland *p. 383* Adrenal gland.

Sutures *p. 103* Joints existing between mature, ossified bones; immovable synarthrodial joints.

Synarthroidal joints *p. 103* Joints lacking a joint cavity, with the bones being tightly united by cartilage or fibrous tissue.

Synarthroses *p. 103* One of two major groups of joints in the body; immovable joints; joints lacking a joint cavity.

Syncytium The fusion of cells that were originally separate.

Synovial cavity *p. 104* A structure formed by a fibrous capsule and lined with membranes that secrete a lubricating fluid enabling smooth movement of bones involved in the joint.

Synovial fluid *p. 104* A lubricating fluid around a joint.

Systole *p. 225* The contraction phase of the cardiac cycle.

Telophase *p. 28* A stage in mitosis during which spindle fibers disappear, cytokinesis occurs, and the nuclear membrane reappears.

Tendon A dense, fibrous connective-tissue band connecting bones to muscles.

Testis *p. 355* A paired primary reproductive organ in the male that produces sperm and testosterone.

Testosterone *p. 360* The primary male sex hormone.

Tetanus *p. 146* A condition in which a muscle is in a state of continual contraction.

Tetany *p. 383* A condition resulting from hypocalcemia.

Thoracic vertebrae *p. 85* Twelve vertebrae inferior to the cervical vertebrae in the vertebral column.

Thorax (thoracic) *p. 156* Chest region.

Threshold stimulus *p. 145* The amount of stimulus necessary to cause a response in a nerve, a muscle, and certain sensory receptors.

Thrombus *p. 221* A blood clot formed within a vessel.

Thyrocalcitonin *p. 383* A hormone secreted by the thyroid gland that lowers serum calcium levels.

Thyroglobulin *p. 383* A glycoprotein that is the storage form of thyroxin.

Thyroid gland *p. 383* An endocrine gland located in the neck, anterior to the trachea, inferior to the larynx.

Thyroid stimulating hormine (TSH) *p. 382* A tropic hormone produced by the adenohypophysis whose target gland is the thyroid.

Thyroxine *p. 383* The primary thyroid hormone.

Tissue *p. 44* A group of cells performing a common function.

Transitional *p. 44* A type of stratified epithelium capable of stretching.

Transverse *p. 5* A plane dividing the body or its parts into superior and inferior portions.

Transverse colon *p. 158* The portion of the large intestine passing horizontally across the abdomen.

Treppe *p. 146* Staircase phenomenon in muscle contraction.

Tricuspid *p. 155* A tooth located posterior to the bicuspids having three cusps; molar.

Tricuspid valve *p. 214* A valve in the heart consisting of three flaps, or cusps, of tissue between the right atrium and right ventricle; prevents the backflow of blood from the ventricle into the atrium during systole.

Trochanter *p. 72* A bone marking—a large irregularly shaped projection to which muscle attaches.

Tropic hormone *p. 382* One of several hormones secreted by the adenohypophysis that stimulates other endocrine glands to secrete hormones.

Trypsin *p. 176* A protease in the pancreas that hydrolyzes proteins to proteoses, peptides, and amino acids under alkaline conditions.

Tubercle *p. 72* A bone marking—a small rounded process.

Tuberosity *p. 72* A bone marking—a large rough process, often smaller than a trochanter.

Tunica adventitia (externa) *p. 196* The outermost coat of blood vessels that contains a well-defined elastic layer with fibrous and areolar tissue supporting the nearby tissue.

Tunica intima *p. 196* The innermost coat of blood vessels composed of endothelium and fibroelastic tissue in longitudinal folds.

Tunica media *p. 196* The medial coat of blood vessels that contains smooth muscle and elastic fibers.

Umbilical *p. 5* The central region of the abdomen; pertaining to the umbilicus.

Ureter *p. 314* A 10- to 12-inch tube used for urine release connected superiorly to the kidneys and inferiorly to the bladder.

Urethra *p. 314* A urinary drainage duct from the bladder to the external orifice in the female, a joint urinary and genital canal in the male; a single tube leading to an external opening used for urine excretion; the terminal canal for urinary excretion.

Urinary bladder *p. 314* A muscular reservoir for urine.

Urinary meatus (urethral orifice) The anterior-most opening of the female vestibule; for urine excretion.

Urine *p. 320* Water excretion containing nitrogenous wastes, urea, ammonia, pigments, and electrolytes.

Uterine tube Fallopian tube; oviduct.

Uterus *p. 350* A female secondary sex organ in which the embryo becomes implanted and develops prior to birth; womb; female receptacle for fetal growth and menstrual function.

Vagina *p. 350* A canal in the female that extends from the cervix to the external genital orifice.

Vasa vasorum Blood vessels supplying blood to the outer layers of the large blood vessels.

Vas deferens *p. 353* An extension of the epididymis through which sperm pass to the ejaculatory ducts; ductus deferens.

Vasopressin *p. 383* Antidiuretic hormone; Pitressin; secreted from the neurohypophysis.

Ventral *p. 2* Of or near the belly.

Ventricles *p. 215* The inferior left and right chambers of the heart.

Vertebral column *p. 82* A collection of vertebrae through which the spinal cord passes.

Vertigo Dizziness.

Vestibule *p. 350* A body cavity that serves as an entrance to another cavity or space.

Villi *p. 169* Elevations of the mucosa of the small intestine.

Viscera (visceral) *p. 156* The internal organs of the abdomen.

Vulva The external female genitalia.

Z band *p. 142* A microscopic band that divides an I band in a myofibril; Z line.

Zona fasciculata *p. 388* The middle zone of the adrenal cortex; secretes glucocorticoids and sex hormones.

Zona glomerulosa *p. 388* The most peripheral portion of the adrenal cortex; secretes aldosterone.

Zona reticularis *p. 388* The inner portion of the adrenal cortex; secretes glucocorticoids and sex hormones.

Solution Appendix

Some of the following solutions are available as commercially prepared solutions:

Barfoed's Solution

4.5 g crystallized neutral cupric acetate
100 ml distilled water
0.12 ml 50% acetic acid

Benedict's Solution

17.3 g $CuSO_4$
173.0 g sodium citrate
100.00 g Na_2CO_3 (anhydrous)
Dissolve citrate and carbonate by heating in 800 ml of distilled water.
Dissolve $CuSO_4$ in 100 ml of distilled water. Add $CuSO_4$ slowly to the first solution, stirring constantly. Add water to make a total of 1000 ml. May be kept indefinitely.

Holtfreter's Solution

Stock Solution:
  3.5 g NaCl
  0.05 g KCl
  0.1 g $CaCl_2$
  0.02 g $NaHCO_3$
Dissolve in 1 liter of H_2O (all glassware must be scrupulously clean).
  When ready for use:
Add one (1) part stock solution to nine (9) parts distilled water.

Lugol's Iodine Solution

4.0 g iodine
6.0 g Kl (Potassium iodide)
Dissolve in 100 ml of distilled water.

Methylene Blue Solution

10 mg methylene blue dissolved in 100 ml of distilled water

Millon's Solution

Dissolve 100 g of Hg in 200 g of concentrated HNO_3.
Dilute solution with two volumes of water.

Molisch Reagent

5% alcoholic solution of alpha naphthol

Ninhydrine Solution (also commercially available as a spray)

0.1 g ninhydrine
100 ml distilled water

Ringer's Solution (for frogs)

6.5 g NaCl
0.2 g $NaHCO_3$
0.1 g $CaCl_2$
0.1 g KCl
Add enough distilled water to make 1000 ml.

Saline Solution (physiological)—humans

9.0 g NaCl dissolved in 1000 ml distilled water

Saline Solution (physiological)—amphibians

7.0 g NaCl dissolved in 1000 ml distilled water

Seliwanow's Reagent

0.5 g resorcinol
100 ml concentrated HCl
Add water to make a total of 300 ml.

Illustration Appendix

FIGURE 7A   Mitosis in onion root cells
1. Late Telophase

FIGURE 7C   Mitosis in onion root cells
2. Anaphase

FIGURE 13   Cuboidal epithelium as found in the collecting tubules of the kidney
1. Cuboidal Cells

FIGURE 31   Thick skin, palm, showing epidermal layers
3. Stratum Granulosum

FIGURE 33   Skin, human scalp, showing hair follicles
1. Stratum Corneum
2. Cortex of Hair

FIGURE 103   Tongue showing taste buds
2. Taste Bud

FIGURE 107   Stomach showing detail of mucosa, submucosa, and muscularis layers
1. Lumen of Stomach
3. Arteriole
5. Submucosa Layer

FIGURE 137   Sheep heart in situ with pericardial sac removed, ventral aspect
1. Trachea
5. Ventricle, Left
6. Apex

FIGURE 138   Sheep heart in situ with pericardial sac removed, dorsal aspect
1. Trachea
5. Left Ventricle

FIGURE 139   Sheep heart with ventral portion reflected to show chamber detail
1. Trachea
7. Left Ventricle
8. Apex

FIGURE 140   Sheep heart showing chamber and valve detail
2. Atrium, Right
4. Ventricle, Right

FIGURE 158   Sheep brain, superior aspect
3. Spinal Cord
4. Longitudinal Cerebral Fissure

FIGURE 159   Sheep brain, inferior aspect, showing gross detail
1. Frontal Lobes
5. Pons
7. Brainstem (Spinal Cord)
9. Temporal Lobe

FIGURE 160   Sheep brain, lateral aspect, showing gross structures
4. Cerebellum
5. Medulla Oblongata
6. Spinal Cord

FIGURE 161   Sheep brain, midsagittal section, showing internal structures
6. Optic Chiasma
7. Pituitary (Hypophysis)
10. Medulla Oblongata
11. Brainstem
15. Gray Cortex of Cerebellum

FIGURE 171   Sheep eye, midsagittal section, showing gross internal detail
3. Lens
4. Cornea
5. Pupil
8. Uvea (Choroid)

FIGURE 172   Sheep eye, coronal section, showing internal detail posterior to lens
1. Sclera
4. Uvea (Choroid)

FIGURE 180   Sheep kidney, midsagittal section, showing internal detail
4. Ureter

1 2 3 4 5 6 7 8 9 10